'It is great to read one of the rare books responding to [illegible] *not with moral reproach, but with a subtle strategy of turning apparent trouble* [illegible] *advantage. Combining urbanism, material flows and democracy might indeed be a way out.'*

Professor Marina Fischer-Kowalski, Director, Institute of Social Ecology, Alpen-Adria Universitaet, Vienna

'The book provides material for understanding the science of sustainability and the practice of sustainable living. It X-rays, in a very lucid way, the complexity of transitions to sustainable economies.'

Professor Kevin Chika Urama, Executive Director, African Technology Policy Studies Network (ATPS), and President, African Society for Ecological Economics (ASEE)

'A stimulating book that thinks critically and reflectively about how we can re-construct socially just transitions in a period marked by deep economic and ecological crises. But what makes this book distinctive is both its professional commitment to intellectual inquiry and the authors' own personal commitment to experimenting and living with (more) just transitions in the Lynedoch EcoVillage. An inspiring thought-provoking read about how alternatives could be reconstructed.'

Professor Simon Marvin, Co-Director, Centre for Sustainable Urban and Regional Futures, University of Salford

'For anyone interested in the necessary transition that awaits us all, this book is testimony to a thoughtful marriage of theory and practice.'

Professor Malcolm McIntosh, Director, Asia Pacific Centre for Sustainable Enterprise, Griffith University, Queensland, Australia and co-author of *SEE Change: Making the Transition to the Sustainable Enterprise Economy*

'Just Transitions *brings together powerful arguments and evidence that a fairer world is not only a moral imperative but an ecological and economic necessity.'*

Ashok Khosla, Chairman of Development Alternatives (India), President of IUCN and Co-president of the Club of Rome

'We all know that the current model of global capitalism and its unjust settlement patterns are wrong, unsustainable, and yet, seemingly interminable. Just Transitions provides an excellent academic and activist resource to simultaneously conduct incisive critique and visionary proposition. It is indispensible reading for anyone with a desire to cast light on our bleak world bereft of ideas.'

Professor Edgar Pieterse, Director of the African Centre for Cities, University of Cape Town and author of *City Futures*

Just Transitions
Explorations of sustainability in an unfair world

Dedicated to our boys—Michael,
Ronen and Xolani—who have seen
the future

Just Transitions
Explorations of sustainability in an unfair world

MARK SWILLING
EVE ANNECKE

TOKYO • NEW YORK • PARIS

Just transitions: Explorations of sustainability in an unfair world

Published in 2012 in South Africa by UCT Press
An imprint of Juta and Company Ltd
First Floor Sunclare Building, 21 Dreyer Street
Claremont, 7708 South Africa
http://www.uctpress.co.za

Published in 2012 in North America, Europe and Asia by
United Nations University Press
United Nations University, 53-70, Jingumae 5-chome,
Shibuya-ku, Tokyo 150-8925, Japan
Tel: +81-3-5467-1212 Fax: +81-3-3406-7345
E-mail: sales@unu.edu general enquiries: press@unu.edu
http://www.unu.edu

United Nations University Office at the United Nations, New York
2 United Nations Plaza, Room DC2-2062, New York, NY 10017, USA
Tel: +1-212-963-6387 Fax: +1-212-371-9454
E-mail: unuony@unu.edu

United Nations University Press is the publishing division of the United Nations University.

ISBN 978-1-91989-523-9 (Southern Africa)
ISBN 978-92-808-1203-9 (North America, Europe and South East Asia)

Library of Congress Cataloging-in-Publication Data

Swilling, Mark.
Just transitions : explorations of sustainability in an unfair world / Mark Swilling, Eve Annecke.
p. cm.
Includes bibliographical references and index.
ISBN 978-9280812039 (pbk.)
1. Sustainable development—Developing countries. 2. Natural resources—Developing countries—Management. 3. Developing countries—Economic conditions. 4. Sustainable development—South Africa. I. Annecke, Eve. II. Title.
HC59.72.E5S96 2012
338.9'27091724—dc23 2011044276

The views expressed in this publication are those of the authors and do not necessarily reflect the views of the publishers.

Cover design: Luke Metelerkamp
Project Manager and Editor: Glenda Younge
Proofreader: Catherine Damerell
Typesetter: Exemplarr
Printed by Permanent Printing Limited, Hong Kong

Typeset in 10.5 pt on 13 pt Minion Pro

This book has been independently peer-reviewed by academics who are experts in the field.

Contents

Acknowledgements

We would like to thank the following for their collegial, institutional and, in some cases, funding support over many years: the Enthoven family, Sustainability Institute, School of Public Leadership at Stellenbosch University, Centre for Renewable and Sustainable Energy Studies at Stellenbosch University, Spier Wine Farm, Lynedoch EcoVillage, Schumacher College, Freiburg University, National Research Foundation and the various actors in the Lynedoch-Eerste Rivier valley. Although the individuals who inspired, encouraged and assisted us are too numerous to name, some of the most significant need to be mentioned: Adi Enthoven, Sally Wilton, Teresa Graham, Lawrence Boya, the late Paul Cilliers, Edgar Pieterse, Gita Goven, Naledi Mabeba, Rosieda Shabodien, Johan Hattingh, Tarak Kate, Malcolm McIntosh, John Benington, Patrick Fitzgerald, Bryce Anderson, Ron Heifetz, Marty Linsky, Fritjof Capra, John van Breda, Martin de Wit, Alan Brent, Gareth Haysom, Candice Kelly, Camaren Peter, Simon Marvin, Pru Ramsey, Mazibuko Jara, Mokena Makeka, Manfred Max-Neef, Chris Brink, Pieter du Toit, Robert Davids, Sharifa Ismail, Tom Darlington, Joel Bolnick, Martin Yodaiken and colleagues involved in the International Resource Panel, namely Ernst von Weizsacker, Marina Fischer-Kowalski, Kevin Urama, Ashok Khosla and Janet Salem. We also want to thank the wonderful staff at the Sustainability Institute, in particular June Stone and Louise Bezuidenhout, and all our students who have over the years taught us so much. For all the assistance in editing and publishing this book we thank Sandy Shepherd, Glenda Younge, Ruenda Odendaal and the anonymous reviewers.

Front cover:

The cover depicts a threatened species – commonly known as the quiver tree. Known as *Choje* to the indigenous San people, the quiver tree (*Aloe dichotoma* or kokerboom) is named from the San practice of hollowing out the tubular branches of this tree to make quivers for their arrows. Quiver trees are a species of aloe which are indigenous to South Africa, specifically to the Northern Cape region as well as and Namibia. These trees, which are dying in the most northern areas as a result of rising temperatures, are our warning signals that transitions are already underway. For some, justice is too late. We acknowledge our interconnectedness with all life, and the gifts Earth has provided.

List of Maps, Figures and Tables

Maps

Figures

Tables

Introduction: On Becoming Visible

It is better to be invisible. His life was better when he was invisible, but he didn't know it at the time.

He was born invisible. His mother was invisible too, and that was why she could see him. His people lived contented lives, working on the farms, under the familiar sunlight. Their lives stretched back into the invisible centuries and all that had come down from those differently coloured ages were legends and rich traditions, unwritten and therefore remembered. They were remembered because they were lived. (Okri 1995. *Astonishing the Gods: 1*)

Context

We live in the Lynedoch EcoVillage located near the historic town of Stellenbosch, a 30-minute drive from Cape Town. This south-western tip of Africa is where the Indian and Atlantic oceans notionally meet and it is the home of *fynbos* which is the smallest and most diverse of Earth's six biomes. As the first hinterland outpost of the Dutch colonial settlement in the mid-1600s in what is now Cape Town, Stellenbosch is often regarded as the intellectual birthplace of apartheid. This legacy is reflected in the harsh inequalities in wealth and land ownership that still characterise this rich agricultural region.

We started building the Lynedoch EcoVillage in 1999 with a group that included a local housing activist, landless farmers, the local school principal and two architects[1] who were prepared to figure out what it means to design for sustainability. We have, in short, lived through profound transitions in our everyday lives as we have struggled towards a dimly understood goal, working things out along the way, often long after they have happened.

We were fortunate: we could experiment and fail because there was the space for innovation in a society desperate to break from its (apartheid) past, but often without a coherent vision of the future. In a society too traumatised to unite around a specific solution, there was enough uncertainty for a small group of us to mount an experiment that was not ridiculed in advance. Luckily we never knew enough to be too clever about the enormity of the challenges ahead.

After all is said and done everyday life in the village goes on. Ranen, our 17-year-old son, strides down the hill to catch the train as the orange dawn light breaks from behind the towering Helderberg Mountains, chatting to Tebo, Naledi's daughter, along the way. Willem rides out of the gate on his bicycle on his way to work in a local wood factory. Eric, the organic farmer, drives out in his small truck for another day on his organic vegetable farm. Two hours later the shriek and chatter of 300 schoolchildren arriving for school

---

1 Gita Goven and Alastair Rendall who now run a large architectural and urban design practice in Cape Town, called ARG Design.

from farms and nearby townships, shatters the quiet of another frosty autumn morning. Then the university students doing their master's degrees in sustainable development arrive — some emerge sleepily from the guest house, others arrive by car and train. They gather in the hall before going out to work on the farm and in the gardens before lectures begin at 09h30. As the day warms, Joseph inspans the six oxen on Eric's farm, coaxing them into another day of ploughing soils that have benefited from a decade of organic farming. In the crèche, 40 small people — some suffering from foetal alcohol syndrome — have finished their daily t'ai chi exercises and are going about their different activities in their carefully crafted Montessori environment.

By the afternoon the pace changes: as the children head home (to some less-than-safe areas), a quieter reflective space opens up. After lunch on the guest house veranda with its spectacular views across the green valley fractured into its patchwork of land uses, the teenagers head up to the 'Changes Youth' clubhouse built into the roof of the crèche. Oblivious to the electricity being generated by solar roof tiles less than two metres above them, they do their homework, plan their sports programmes and chat in a safe environment, far removed from the predatory threats that await them in the few hours between the end of school and the arrival of parents from work.

In the near distance, the sound of building is an ever-present reminder of big things happening — someone out there has placed another brick on a rising wall, hammered home another nail, or lifted another roof truss into place. With every shrill whine of a power tool and dull thud of a hammer, someone gets closer to completing the home of their future memories.

As the light fades and the chill returns, neighbours head back from the vegetable gardens, greeting others along the way, tending their own allotments. It is a routine greeting — a smile, a nod — but the connection is made as if acknowledging yet another moment has passed in that excruciatingly slow process of becoming familiar.

As the light disappears, the solar street light switches on. The worms in the biolytix sewage treatment plant wriggle their way up to the darkening surface to access more oxygen without encountering the light that they seem to loathe, and the pumps switch on to replenish the water tanks with the treated effluent of the day.

Life goes on — oblivious to the hundreds of documents that lie gathering dust in dozens of lever-arch files; documents needed to get all the necessary approvals from government to raise the funds, to appoint the contractors, to constitute the home owners' association, to get housing subsidies from government, and to define with false certainty the semblance of a believable story (or what, in more conventional language, would be called a 'strategic plan').

This book, in many ways, is a tribute to this lived transition. In being part of becoming visible in a place, we were able to bring up our sons while exploring new ways of seeing and being in our troubled but rapidly changing world. The more we endeavoured to teach people about sustainability, as our master's programme matured, the more we learnt from our students. They pushed us to look beyond the dismal platitudes of comfortable critique. They were impatient with the self-satisfying banalities about how bad the world is, especially if the underlying suggestion was that they could do nothing to change things. They were particularly frustrated with ecologists who talked about how the environment was being destroyed before concluding that if more people were aware of this destruction, things would change. How would things change, the students asked? Many people *are* aware, but what can they do if they are locked into urban systems that condition environmentally destructive behaviour? Were there any practical examples of alternative ways of living? The problem is that most ecologists cannot answer these questions because they have a limited understanding of the dynamics of institutional power and cultural change. Nor do social scientists have answers, because they ignore the ecological context (and often causes) of the socio-economic challenges on which they prefer to focus. Interestingly, it was not solutions as such that the students demanded, but *patterns of thinking* informed by what Gibbons *et al.* have usefully called 'socially robust' knowledge about future orientations at different scales (Gibbons *et al.* 1994). This book is really for these students and their hungry imaginations, but it is also for anyone wanting to share this search for actionable imaginaries.

The perspective and argument in this book is inescapably shaped by the fact that we live and come from the most unequal society in the world. As will be clear in the chapters that follow, what is at stake is not simply a transition to a mode of production and consumption that is not dependent on resource depletion and environmental degradation, but as important is the challenge of a *just transition* that addresses the widening inequalities between the approximately one billion people who live on or below the poverty line and the billion or so who are responsible for over 80 per cent of consumption expenditure. The very rich who comprise a small fraction of this class of over-consumers enjoy enormous political, economic and cultural power. A transition to more sustainable forms of development that leaves these socio-economic inequalities intact will not, in our view, deliver an end result that can be called *sustainable.* A *just transition,* therefore, must be a transition that reconciles sustainable use of natural resources with a pervasive commitment to what is increasingly being referred to as *sufficiency* (that is, where over-consumers are satisfied with less so that under-consumers can secure enough, without aspiring for more than their fair share). This, however, will involve deep structural changes that will require extensive interventions by capable developmental states, active commitments by progressive business coalitions, and a mobilised civil society rooted in experiments that demonstrate in practice what the future could look like.

Our experience as co-creators of the Lynedoch EcoVillage and as engaged citizens of the new South Africa fused two seemingly contradictory sensibilities. On the one hand, like many South Africans, we have watched with mounting despair how the great promise of the democratic transition of 1990–1994 has failed to translate into a more just, equitable and sustainable South Africa. Instead, the patterns of elite consumption, intense resource exploitation and grinding poverty that were brutally enforced during the apartheid era have persisted in a different form within an open, democratic and culturally expressive society. What changed was the broadening of the class who now benefit from the wealth to include the new black elites. Our despair was balanced out however — sometimes even neutralised — by the inspiration that came from the gradual process of building South Africa's first socially mixed, ecologically designed community.

Suspended in this paradox of despair and inspiration, we felt compelled to explore both the micro-details of alternative ways of living and alternative ways of understanding the big developments and ecological challenges of our times.

We do not prescribe a specific technical alternative, nor do we satisfy the popular appetite for a simple set of steps for replication. This book is self-consciously and unapologetically 'academic' in the way in which it explores the perplexing logics of a range of different literatures (complete with references so that researchers/practitioners can follow up on the threads of thinking that shaped our own). Each chapter in one way or another synthesises related but hitherto disconnected literatures in order to illuminate new ways of thinking about that particular subject from a sustainability perspective. Although academic writing is often dismissed because it seems overly concerned with its own internal (often arcane) debates, the advantage of the academic discipline is that it maps out the complex overlapping trajectories of thought that shape the assumptions at the centre of the (often partially understood) common-sense ideas that condition everyday life. It is not possible to understand our world — to become visible — without understanding the language we use and the origins of the concepts that are embedded in the common sense ideas that get mobilised in everyday conversations, in the media and elsewhere, all the time. We choose to see academic writing as a 'global positioning system' in a conceptually cluttered world.

This book is also more normative than the average academic text. For us, critical analysis is vital, but insufficient on its own. Drawing from experience, research and inspirations from around the world, we suggest ways of thinking about the future that may assist in building a more sustainable and just world. To this end we strongly agree with Costanza who argues that we need to restore the 'balance between synthesis and analysis'. In his words:

> *Science, as an activity, requires a balance between two quite dissimilar activities. One is analysis — the ability to break down a problem into its component parts and understand how they function. The second is synthesis — the ability to put the pieces back together in a creative way in order to solve the problems. In most of our current university research and education, these capabilities are not developed in a balanced, integrated way.* (Costanza 2009: 359)

So why have we entitled this book *Just Transitions*? We are not offering another grand theory of transition. Nor do we prescribe a particular programme or type of transition. The notion of a 'transition' has a specific meaning and history in South Africa. It is hard to imagine these days, but as recently as 1990 South Africa was ruled by a harsh militaristic regime that was able to deploy a vast modern police force and army to protect the power and privilege of a white capitalist elite that appeared reluctant to face the prospect of a democratic future. The opposition forces had no significant military power, yet subscribed to the view that change would come about by way of a revolutionary rupture. Ironically, the regime and the opposition shared the same theory of change, namely that change is systemic and the outcome is the function of the balance of military power. For the regime, brute force was needed to stop systemic change, and for the opposition change would come about when the oppressed majority rose up in sufficient numbers to forcefully detonate the collapse of the regime's capacity to govern. In reality, both the regime and the opposition spent at least four years between 1990 and 1994 negotiating a transition that resulted in systemic change but without a revolutionary rupture (Marais 2011; Swilling *et al.* 1988; Swilling & Phillips 1989). The founding democratic elections in 1994 ushered in a new era of unprecedented democratic space. Although many of the fundamental economic problems of inequality and resource exploitation have remained intact, this democratic space has made possible countless innovations and changes across the social, economic and environmental spectrum. They have yet to accumulate into a qualitative shift that will contribute to substantive solutions to the problems of poverty, inequality and resource depletion.

To explain the South African transition it was necessary to turn to a body of literature that took into account more than just the balance of force. This was provided by the remarkable work on transitions to democracy assembled by political scientists Philippe Schmitter and colleagues (Schmitter *et al.* 1986). This highly influential work revealed that South Africa was not unique. Indeed, it became clear that South Africa was part of a trend that began in Southern Europe in the late 1970s/80s (Portugal, Spain and Greece), spread across Latin America through the 1980s, and extended into Eastern Europe from the late 1980s. Instead of big lumpy categories such as 'the state', 'power', 'violence', 'revolution' and 'class', Schmitter and his colleagues zoomed in on more granular institutional processes, personalities, complex interest groups and relational dynamics in order to reveal unexpected drivers of these 'non-revolutionary regime transitions' that are not easily understood from a traditional 'balance of force' perspective. So while on the surface a stalemate persisted in all these cases for extended periods of time, and although neither side had the combined military and political force to defeat the other, various things started to happen that eventually prepared the way for negotiations and — following all sorts of setbacks and reversals — a political settlement that satisfied no-one completely, but which was seen by a cross-section of key actors as the best available option at that particular moment. These 'non-revolutionary transitions' were not regime change in the classic sense ('seizure of state power', *coup d'état* or military invasion), nor popular revolution, nor 'decolonisation', nor merely political compromise to maintain the status quo. Something fundamental eventually happened to dismantle the 'colonial' nature of the apartheid state, but in ways that were often quite obscure and unexpected.

At a deeper level, the lesson from the South African transition is that change is divisible: many small changes at different scales (local, national, global) started adding up in a way that created conditions for a 'tipping point' at regime level. The timing of this tipping point and the nature of the outcome were, however, both unpredictable: seemingly tiny factors had disproportionately large impacts within particular contexts in ways that would have been impossible if the context had been different.[2] But up until this tipping

[2] For example, the South African transition was affected critically by the fact that in 1989 President P.W. Botha (the chief 'securocrat') had a stroke that incapacitated him, and political prisoners went on a hunger strike more or less at the same time.

point everything seemed implacably unchanged. So much so, that many grew impatient and resorted to more extreme violent action to force faster change — a response that triggered equally violent counter-actions to maintain the status quo. Extreme actions had the counter-intuitive impact of reinforcing the resilience of the system (at least temporarily). For the transition to run its course a core adaptive leadership (Heifetz 1994) comprising key movers from both camps was required to 'hold the centre' of an unpredictable process that the word 'transition' seemed to capture. Since 1994, however, the quest for certainty and the tolerance of poverty has threatened the democratic space created by this remarkable transition.

On transitions

Unsurprisingly, the language of 'transition' has started to penetrate the sustainability literature in ways that have influenced our understanding of global change at different scales (for some key examples from different traditions see Brown 2008; Fischer-Kowalski & Haberl 2007; Grin *et al.* 2010; Guy *et al.* 2001; Hodson & Marvin 2010; Korten 2006; Madeley 2002; Rotmans & Loorbach 2009; Scheffer 2009; Smith *et al.* 2005; Van den Berg *et al.* 2011). They are all motivated, in one way or another, by the search for clues using a systems perspective that might reveal insights into how the transition to a more sustainable order might come about. They analyse transition dynamics at different spatial and temporal scales, and they all (either explicitly or implicitly) suggest strategic roles for key agents of change. In our view there is still a way to go before we have a substantial body of literature on transitions to a more sustainable future that reflects a degree of consensus on what needs to happen and who the key change agents really are.

A group of Dutch researchers working within the wider European 'sustainability science' tradition have, in recent years, synthesised technology science, evolutionary economics and structuration theory in order to develop what may well be the first systematic attempt to develop a comprehensive 'theory of transition to sustainable development' — or what is commonly referred to now as the Multi-Level Perspective or MLP (Grin *et al.* 2010; Smith *et al.* 2010). Like this book, they understand the current crisis as playing itself out across the nexus between the global financial systems, market-governance-society relations and the values as expressed in lifestyle and consumption. They define transitions (both in the past, with respect to industrial change, and possibly in future, to a more sustainable order) in ways that we would agree with, namely they are processes characterised by the following:

- They are the *co-evolution* of technological change, consumption behaviour and the institutional reforms that are required to embed the new technologies in society
- Transitions are *multi-actor* processes that engage actors in unpredictable ways from all sectors (public, private and non-profit)
- Transitions are *long-term* processes, often 40–50 years, with distinct internal phases (from initiation to maturation)
- Transitions are about the *reconfiguration* of the institutional and organisational structures and systems of society.

We have not, however, adopted the Multi-Level Perspective nor any other general theory of transition to make sense of the issues addressed in the various chapters. We have chosen to explore the dynamics of transition empirically in contextually specific ways rather than depict them in generic terms that could, over time, create the misleading impression that there is a particular transition pathway that is relevant for all contexts. We remain open to the possibility of an appropriate general theory of transition to what we have conceptualised in Chapter 3 as a *sustainable long-term development cycle.* As we show in Chapters 3 and 4, this needs to build on a synthesis of the work of the Dutch school, the Vienna school centred around the work of Marina Fischer-Kowalski and colleagues, the work on technological change by Perez (Perez 2002) and the development challenges addressed by the work of Peter Evans, Gilberto Gallopin and Charles Gore (Evans 2010; Gallopin 2003; Gore 2010). Indeed, when African realities are factored into the equation, what may be required is not merely a 'general theory of transition', but rather a 'general theory of transition *and collapse*'. Whereas the European discussion is largely about *low-carbon transition* as an alternative to preserving the status quo, in many other parts of the world that are exploited for their resources the alternative to transition may well be collapse. Although European writers on transition have paid far more attention to governance and the role of the state than their North American colleagues,[3] both make no significant references to contexts in which states are incapable, for various institutional and political reasons, of intervening effectively to stop what Gallopin calls 'maldevelopment' (Gallopin 2003), to say nothing of the complex challenges of a transition to more sustainable modes of development. Indeed, history is replete with examples of societies that failed to make a transition and eventually collapsed (Diamond 2005). This is why this book includes a chapter on resource wars with Sudan as a case study (Chapter 7).

The logic of this book

Inspired by our experience in building the Sustainability Institute within the Lynedoch EcoVillage, this book addresses what we regard as the five key transitions that will in one way or another intersect and shape the next half century. As discussed in more detail later, these are what we have called the *epochal, industrial, urban, agro-ecological* and *cultural* transitions. We suggest that if we want the *next long-term development cycle* to be *sustainable,* we need to pay attention to the rapid and seemingly disconnected transitions taking place at these different spatial and temporal scales. This will help us to understand how best to promote a more *just transition.*

The book is divided into three parts. In Part I we address in three chapters the broad general conceptual themes of the book, namely the need to break from reductionism (Chapter 1), what is so unsustainable about the world today (Chapter 2), and how we can begin to understand the dynamics of transition (Chapter 3). In Part II we argue for

[3] Contrast, for example, the work by Clark *et al.* (Clark *et al.* 2005) and the work by Smith *et al.* (Smith *et al.* 2005).

a rethinking of development with special reference to the greening of the developmental state (Chapter 4), the key role that cities could play in the transition to a more sustainable largely urbanised world (Chapter 5), and the neglect of soils in the global discussions about the potential of sustainable agriculture to feed the world (Chapter 6). Part III comprises a set of case studies drawn from the African context, namely the nexus between failed states and resource conflicts with special reference to the case of Sudan (Chapter 7), the challenge of transcending resource and energy intensive growth paths in modernising developing countries using South Africa as a case study (Chapter 8), and finally an exploration of what sustainability and liveability means in a rapidly urbanising world by way of case studies of Cape Town (Chapter 9) and the building of the Lynedoch EcoVillage (Chapter 10).

We lay out below brief summaries of the logic of each chapter that substantiate our core argument that a *just transition* must be a transition that reconciles sustainable use of natural resources with a pervasive and meaningful commitment to *sufficiency.* A just transition to a more sustainable long-term development cycle is what this book is about.

Part I: Complexity, Sustainability and Transition

Chapter 1 argues that sustainability is a challenge that is difficult to comprehend through disciplinary lenses. Towards the end of his magisterial account of biological and cultural history entitled *Human Natures: Genes, Cultures and the Human Prospect,* renowned Stanford University biologist, Paul Ehrlich, prophetically concluded that a sustainable and more just future will depend on whether the human species can develop the capability for 'conscious evolution' (Ehrlich 2002). This call for cultural transformation to inspire the need for a more sustainable world has since been echoed many times over (for particularly influential publications see Capra 1996; Hawken 2007; Kauffman 2008; Korten 2006; Okri 1999). We concur with Kauffman that unless we break from the 'injuries of reductionism', we will never discover the contextually specific cultural sensibilities for living out the 'fullness of human life' (Kauffman 2008: 7).

We argue that the significance of complexity theory lies in the fact that it shows how to make this break from reductionism. Whether it goes far enough when it comes to imagining an alternative world is explored.

Chapter 2 provides a summary of well-known mainstream documents that, when read together, provide an understanding of what is so unsustainable about the world. However, we concur with the contributors to *What Next?* that an *unjust* transition is a distinct possibility (Dag Hammarskjöld Foundation 2007). We discuss this possibility with respect to the widely held notion of *ecological modernisation* (Korhonen 2008). For us, ecological modernisation and an unjust transition may well mean the same thing in certain contexts—the greening of existing modes of consumption and production without addressing the challenge of (global) inequality. There is mounting evidence that an unjust transition would involve massive private sector investments to build low-carbon, resource-efficient economies with reduced environmental impacts, while leaving intact existing inequalities. Public sector subsidies to reduce the risk of these investments would be justified using the compelling logics of climate science and, to some extent, environmental conservation. A divided, poverty-stricken, conflictual and socially unsustainable *low carbon* world would then be the outcome of an unjust transition.

Chapter 3 addresses the nexus between the epochal and industrial transitions in order to conceptualise possible ways in which the next long-term development cycle will unfold. We are convinced that there is much to be learnt from previous epochal transitions, in particular the transition that began some 13,000 years ago with the birth of agricultural systems in the Fertile Crescent, and the (partially completed) transition from the agricultural to the industrial epoch that started over 250 years ago. As we hit the limits of the resource requirements of the industrial epoch (discussed in Chapter 2), so we witness global discussions that explore the modalities and temporalities of the next epochal transition, namely the transition from an 'industrial' to a 'sustainable' socio-ecological regime and what this means for developed and developing countries. The rapid rise since 2009 of the notion of a 'green economy' within the UN system captures this dawning realisation, albeit in ways that could potentially undermine a just transition. As argued in this chapter, the contemporary economic crisis (which began in 2007) marks the mid-point of the digital transition (the fifth industrial transition). What is significant, though, is that this particular crisis dovetails with the wider epochal crisis of the industrial era as a whole. Thus there is sufficient evidence that we may be experiencing an unevenly developed dual transition at the *epochal* and *industrial* levels which could constitute the basis for what we have called a *sustainable long-term development cycle.*

Part II: Rethinking Development

In Chapter 4 we argue that we need to rethink the developmental state in light of the transition to the next *long-term development cycle*, no matter how more or less sustainable this will be. The classical conceptions of the developmental state were formulated to make sense of the role that states played in the modernisation processes that resulted in the transition from agrarian to industrial economies. This became the prime focus of development economics. We foresee an interventionist role for states, but now with respect to both developmental and ecological challenges. For this to be possible we argue that it will be necessary to synthesise development economics and ecological economic theory in order to conceptualise a developmental state that can foster sustainability-oriented innovations. This means going beyond the traditional socio-economic boundaries of mainstream Keynesian and Marxist economics.

Chapter 5 argues that the epochal, industrial and developmental trends take place within a rapidly changing spatial context that is most clearly marked by two simple but overwhelming facts: some time in mid-2007 there were, for the first time ever, more people living in cities than outside cities, and over the next four decades as the global population grows from 7 to 9 billion people (or more) another 3 billion will land up in African and Asian cities. As explored in Chapter 5, the city becomes the key locus for making sense of the epochal and industrial transitions that are transforming all cities in profoundly different ways. To prepare cities for the next long-term development cycle, we need new methodologies to understand the metabolisms of cities, with special reference to the networked urban infrastructures that conduct the flow of resources upon which (nearly) all city dwellers are dependent. We review the history of urbanism, tracing the evolution of inclusive urbanism, splintered urbanism, green urbanism, slum urbanism and end up proposing a radical developmental alternative that we call 'liveable urbanism'.

Chapter 6 addresses the challenge of sustainable agricultural development by arguing that insufficient attention is paid to the problem of degraded soils. Without food, we perish. Without biologically healthy eco-systems such as soils, stable climates, sufficient water and viable nutrient cycles, there will be no food. Yet few city dwellers realise how vulnerable their food supplies really are as they seldom think about these degrading and collapsing eco-systems, often located thousands of miles away. As discussed in this chapter, there is an agro-ecological transition underway that is one response to a very particular, but often misunderstood, crisis that threatens the conditions for global food security. At the centre of this transition is the much-neglected subject of our soils. Even in some of the most advanced assessments of agricultural science and practice, soils are neglected. In almost every rural development strategy in the world, soils are forgotten. We argue that as long as we take soils for granted, the chances of finding ways of feeding the world sustainably will remain very small indeed.

Part III: From Resource Wars to Sustainable Living

To counter the spectre of an unjust transition, Chapter 7 addresses the vexed issue of resource wars, focusing in particular on Sudan. Sudan tops two of the most depressing lists in the world: the list of 'failed states' and the list of 'resource wars'. For us it is a tragedy that provides everyone with an insight into what could happen if we have an *unjust transition*. Sudan is a resource war and failed state because it has resources that others want and elites prepared to fight over the spoils. A world with many countries destroyed by resource wars is a distinct possibility if the billion over-consumers that live out urbane, sophisticated lives continue to assume that everything will be all right if minor adjustments are made, such as fitting solar panels and eating organic food. These are important for individuals, but insignificant in the greater scheme of things. A just transition will have to be far more fundamental and globally relevant, especially as we hit global resource limits. Similarly, the business elites and governments in resource-rich countries in Africa (including South Africa) who sell off the natural capital of their countries in return for personal enrichment must realise that they are just as guilty of creating a world of endemic resource wars for their children.

Using South Africa as a case study, Chapter 8 addresses the challenge that faces many resource-rich countries that have relatively capable states which are committed to rapid industrial modernisation. We question how long South Africa's image as an African success story will last because its growth model is based on elite consumption and extreme levels of resource exploitation. As the country hits many resource limits, it will have to manage a transition to a more sustainable and equitable economy, which will come up against a powerful set of vested interests in the mineral and energy sectors. We use South Africa as a case study of a developing economy which cannot assume that it is possible to develop first to eradicate poverty, followed by an environmental clean-up later. This development paradigm is intellectually bankrupt. South Africa is a robust democracy which has a Constitution that obliges the state to eradicate poverty and racism in a way that is 'ecologically sustainable' (Section 24(b) of the Constitution). Until recently very little attention was paid to this mandate. However, as the economic consequences of dependence on resource- and energy-intensive growth become apparent, South African debates about building a 'green economy' may be instructive for other fast-growing developing countries. We critically examine these claims.

With linkages back to themes addressed in Chapters 4, 7 and 8, in Chapter 9 we use Cape Town to demonstrate what it means to dissect and reassemble the flows and networks of a city. Our conclusion is that the transition to a more sustainable long-term development cycle may well depend on the way in which cities take the initiative to figure out what this means in practice. Our argument is that urban infrastructures sit at the nexus between the spatial reconfiguration of resource flows through cities and macroeconomic expenditures aimed at countering the global recession. How these are aligned will determine the future of urbanism.

Instead of adding our voice to the cacophony of calls for global mindshifts and/or grassroots action (Hawken 2007), in Chapter 10 we explore our own 10-year experience of building a community called the Lynedoch EcoVillage, which aspires to more sustainable living. We have reflected here on what has emerged from our experience. Drawing on theories of adaptive leadership and ecological design, we suggest that it is no longer good enough to merely minimise environmental damage (which is what green urbanism is largely about). For us, liveable urbanism is about restoration of eco-systems and resources. We concur with Hawken when he praises the 'infinite game, the endless expression of generosity on behalf of all' (Hawken 2007: 187). Chapters 1 and 10 are, appropriately, the 'book-ends' of this volume: we start by making the case for a break from reductionism, and we end with an exploration of living and learning a generous and restorative life.

Values

Taking a lead from Hawken's conclusion about what motivates activists who contribute in their localised ways to epochal transition (Hawken 2007), we believe that there are two ancient values that have re-emerged at the heart of this new movement of conscious evolution:

1. 'I am because we are',[4] or more specifically, never let happen to anyone else what you would not have happen to yourself
2. 'All life is precious', thus all action must be judged by whether it restores or destroys the greater web of life.

[4] There is a subtle but important distinction between this notion and a variant that is expressed as 'I am because you are'—this latter notion, which is sometimes used to define what is meant by the African concept of ubuntu, is still rooted in an individualist sensibility.

These values reflect what this book is about: generosity and restoration. The first implies an ethics of sufficiency and cooperation which undercut the capitalist values of individualism and crass materialism that are of little use when it comes to considering alternatives to the polycrises we now face.

'All life is precious' is a value which questions the anthropocentric perspective that reduces nature to a set of resources and eco-system services that humans can use as they want. We recognise, though, that for the sake of analysis and to engage the prevailing literature, we have used this anthropocentric language in this book. Nevertheless, our inspiration stems from much of the deep ecology literature which advocates that, ultimately, sustainability will depend on the diverse ways humans have available to them to reconnect with the various dimensions of nature (Harding 2006). We explore this idea towards the end of the book, when we suggest that sustainability will not result from doing less damage over time, but rather by finding ways of living that restore the eco-systems upon which we depend.

To conclude, we reiterate the core argument of this book. We argue that various transitions are already underway in response to resource depletion and negative environmental impacts. Unlike previous global economic crises, economic recovery will not be able to depend on cheap resources extracted from some outlying region, nor will prices be lowered by applying new technologies that will somehow magically deliver ways of transcending geophysical and biological limits. Something fundamental must change in the way in which economies relate to their environments. These transitional dynamics, however, are taking place in a world characterised by population growth, severe inequalities and rapid urbanisation. The challenge is to ensure that these transitions are *just* transitions. Although the world's poor are most affected by resource depletion and negative environmental impacts, it is possible to envisage investments and interventions that result in a more sustainable use of resources and reduced impacts, without fundamentally altering the balance of power and distribution of resources between richer and poorer sectors of global society. A *just* transition, by contrast, will regard the innovations, investments and interventions required to address resource depletion and impacts as unique opportunities to simultaneously address the wide range of fundamental needs of everyone, but in particular the world's poor—most of whom are concentrated in the global South. These kinds of *just transitions* can, of course, occur across many contexts, from the smallest eco-village right up to the nation-state and the global stage. The emergence of new indicators, such as the 'extended Human Development Index' proposed in the 2010 Human Development Report and the idea of a Happiness Index proposed by Nobel Prize winners Amartya Sen and Joseph Stiglitz (Stiglitz *et al.* 2009), suggests that ways may indeed emerge to validate qualitative rather than quantitative measurements of progress. We need more of this kind of thinking, and many more experiments across diverse scales which demonstrate in practice that a *just transition* is both desirable and feasible.

We conclude by paraphrasing Mahatma Gandhi who once said that there is enough in the world to provide for what we need, but not enough to satisfy our greed. For us, generosity and restoration are the inseparable values that seem to be foundation stones for the messy, unevenly developed epochal transition to a more just and ecologically sustainable *long-term development cycle*. But this transition is by no means inevitable. Indeed, resisting this transition in some of the richer parts of the world may well entail collapses in other poorer parts of the world, especially if there are much-wanted concentrations of primary resources in these regions. The complex trajectories, convergences and disjunctures of the epochal and related transitions underway at the industrial, agro-ecological, urban and cultural levels need to be understood by those who share our commitment to this sustainable future if they want to figure out how best to make it happen. We hope this book assists in this endeavour.

PART I

Complexity, Sustainability and Transition

Chapter One

Complexity and Sustainability[1]

Dedicated to the late Paul Cilliers, in deepest appreciation

Introduction

By the end of this book, there should be little doubt that imagining and implementing more sustainable futures is the greatest challenge that our generation faces. To do this we not only need new ways of thinking, but we need to understand the history of patterns of thinking that fail to appreciate the evolutionary significance of our incontrovertible dependence on other living species and nature in general.

Evolutionary thinking pervades the discussions in this book. We simply cannot fathom how it will be possible to contemplate building a global transdisciplinary commitment to sustainability without an appreciative understanding of the dynamics of evolution at the micro- and macro-scales of life. At the same time we question whether reductionism will assist with the building of this transdisciplinary commitment because we may well need to think of the passage from beginnings to ends in terms of complex patterns, not building blocks. This may, for example, entail a rethink not of evolution per se, but rather of what drives the evolutionary process and, by the same token, the quest for more sustainable futures.

Imagining modern society

Finding a new collaboration between the sciences and humanities which is appropriate to our specific context must entail both coming to terms with our past and synthesising something new. We argue, however, that this will not be achieved via the now faded promise of Newtonian science, which has not only collapsed as an all-embracing scientific paradigm appropriate for analysing all reality, but that the universal developmental modernity that this kind of science underpinned has been a cataclysmic disappointment. Instead, the challenge today is not the unleashing and application of modernist science, but rather its containment for the sake of a sustainable and more equitable and just future.

Instead of the First World model of industrial modernity as the primary driver of growth, the contemporary challenge is equity (specifically, poverty eradication) via the search for more sustainable livelihoods in which industrialisation and urbanisation are constrained by limits to the carrying capacity of the natural systems within which these human systems are embedded (see Chapter 2).

---

1 This chapter is based on an opening address prepared for the workshop on the Origins of Humanity and the Diffusion of Human Populations in Africa, 17–19 September, Lanzerac Estate, Stellenbosch, convened by the Human Sciences Research Council. A version of this paper was published in: E. Pieterse & F. Meintjies (eds) (2004) *Voices of Transition: The Politics, Poetics and Practices of Development*. Johannesburg: Heinemann.

Instead of the Eurocentric cultural synthesis of English classical literature and Newtonian science which a previous generation longed for (Snow 1993), we propose that a theory of complex adaptive systems helps to create the basis for the kind of sustainability science that can cope with uncertainty without obliterating hope.

In short, to break out of a 'Western perspective' on potential transitions to more sustainable futures will mean embracing bodies of knowledge which recognise what Margulis and Sagan call the 'incontrovertible partnership' (Margulis & Sagan 1997) between human and natural systems, and which also recognise the enormous diversity of contexts that will shape the way the future is imagined.

On reductionism, and horror

As African social scientists, we often contemplate a journey which many African social scientists have travelled since the mid-twentieth century. We have witnessed the breakdown of many of the major twentieth-century social projects that in one way or another were premised on the modernist promise of the 'good society'. It was this promise that constituted the basis for hope by many who founded these social projects. This included the Western European social democracies, the Soviet-style socialist societies, the national democracies in post-colonial Africa and Asia, and the various revolutionary options such as Cuba, Nicaragua, Mozambique, Algeria and Vietnam. Out of the collapse of Soviet socialism and many left-wing experiments has arisen an aggressive neoliberalism with a breathtakingly ambitious twenty-first century global agenda to bring politics to an end via a complete consensus on the virtues of the market. Although Soviet socialism was the formal enemy of the neoliberals, their real enemy was the social democratic aspirations of the post-World War II generation with, of course, John Maynard Keynes, who attempted to reconcile capitalism and social justice.

Many social scientists have tried to dig deep into the basic structure of social theory to find explanations for both the rise and fall of the twentieth-century social projects and the Godzillan spectre of globalised neoliberalism that was so hegemonic from the late 1980s through to the financial crash of 2008. Both in many ways depended on the possibility of certainty derived from an understanding of what were deemed to be the basic laws of social progress. For twentieth-century social projects, it was Marxian class analysis, or nationalist anti-colonialism, or the rationally planned social market of Keynesian economics. But also present throughout the twentieth century (with deep roots in the nineteenth century) was the certainty of the market mechanism inspired by economic liberals such as Milton Friedman and Friedrich Hayek.

What many social scientists wanted to know is whether the epistemology of the social sciences contributed in any way to the different ideologies that helped legitimise the mass mobilisation of great hopes, but also the perpetration of great horrors. These horrors apply not only to the obvious such as colonialism, dekulakisation, genocide, apartheid, and the numerous overt and covert military attacks on developing countries by the USA since World War II, but also to poverty, racism, ethnic cleansing and gender oppression. The origin of this inward search was, of course, an examination of fascism, starting with the work of the Frankfurt School. Fascism was an unsurprising starting point because it represented the ultimate certainty, rooted as it was in a brutally reductionist social theory that reconciled atavistic impulses and modernity. But this just provided the basis for a generalised critique of reductionism where it appeared in social theory. Over time, this critique came to be applied to a wide range of social projects no matter what their ideological orientation. This method eventually came to underpin the deep scepticism that existed in many branches of the social sciences by the late 1990s and which ran contrary to the rigid certainties of the neoliberal discourse, articulated most forcefully in Fukuyama's triumphant book, *The End of History and the Last Man* (Fukuyama 1992).

A reductionist analysis is basically an explanation of a complex reality which depends on the reducibility of the multiplicity of components of this reality to a few basic elements which are deemed *a priori* to hold greater explanatory weight than any others in the system. This becomes a powerful tool for predicting the future. Following the work of Foucault (Foucault 1998), this analytical logic makes it possible to claim with great certainty what is knowable about the present which, in turn, makes possible predictions about the future which can be equally validated by high levels of certainty. By applying the Newtonian scientific method to the social sciences, it was thus possible to claim to know the basic laws of motion in society, and therefore also claim to be able to predict the future. This, in turn, made it possible for those with power to legitimise their actions: by claiming to know the objective reality they were able to justify actions based on the supposed predictive capacity of social theory. When these actions were deemed to be constructive and good, no one seemed to mind. But when the opposite applied, the ability to resort to 'objective truth' was used to justify the horrors as well. Hence, many post-modern social theorists wish to obliterate the possibility of repeating these horrors by negating the possibility of the existence of objective truth. They may be going too far, but the motive is understandable.

For African intellectuals, this journey towards uncertainty was particularly brutal. Decolonisation was celebrated with great dreams, but the disappointments have been bloody and the price paid in mass misery and suffering has been so awesome that many intellectuals have fled the continent in despair. It is not surprising that the most passionate and compelling critiques of certainty have emanated from these intellectuals. Whether it is Ben Okri in literature (Okri 1995; Okri 1996; Okri 1999) or Achille Mbembe in the social sciences (Mbembe 2002; Mbembe 2004), their fear of certainty is tangible in every word — just raw bloody nerves between them and their experience of reality. For them, there is a direct and short march between reductionism and authoritarianism. Equally, therefore, is their desire to increase the distance between what constitutes knowing and the use of power. Given the increasing centralisation of power in contemporary South Africa and what Mbembe calls the 'necropolitics' of Africa, this is a desire which has its merits.

Edgar Morin, a renowned French social theorist who has pondered these questions for decades, does not go this far. He represents a tradition which wants a better understanding of reality than reductionism has to offer, but he stops short of extreme deconstructionism:

> *Intelligence that is fragmented, compartmentalised, mechanistic, disjunctive, and reductionistic breaks the complexity of the world into disjointed pieces, splits up problems, separates that which is linked together, and renders unidimensional the multidimensional. It is an intelligence that is at once myopic, colour blind, and without perspective; more often than not it ends up blind. It nips in the bud all opportunities for comprehension and reflection, eliminating at the same time all chances for corrective judgement or a long-term view. Thus, the more problems become multidimensional, the less chance there is to grasp their multidimensionality. The more things reach crisis proportions, the less chance there is to grasp the crisis. The more problems become planetary, the more unthinkable they become. Incapable of seeing the planetary context in all its complexity, blind intelligence fosters unconsciousness and irresponsibility.* (Morin 1999: Ch. 7)

For Morin, the problem with reductionism is not certainty, but its failure to generate a reliable enough understanding of reality as the basis for imagining transitions to more sustainable futures. By contrast, many African intellectuals are so terrified of any intent to change things for the better because of the horrors these intentions have generated in the past, any claim to understand reality in order to change it is automatically regarded with intense suspicion. Strangely, therefore, this fear of change can unintentionally reinforce the status quo by raising doubts about *any* claim to be able to know or create a better future.

The journey beyond certainty has not resulted in a new consensus which could become the basis for a new mobilisation and new hope. For some, this is a good thing — we are not so sure. For Manuel Castells, the towering contemporary philosopher whose three-volume work on the state of the world after the Information Revolution defined the analytical perspective of a generation, argues that at best we will enter the twenty-first century in a state of 'informed bewilderment' (Castells 1997). Much of social theory goes further: it has degenerated either into deconstructionist postmodernism, or a cynical belief in a very crude form of economic liberalism which has forgotten about the ethical foundations emphasised by the classical liberals. For the neoliberals, virtue flows from the maximisation of advantage in the market. Postmodernism has, of course, prompted a reaction, captured most aptly by Raymond Tallis (Tallis 1997) in his book appropriately entitled *Enemies of Hope*, which is a diatribe against the 'semantic turn' in the social sciences and a call for a return to methodological individualism and Cartesian rationality. Neoliberalism, on the other hand, has been the target of sustained attacks by Marxists and social democrats for some time for similar reasons, namely the claim by neoliberals that — in the words of Margaret Thatcher — 'there is no alternative'.

Remarkably many contemporary social scientists have a limited appreciation of the impact of complexity theory in the natural sciences. Castells, for example, argues that the global economy and society have been reorganised as a grand set of networks, but he does not refer once to the huge body of theory about networks in the natural sciences. This, in our view, seriously hampers the search for a way out, which avoids hopelessness without returning to the false promise of certainty. Indeed, there are profound ethical implications for the notion that the quest for certainty is the greatest threat to democracy. We need to go further than many social theorists have gone — beyond the traditional boundaries set by the social sciences. This means taking a lead from contemporary theorists such as the French socio-philosopher Edgar Morin (Morin 1992), Chilean economist Manfred Max-Neef (Max-Neef 2005), the Indian eco-feminist Vandana Shiva (Shiva 2005), the work emanating from the Institute for Social Ecology in Vienna (Fischer-Kowalski 1998; Fischer-Kowalski 1999; Fischer-Kowalski & Haberl 2007), the work of the Chilean neuro-scientists Humberto Maturana and Francisco Varela (Maturana & Varela 1987), the critical complexity of Paul Cilliers (Cilliers 1998) and even the expanding scholarship associated with the prodigious work of Niklas Luhmann (Luhmann 1996).

Revisiting the scientific revolution

The most distinctive feature of the European Scientific Revolution of the sixteenth and seventeenth centuries, and of the subsequent eighteenth-century Enlightenment, was the split, within what up until then was generally called Philosophy, between theories of nature and theories of society. Building on the intellectual and cultural ramifications of the fifteenth-century Renaissance and sixteenth-century Reformation, the Scientific Revolution and eventually the Enlightenment established the foundations for what has generally come to be called the culture of modernity. Some key principles have emerged as central tenets of the culture of modernity, specifically a belief in progress, the power of reason, the primacy of the individual, the sanctity of empiricism, the unlimited universalism of scientific knowledge, and the virtues of secularism. Central to classical modernism was the great legacy of Galileo (1564–1642) — the notion that valid knowledge can be quantified and therefore measured.

The revolutionary Enlightenment intellectuals (with origins in Galileo's battle against the Catholic Church in defence of Copernicus) were struggling against a universalising cosmological order which was legitimised and institutionalised by the Church. Starting with Descartes (1596–1650) and ending with Habermas (1929–) today, they needed to assert that the mere act of thinking constituted an act of rebellion and therefore of self-identity — the famous 'I think, therefore I am'. This liberatory aspiration, centred in the endogenous rights that stem from being human, was captured in the African context by Kwame Nkrumah's *I Speak of Freedom* (1961) and Steve Biko's *I Write What I Like* (1972).

It was, however, Sir Isaac Newton who stitched together the various elements of the secular scientific understanding of natural phenomena. His basic and most profound claim was that it is possible to understand a given complex physical phenomenon by discovering — and then reducing its content to — the phenomenon's quantifiable component parts. A valid scientific explanation, in other words, was one that established the parts, distinguished them from one another, and then explained the whole in terms of these identifiable and quantifiable components. From a huge number of cases which confirmed the basic method, he was able to finally claim that all physical reality could, in the final instance, be reducible to three factors, namely gravity, molecules and time. Like the weights, hands and cogs of a clock, all three, for Newton, were constants. And because these were the only constants, everything else was explicable in terms of some manifestation of these three factors. Modern science came of age.

This grand act of reduction unleashed a phenomenal energy because it made possible a wide range of scientific endeavours which were able to assume that the basic assumptions of the paradigm were valid. In Kuhnian terms, as long as these assumptions remained 'generally accepted', scientific inquiry could proceed to focus on method and substance without too much concern about the validity of these underlying epistemological and ontological assumptions (Kuhn 1962). The results are well known and need no repeating here other than to say that this made modernity—and its associated revolutions, transformations and progressions—possible. These results include, of course, the entire body of political and ethical theory that underpins our collective commitments to human rights, democratic governance, rule of law and self-determination. It is these remarkable achievements that have made it so difficult to reflect on the consequences of the dark side of modernity.

Once the break from the iron fist of the Church and the definition of the scientific method had been established in the natural sciences, there was a certain inevitability that the most influential social theories would emulate this basic approach. Although there were competing scientific traditions that were open to other ways of knowing (for example, Goethe and phenomenology), the social sciences emulated the dominant tradition in the natural sciences via metaphor. The classic metaphorical transplant from Newtonian physics to social reality was achieved by Adam Smith: gravity became the market, molecules became individuals, and the constancy (and reversibility) of time became value. But the same applied to the ideological polar opposite, namely Karl Marx: class struggle instead of gravity, capital and labour in place of molecules, and the dialectic replacing time. And similar reductions applied for others: Geist through history for Hegel; the 'Leviathan' as the antidote for the Hobbesian notion that life in its natural state was 'nasty, short and brutish'; citizens and the social contract for Locke; bureaucratisation for Weber; monarchy and liberty for the anti-feudal revolutionaries, and so on. Finding a pattern of thinking about social reality which was similar to the pattern set by the natural sciences gradually became the primary intellectual ambition of social scientists because it was assumed that just as nature operated in accordance with a defined set of basic laws, so too could society be seen to operate in accordance with a defined set of basic laws. Without this ambition there would be no *social* science.

What was common to both the natural and social sciences was the assumption that physical or social reality could be comprehended only via the application of the (trained) rational mind, that language was constructed to reflect this reality, and that once the basic laws of a given phenomenon were discovered, all else could be explained. Although modern science cannot be reduced to scientific method, it is possible to argue that the evolution of these scientific methods determined the boundaries of what came to be regarded as scientifically knowable. Of course it was accepted that knowledge existed outside these boundaries, but this knowledge did not conform to the (largely quantitative) strictures of the scientific method.

The problem was that Newton's constants did not stay constant. Social science was betrayed by its role model. Thanks to quantum physics dating back to Einstein's time (but with earlier antecedents), Newton's basic building blocks turned out to be rather malleable and indeterminate. There was the famous experiment in which two electrons with opposite spins (positive and negative) were taken from the same atom, separated in space, and the spin on one reversed. When the spin on the other automatically and simultaneously reversed itself, scientists were faced with a profound mystery because the possibility of cause and effect mediated by time had disappeared. For Einstein, this was a logical impossibility and he insisted the explanation would be found in time. Einstein's faith in physics has been carried into the contemporary era by world-renowned ecologist, E.O. Wilson, whose critique of the split between the social and natural sciences ends by arguing that ultimately all natural and social phenomena are 'ultimately reducible' to the 'laws of physics' (Wilson 1998: 266). Wilson's text is the most sophisticated contemporary defence of reductionism as the basis for what he calls 'consilience' — a unified theory of nature and society as the basis for a sustainable future.

Others followed physicists such as Neils Bohr and David Bohm out of a reductionist world and came to accept inter-dependency as an inherent property of matter (for an overview of this period in intellectual history see Peat 2002). From this has flowed an extraordinary literature with key terms becoming increasingly popular today, such as non-linear dynamics, chaos, complexity, systems thinking, strange attractors, implicate order, quantum thinking and the butterfly effect. In essence, quantum physics gave to the world an image of reality that was about relationships rather than building blocks, and qualitative explanations of phenomena in terms of the interactions between the component parts rather than linear causal relations between a few primary parts and the rest (determinism). The entry of this literature into the humanities has been largely via management theory because it has been recognised for decades that reductionism is not at all useful when it comes to analysing the internal dynamics of human organisations. Quantum physics provided the metaphors for an anti-reductionist theory of organisational behaviour. It remains to be debated whether this metaphorical foundation for an entire field of study can continue to be viable. But this is a matter for a different discussion.

The living cell holds clues to so many contemporary dilemmas because it sits at a kind of epistemological intersection point. It holds the key to an understanding of evolution, and via bio-mimicry it may also hold the key to a sustainable future (Benyus 1997). But how it is conceptualised also has much to add to our search for transdisciplinary ways of thinking about transitions to more sustainable futures. For a long time after the discovery and elaboration of DNA in the 1950s, molecular biology was caught in a reductionist vice grip. The awesome explanatory power of DNA seemed to obliterate the need to even ask questions that a DNA-centred approach was not geared to answer, for example, about the role of enzymes and metabolic regulation. However, this too has begun to change. The renowned Warwick University mathematician, Professor Ian

Stewart, wrote a book with the telling title, *Life's Other Secret: The New Mathematics of the Living World* (Stewart 1998). What then is this 'other secret'? His response to the DNA-centred approach is as follows:

> *As a consequence [of this approach], we are in danger of losing sight of an important fact: There is more to life than genes. That is, life operates within the rich texture of the physical universe and its deep laws, patterns, forms, structures, processes, and systems... Genes nudge the physical universe in specific directions, to choose this chemical, this pattern, this process, rather than that one, but the mathematical laws of physics and chemistry control the growing organism's response to its genetic instructions.*
>
> *The mathematical control of the growing organism is the other secret — the second secret, if you will — of life. Without it, we will never solve the deeper mysteries of the living world — for life is a partnership between genes and mathematics, and we must take proper account of the role of both partners.* (Stewart 1998)

For many, the notion that a hard square science like mathematics is reconcilable with the soft gooey things that biologists deal with seems bizarre. However, Stewart is quick to point out:

> *If we are going to understand the second secret, we must begin by recognizing that biology is not the only science that has undergone a revolution... Physics and mathematics have also changed beyond recognition, becoming more powerful, more general, more flexible, and* a lot closer to the intricacies of life. *These advances offer radical new opportunities for uniting the biological and mathematical worldviews, at a time when there is a renewed and urgent need for just such a unification.* (Stewart 1998; emphasis added)

Stewart devotes the rest of his impressive book to a detailed analysis of the operation of the 'second secret' via the application of complexity theory, and succeeds in providing a mathematical description of the physical and chemical context of cellular processes.

It is significant, however, that one of the world's leading scientific initiatives aimed at replacing a DNA-centred with a cell-centred approach is based here at the University of Stellenbosch. Under the leadership of Professor Jannie Hofmeyr in the Biochemistry Department, the so-called Triple-J Group for Molecular Cell Physiology is engaged in a project which aims to build an 'integrative theory of how the molecular economy in living cells is organised, controlled and regulated'. By breaking from DNA reductionism, this group aims to construct an explanation of cellular behaviour which is rooted in complexity. As Professor Steven Oliver of Manchester University argued in a review of their work in *Nature*, if they succeed this 'could mean that biologists in the twenty-first century need a rethink of their view of cellular economy that is every bit as radical as that initiated for political economy by John Stuart Mill and William Stanley Jevons in the nineteenth century.' (Oliver 2002)

In short, like physics in the 1930s, genomic science as the premier science of the early twenty-first century may be facing its moment of truth: either remain wedded to reductionism and blindly pursue the false promises of genetic determinism, or open up to a complex systems approach which not only makes the whole cell the focus of research, but connects this knowledge of the secret to all life to both evolution (as the history of life) and sustainability (as memory of a future for life). How the Kuhnian revolution in biology pans out will have a direct impact on how we understand ecosystems and life, and therefore how we negotiate the transitions to more sustainable futures across different contexts (for a review of these paradigmatic shifts in biology, see Capra 2002; Keller 2000).

Learning from complexity theory

We have tracked the story of reductionism as it has played itself out in the natural and social sciences and suggested that the alternative lies in an appreciation of complexity. We have not, until now, made explicit the theoretical dimensions of complexity thinking. We argue that it provides social science with a way forward that avoids the extremes of hopelessness and certainty, and brings the natural sciences closer to what Ian Stewart so delicately described as 'the intricacies of life'. But these are not separate movements — they depend on and feed off each other. As social scientists are realising that future social transformations will be determined and constrained by sustainability challenges, they are learning about the dynamics of natural systems (including evolution) from their colleagues in the natural sciences. As they do, they are discovering a new language for comprehending social reality which may revitalise their disciplines (as has already begun). Equally, as the impact of unsustainable practices starts to affect the bulk of humanity as negatively as it does the poorest two billion, the work of social scientists becomes key to the survival of all species, because it is the social scientists who have accumulated knowledge about cultural change, social transformation, collective behaviour, organisational change, conflict and socio-psychological responses. However, the chances of these mutual syntheses occurring in an ethical way are substantially reduced if this transdisciplinary endeavour is not rooted in an appreciation of the deep connections between the human species and natural systems (Costanza 2009; Costanza 2003). Evolutionary thinking creates a profound sense of the long-term histories of complex systems, how they adapt and how fragile they are (Margulis & Sagan 1997). An understanding of evolution gives humanity its memory of potential futures and shapes the way we manage ourselves into the future (Costanza 2003).

Complexity theory is not a single body of thought which stems from a clearly identifiable central source. Morin refers to two twentieth-century 'scientific revolutions' that laid the foundations of complexity science. The first is the Second Law of Thermodynamics that successfully challenged the absence of time in Newtonian science. Over time, as entropy increases, energy is dissipated which means the 'arrow of time' is irreversible. This — in the language of social scientists — is effectively about recognising that context (that is, the operation of a law within specific conditions) is not extraneous, but intrinsic to the operation of a system. From this flows the logic of the quantum world and the emergence of a new form of scientific explanation in terms of irreducible relations between the component parts of a system, as an alternative to reductionism (that is, explaining the whole in terms of the properties of a few selected components). The second revolution was the regathering of disciplines to understand the dynamics of the Earth and eco-systems when it became increasingly apparent from the 1970s onwards that unsustainable resource use, driven by reductionist disciplinary science, was threatening human and other life (Morin 2005).

Like the process it is best at explaining, complex adaptive systems are themselves an emergent condition. This is also often confused with chaos theory. However, we agree with our colleague, the late Paul Cilliers (from the Centre for Complexity Studies and the Philosophy Department at Stellenbosch University) who argues that chaos theory and complexity theory may overlap, but at the core they are fundamentally different. Chaos theory still tends to look for a patterned 'order' inherent in the effects of seemingly repetitive functions and, therefore, has not fully broken from the reductionist aspiration to find primary determinants. We have used Cilliers' internationally renowned work to elaborate some of the key elements of complexity theory (Cilliers 1998).[2] These are the following:

1. Complex systems comprise a large number of diverse elements that in themselves can be simple. Put another way, many seemingly simple elements or transactions can interact in ways that generate an extremely complex system. The whole is, therefore, more than the sum of its parts.
2. The interactions between the elements are non-linear. This means they interact dynamically in richly textured patterns by exchanging energy or information. Even if only some of the elements interact with others, the effects are propagated throughout the system. The results are non-linear because the dimensions of these effects cannot be predicted with certainty, with distinct possibilities that the effects of the same causes may differ from context to context, and that even in the same context, consecutive effects generated by the same causes may suddenly differ and appear to be completely different or out of all proportion to the effects that came before.
3. There are many direct and indirect feedback loops operating simultaneously all the time. This makes it impossible to identify a simple linear cause-and-effect relationship. When multiple effects become multiple causes, it becomes impossible to assume *a priori* that any one cause has greater explanatory weight than any other. This is why the specificities of context become so important. It is only within a particular context that there can be greater certainty (but never absolute certainty) about the number — and nature — of feedback loops that may be in operation at any point in time.
4. Complex systems are open systems. This means they continuously exchange energy or information with other systems located in the external environment. This continuous throughput of energy and information entering the system from

[2] Cilliers identified 10 principles of complexity whereas we have described seven. We have taken the liberty of condensing his 10 principles into seven, which we believe still captures the essence of his argument.

external sources means that complex systems operate at conditions that can be described as 'far from equilibrium'. A system in equilibrium is dead because there is no energy or information throughput. It is alive when throughputs are active, and because it is alive it is said to be far from equilibrium yet simultaneously remarkably stable.

5. Complex systems have a memory which is held by the system as a whole. No single element of the system has exclusive control of or access to the memory. It is this distributed memory that makes it possible for complex systems to have a history, which, in turn, is a critical determinant of the system's future behaviour.
6. The nature and behaviour of the system is determined by the quality of the interactions between the elements, and not by the properties of any one or more of the elements. Because these interactions are dynamic, fed back, rich, embedded in memory and, above all, non-linear, the behaviour of the system cannot be predicted by reference to the nature of any of its elements. This is why outcomes that do result from these interactions are referred to as emergent properties and the process of getting to these outcomes as emergence. Water, for example, is not made simply by adding hydrogen and oxygen together — how this is done (that is, context) is what determines the outcome. Similarly, democracy is not about regular elections, but rather the quality of all the social and political interactions between elections. Although this way of thinking disallows reductionism, and therefore deterministic forms of prediction, causality still exists but in this case as sets of probabilities with actual outcomes dependent almost entirely on context.
7. Complex systems are inherently adaptive. They can organise and reorganise their internal structures and operations without the intervention of an external agent. Invariably, however, they do reorganise in accordance with their own dynamics and logics in response to an external intervention, but not necessarily aligned to the intentions or logic of the external intervention.

The impact of this core body of theory has had a major impact across a wide range of social and natural science disciplines (Byrne 1998; Cilliers 1998; Costanza 2009), including genetics (Keller 2000), biology (Kauffman 1995), ecology (Gunderson & Holling 2002; Kay *et al.* 1999; Scheffer 2009), organisational science (Maguire & McKelvey 1999), economics (Arthur *et al.* 1997), and public management (Kooiman 1993; McCarthy *et al.* 2004).[3] We have used complexity perspectives to reconceptualise the dynamics of African cities (Swilling *et al.* 2003) while others have used it to rethink cities in general (Allen 1997; Byrne 2001; Uprichard & Byrne 2006). Leading scientists from South Africa's Council for Scientific and Industrial Research (CSIR) in collaboration with academics from the University of Stellenbosch have used complexity theory to rethink various dimensions of the scientific endeavour within a Southern African context (Burns *et al.* 2006; Burns & Weaver 2008). Others have used it to rethink the dynamics of the

[3] See also the special edition of the journal, *Public Management Review* (2008) on the theme Complexity theory and public management, 10: 3.

entire global eco-system that gives rise to Gaia Theory (Lovelock 1979) and the notion of a 'micro-cosmos' (Margulis & Sagan 1997). Social scientists have used complexity to re-imagine a post-corporate developmentalism (Korten 1995; Korten 2006), propose a future for human development in relationship with planetary systems as the basis for a new 'politics of humanity' (Morin 1999), reconceptualise a sociology of communicative structures and processes (Luhman, 1995), and develop a new theory of socio-ecological transition (Fischer-Kowalski & Haberl 2007). Complexity science plays a central role in most theories of transdisciplinarity (Nicolescu 2002) and it is also central to a large volume of influential science writing about ecological sustainability (Capra 1996; Capra 2002).

As complexity thinking permeates more and more disciplines, it can become a new orthodoxy. Reflecting on this problem, renowned complexity thinker Isabelle Stengers concludes:

> *Finally, I would stress that the science of complexity also needs rainbows, in order to escape the temptation to power, in order to keep alive the learning process and the capacity not to turn a surprise into a triumph.* (Stengers n.d.)

Like Bruno Latour, Stengers says it means nothing to describe a given system as 'complex'. What matters is the empirical challenge of finding the actual connections within a system, examining them up close to find out how they relate to one another within a particular context, and working out what effects these interactions have. There is no substitute, in other words, for hard work. Taken to its extreme, this means that one cannot assume in advance of hard work that a system is complex. What complexity thinking does is invite the formulation of questions about specific contexts that previously could not be asked. As soon as complexity thinking is used to generate the answers in advance, it will have succumbed to the 'temptation to power'.

Evolution and sustainability

Many textbook histories of science will place Charles Darwin up there at the start of it all, sitting comfortably on the pedestal with Sir Isaac Newton. Somehow Newton's revolutionary secular image of the universe as a vast mechanical clock was confirmed by the notion that the literal biblical image of creation was a myth. The Newtonian scientific method also seemed to be confirmed because there were obvious constants that could serve as primary determinants in a grand reductionist theory of the evolution of all species: molecules became species, gravity became natural selection, and time became this grand epochal evolutionary progression from the least to the most evolved species, namely 'man'.

There was, however, an addition to this basic schema which substantially crudified the Darwinian image of evolution, and this has got much to do with the influence of Hobbesian social theory and the tradition of methodological individualism to which this gave rise. The addition to which we have referred is of course the popular distortion of the notion 'survival of the fittest'. This was grabbed and used by the new believers in the virtues of the market, and by racial supremacists to prove that their conception of rabid individualism and selfish competition was perfectly 'natural'. For social Darwinism, the wealthy capitalist or (white) colonial master was deemed to be the outcome of an inherently natural process. This resort to the 'natural order of things' survives today as a mass cultural belief that free market capitalism is a perfectly natural form of social organisation superior to any other and in the obstinate persistence of most forms of racism and gender oppression.

In 1894 Thomas Huxley delivered his famous lecture entitled *Evolution and Ethics*. Huxley articulated the classical view that remains pervasive to this day that human nature is essentially evil. His argument was that human nature is the product of evolution, which he assumed was a natural survival-driven process which conditioned all species to be inherently selfish, competitive and nasty (that is, incapable of acting for a greater communal good). In short, there is no morality as we know it in nature. Instead morality and ethics were constructed by humans (read 'white Anglo-Saxon males') to combat natural selfishness for the sake of a wider social good. The result of this line of logic was that ethics has nothing to do with evolution — indeed, ethics arose to save humans from their essentially evil natural selves (as represented by the 'primitive civilisations' that pre-colonial and colonial travelogues and stories of conquest were revealing from all around the world). Richard Dawkins, with his image of the 'selfish gene', is the contemporary spokesperson for this view, albeit stripped of its distasteful racist history and legitimised by sophisticated contemporary research which spans the scales of analysis from cells to global cultures (Dawkins 1976).

From an African perspective, Huxley's views had immensely destructive consequences. They reinforced the already existing presentation of African societies in colonial literature as 'still in a state of nature' and therefore lacking the capacity for self-governance, based on the kind of ethical systems that were deemed 'civilised' from the perspective of the coloniser. This, in turn, justified treating Africans like animals who (like the rest of 'nature') were equally worthy of torture (echoing Francis Bacon's description of the purpose of science, which is to 'torture nature's secrets out of *her*') and destroying what were, in fact, sophisticated ethical systems for effectively managing human affairs and the relationships between humans and nature. Like all human systems, they never worked all the time and they could be corrupted, but the point is that they existed and were often completely ignored by the colonisers (for a discussion of this theme with respect to the impact of European settlers on the amaXhosa in the Eastern Cape, see Swartz 2010).

The classical Huxleyian view will collapse if it is possible to demonstrate three things: firstly, that nature and its evolution over time is not simply about selfish competition but also about creative co-operation at both the microbial and social levels; secondly, that there is no single 'human nature' with a fixed set of properties which can be explained exclusively in terms of genetic evolution, but rather a multiplicity of human *natures* which are the product of genetic *and* cultural co-evolution; and thirdly, that human morality and ethics in general may well have evolutionary roots in nature — which means nature and humans cannot be depicted as inherently separate from one another.

Professor Lynn Margulis from the University of Massachusetts and her colleague, Dorion Sagan, have forcefully challenged the traditional competitive theory of evolution in their book entitled *Microcosmos: Four Billion Years of Microbial Evolution* (Margulis & Sagan 1997). Informed by a complexity perspective, these biologists have tried to rethink evolution by synthesising what is known about the living cell in its various microbial forms with what is known about the historic origins of all living species, in particular the hominids and their various predecessors. The result is a detailed, empirically justified account of evolution from the perspective of the microbial world, where evolution is not the story of a linear progression from the first bacteria four billion years ago to 'man' as nature's highest creation, via a brutal competitive struggle for survival, but rather it is the story of an emergent mosaic of all life forms that exist in partnership and interdependence with one another — a cosmic conception not dissimilar to representations of universal history that exist in numerous pre-colonial African cosmologies, Asian religions (Taoism, Buddhism) and various so-called 'indigenous cultures' (in Australia, North America and Latin America). This new scientific image allows them to argue that:

> *... the view of evolution as chronic bloody competition among individuals and species, a popular distortion of Darwin's notion of 'survival of the fittest', dissolves before a new view of continual co-operation, strong interaction, and mutual dependence among life forms. Life did not take over the globe by combat, but by networking. Life forms multiplied and complexified by co-opting others, not just killing them.* (Margulis & Sagan 1997: 29)

Margulis and Sagan argue that this reconceptualisation of the underlying dynamic of evolution is made possible by three scientific breakthroughs. Firstly, the discovery of DNA and how it replicated. Secondly, the discovery that natural genetic engineering has been happening for billions of years. Cells can transfer bits and pieces of genetic material from one gene into the other which, in turn, allows the cell to do things that it would not otherwise have been able to do — this being particularly useful when it becomes necessary to respond to new environments. Thirdly, the discovery that mitochondria within the cell, which have their own DNA, exist outside the nucleus of the cell, reproduce at different times to the rest of the cell, and enable the cell to use oxygen in ways that would otherwise make it impossible for the cell to live. In short, mitochondria were once separate oxygen-breathing bacteria that became part of new cell formations via an evolutionary process which Margulis and Sagan call symbiosis. As they conclude: 'Symbiosis, the merging of organisms into new collectives, proves to be a major power of change on Earth' (Margulis & Sagan 1997: 32).

DNA, recombination and symbiosis make it possible for Margulis and Sagan to apply their conception of co-operative microbial evolution to the evolution of all species up to the present. To this extent they have pioneered a rewriting of evolution which depicts nature as a vast, creative, pulsating and complex network which looks very different to the image of selfish competition that Huxley and the contemporary neo-Darwinists have in mind.

To examine the second assumption in Huxley's worldview, namely that there is a single human nature created by natural evolution, we need to turn to another renowned biologist, namely Stanford University's Professor Paul Ehrlich. Ehrlich's book, *Human Natures: Genes, Cultures and the Human Prospect,* sets out to prove two things: firstly, that there is no such thing as a single human nature, but rather a multiplicity of geographically dispersed human natures; and secondly, that there is no evidence that these human natures can be explained exclusively in terms of genetic evolution (Ehrlich 2002). His primary target is genetic determinism and, in particular, the popular image that genes make us what we are. Like Margulis and Sagan he rewrites the story of evolution, but this time by tracing both the evolution of the natural species and the cultural evolution of the pre-human and human species. His aim is to demonstrate that there is now a multiplicity of different human natures that are a product of genetic and natural 'co-evolution' processes which manifested themselves in different ways in different localities and at different times. He demonstrates that there were moments when genetic evolution drove cultural evolution (for example, in the transition to *homo sapiens sapiens),* and when cultural evolution has resulted in a fundamental reshaping of the physical environment (for instance, the agricultural revolution after the last ice age over 10,000 years ago). He does not insist that the two proceeded in perfect tandem, but rather traces how they matched and mismatched in time and geographic space. His conclusion, however, is that cultural evolution is lagging now, with negative consequences for the sustainability of human society and the natural system. In his words: 'The increasing human ability to *do* things has outstripped the evolution of our ability to *understand* both what we should be doing and the full implications of what we are now doing' (Ehrlich 2002: 281).

Ehrlich has therefore eliminated the notion of a single human nature and softened the division between ourselves and nature by suggesting that some of our social and anti-social cultural traits have evolved from species that preceded us. To this extent he has placed our ethical choices as a species firmly back into the context of evolution, conceived as both a genetic and cultural process. He has also echoed the debate on the same subject in a collection of papers published in the *Journal of Consciousness Studies.*[4] In the primary contribution to this collection, Flack and De Waal argue that there is now evidence from observations of primate behaviour that there exists within these primate communities a sense of reciprocity through food sharing, conflict resolution capabilities, and a capacity for empathy, sympathy and consolation. If our primate relatives could choose to be co-operative (or not), how can we possibly assume that co-operative morality is a purely human invention? This suggests that modern rationality is not the only source of relational morality.

Ehrlich concludes by calling for a process of 'conscious evolution' which must entail 'interdisciplinary scholarship ... [so that] those who choose to tackle problems that cross the boundaries of the moment should not be punished, as they often are in academia today. The conservatism that was useful in the past is a luxury that society can no longer afford. Society also can no longer afford the split between the humanities and the

[4] See *Journal of Consciousness Studies* (2000) 7: 1–2.

sciences (the 'two cultures' of physicist C.P. Snow) or the marginalisation of philosophy' (Ehrlich 2002: 326; Snow 1993).

Reconceptualising evolution in a way that returns humans to their evolutionary place brings us back to the challenge of sustainability. As Margulis and Sagan argue:

> *... we can rescue for ourselves some of our old evolutionary grandeur when we recognise our species not as lords but as partners: we are in mute, incontrovertible partnership with the photosynthetic organisms that feed us, the gas producers that provide oxygen, and the heterotrophic bacteria and fungi that remove and convert our waste. No political will or technological advance can dissolve that partnership.* (Margulis & Sagan 1997: 16)

We have been, and always will be (if we survive), embedded in these natural systems. But we are also slowly destroying these life-support systems through human actions which cause ecological disasters, such as climate change, desertification, soil degradation, deforestation, biodiversity destruction, genetic modification and the misuse of scarce water resources.

As Chapter 2 will demonstrate, our development paradigms ignore this precious 'incontrovertible partnership' because we are exploiting natural resources at a rate that is greater than the capacity of natural systems to regenerate these resources, and we are dumping waste into natural systems at a rate with which they cannot cope. The blind persistence of both these trends is already breaking down the evolutionary partnership that Margulis and Sagan describe, at a cost which is, at this stage, carried largely by nature, the world's poor (who depend on this partnership for daily survival to a far greater extent than middle-class urbanites), and also increasingly by the middle class (who suffer the health dysfunctions of a poisoned environment and gluttonous diet). Erhlich is right, however: a genetically determined response to this crisis might be too slow in coming, and so it is up to the social sciences to rethink the cultural norms of contemporary societies so that they can be weaned off their dependence on life-destroying support systems. If this fails to happen, the richer countries (and richer groups in poorer countries) will use greater and greater force to gain more secure control of diminishing natural resources (see Chapter 7), and the suffering of the poor will inevitably get worse than it already is, which, in turn, will further destabilise global social and political systems. A sustainable society must be both more equitable in social terms and more respectful of the fact that human survival depends on the natural systems that have emerged from the evolutionary process. It is now increasingly obvious that this is not a trade-off: a more equitable world cannot possibly also be a less sustainable one. This brings us to a key question: can this 'incontrovertible partnership' become the basis for a new conception of progress?

Progress versus sustainability

One of the most vexed questions today is whether the modernist belief in progress is reconcilable with the vision of a more sustainable world (for a take on this debate, see Latour 2008). In order to rebuild the cultural foundations for a more sustainable global civilisation, social scientists need to rethink progress and the future goal of development as a set of human endeavours which are inescapably constrained by the finite nature of the ecological systems within which these endeavours are embedded. This is what sustainability science is all about. In the words of Wolfgang Sachs and an eminent group of analysts:

> *Without ecology there will be no equity in the world. Otherwise, the biosphere will be thrown into turbulence. The insight that the globally available environmental space is finite, albeit within flexible boundaries, has added a new dimension to justice. The quest for greater justice has, from time immemorial, required us to contain the use of power in society, but now it also requires us to contain the use of nature. The powerful have to yield both political and environmental space to the powerless, if justice is to have a chance.* (Sachs 2002)

Broadening out our conception of justice in this way sets up the framework for a new politics of eco-social (r)evolution. This section explores the implications of this statement by extending the discussion thus far about non-reductionist conceptions of science and evolution to an argument about the future of society and modernity within a wider commitment to a more sustainable future.

The twentieth century will go down in history as the century of high modernism, with the primary ideological contestations being the purpose and beneficiaries of progress — the ultimate *grundnorm* of modernity. Modernity refers to the various interpretations of human destiny and progress that emerged from the fifteenth- to sixteenth-century Scientific Revolution, and the subsequent eighteenth-century Enlightenment project that were premised on the assumption that science, rationality, individual liberty and the conquest of nature via science were sufficient to imagine, construct and make steady linear progress towards the ultimate promise of the Enlightenment, namely the 'good society'. '*Liberté, egalité, fraternité*'[5] were the great slogans of the French Revolution (1789–1799) — that epochal moment at which the monarchy, aristocracy and politically institutionalised religion were brought down in the name of a free citizenry.

For liberals and socialists, *egalité* was the ultimate end, but they differed fundamentally with respect to the scope and depth of equality across the political and economic spectrum. For liberals, *liberté* was primary, and *fraternité* a logical derivative, the sum — so to speak — of the political freedoms of the individual citizens that make up community. For socialists, *fraternité* became primary because, in their view, it was collectives which made history, thus constituting individuals in the name of the great historical collectives of the twentieth century (that is, nations, classes). From this divergence follows the great ideologies that shaped the twentieth century. *Liberté* runs like a golden thread through the founding theories (from John Locke to Jean-Jacques Rousseau, and John Stuart Mill to Adam Smith) and Constitutions that defined the so-

[5] Generally translated into contemporary popular English as liberty, equality and community (derived from 'fraternity').

called Western Liberal Democracies that emerged in the 300 years leading up to the end of the second millenium. *Fraternité* inspired generations of socialists, who either followed the social democratic way by emphasising collective interest, but without displacing the logic of the market, or agreed with the 1848 *Communist Manifesto* authored by Marx and Engels, which envisaged a post-capitalist world premised entirely on the values of the collective. For both social democrats and communists, it is collectives which make history, and therefore it is the collective interest that must be protected and advanced within the public sphere.

If modernism is the secularised cultural belief that society can make progress towards something better for all, then modernity is the practical manifestation of this belief. In the introduction to the first volume of the classic three-volume work on modernity edited by the British social theorist, Stuart Hall, it is argued that there are six characteristics which, in combination, define what is meant by a 'modern society', or 'modernity' (Hall *et al.* 1996):

- Secular forms of political authority that manifest themselves in the form of a territorially defined nation-state
- A 'monetarised exchange economy' based on private ownership of property, large-scale market-driven production and consumption of commodities, and extended accumulation of capital in private hands (in liberal democracies) or the state (in actually existing socialist societies)
- Decline of the traditional social hierarchies and the creation of a new class-based division of labour, cross-cut by reconstituted patriarchal relations
- Replacement of the religious worldview with a 'secular and materialist culture, exhibiting those individualistic, rationalist and instrumental impulses now so familiar to us'
- The rise of scientific knowledge as a new form of knowledge, managed by a professional elite, who claimed the ability to rationally classify, quantify and, therefore, comprehend all reality
- The emergence of a distinct conception of 'the social' as a domain of cultural life, existence and meaning that made it possible to construct 'imagined communities' (such as, a particular 'nation' — for example, the French nation or the Kenyan nation, as well as subnational group and transnational class identities).

Reflecting the paradigm of the social scientists who developed this text, what was missing was an additional criterion which could have been articulated as follows:

- The separation of nature and eco-systems from the logic of socio-economic development, with nature and eco-systems regarded as sources of unlimited resources which will always be extractable by modern technologies derived from scientific advances.

The absence of this additional criterion for defining modernity reflected the fact that *durabilité* (sustainability) was not the fourth slogan of the French Revolution. This, in turn, was possible because the founders of modern science could never have conceived that the seemingly unlimited bounties of nature could be overwhelmed by the forces of modernity they helped unleash. Liberals and socialists shared the modernist belief in progress and the right of humans to exploit nature, but differed fundamentally when it came to organising modernity. For the former, limiting the social contract to the political sphere and leaving the economy to be organised by the market were the conditions for a successful liberal democracy. For the latter, the social contract needed to penetrate beyond the political into the economic sphere to either socially coordinate the market (for social democrats) or to eliminate the market entirely (for classical communists). All socialists, however, shared the view that ownership of the means of production by the collective interest of the capitalist class needed to be recognised and exposed, rather than hidden behind the façade of individual political 'liberty' and 'equality'. Liberals and socialists embedded their belief in modernism and their conceptions of modernity within boundaries prescribed by the emergent nation-state (Anderson 1991). Notwithstanding the internationalist aspirations of many movements, from the anti-colonial revolutions mounted by the settler classes in the Americas in the 1700s, to the anti-feudal revolutions in Europe from the late 1700s, and onwards to the anti-colonial revolutions of the 1900s, starting with India (1947) and ending with South Africa (1994), the nation-state has been the preferred framework for contesting the content and purpose of modernity. For some, this founding norm has started to crumble in the face of an economic process that is rapidly reconstituting the role of the nation-state in the regulation of globally embedded national economies. But it would be a mistake to assume that the nation-state has been transcended as a key institutional reference point for guiding the future course of natural and social history (Hobson & Ramesh 2002).

The African anti-colonial movements of the twentieth century culminated in self-determination (although not stable liberal or socialist democracies) for most British and French colonies by the 1960s, the Portuguese colonies by the mid-1970s, Zimbabwe by 1980, and South Africa in 1994. The African intellectuals who shaped these movements and crafted the post-colonial visions of democracy and development were united by modernism and their aspiration to replicate one or other version of modernity on the African continent. Their claims to self-determination were continuously legitimated by appeals to the United Nations Charter, and with reference to the founding intellectuals of the Enlightenment. Virtually without exception, intellectuals-cum-political leaders such as Nkrumah, Nyerere, Sékou-Touré, Azikiwe, Nasser, Cabral, Machel and various generations of leaders in South Africa have merged strands of modernism with particular conceptions of Africanism and universal commitments to self-determination.

Given that Africa has been consistently pillaged for its natural treasures for over 500 years by Arabs, Europeans, Americans and African elites, starting with slavery followed by formal colonial annexation for the purpose of raw materials extraction and markets for consumption, it is not surprising that African intellectuals gave primacy to self-determination within a nation-state framework. Thus, for them, their struggles were akin to the anti-colonial movements that resulted in the establishment of independent republics in South and North America from the seventeenth century. The boundaries imposed by the colonial powers were the most significant defining feature of these nation-states. Within the parameters of African Nationalism inscribed within these boundaries, African intellectuals and leaders articulated various permutations of modernity, from Nyerere's 'African Socialism', to Nkrumah's 'Pan-Africanism', to Machel's state-centric 'people's democracy', and the rights-based liberal democracies created in Botswana and South Africa. The founding values of the liberated 'African Nations' rested on an all-pervasive belief in the possibility of progress determined via reason and achieved via the tried-and-tested instruments of liberal or socialist governance. These values were Africanised to an extent via appeals to 'Pan-Africanism', which, in turn, was rooted in an overwhelming sense of a shared continental destiny and history, and a common experience of everyday life. This occurred despite vast differences in culture, religious beliefs, political traditions and the accumulation strategies of local elites. The formation of the African Union after South Africa's democratic transition in 1994 suggests that the integrative 'Pan-African' discourse has survived politically in spite of the disintegrative dynamics that pervade African societies and economies. Both discourses co-exist now, forever intertwined in remembrance of the inherent limits of the promise of modernity.

Of course, '*durabilité*' was not the fourth slogan of the French revolutionaries, nor was it a central concern of the African revolutionaries 200 years later. Nature and eco-systems for most of them — and for intellectuals who have articulated the modernist vision across diverse contexts — were regarded as unlimited resources and the domain of the 'natural scientists', whose job it was to construct the knowledge required to realise the modernist vision. The so-called 'progress of science' became synonymous with the 'progress of humankind'. Maybe it was William Blake and the romantics who held out a lonely voice for some sort of nostalgic connectionism to nature and creation. As Paul Ehrlich so aptly observes, as a species we evolved to see, smell, hunt and kill an agile and strong quarry within a fixed unchanging environment that our cosmologies told us was God-given and permanent (Ehrlich 2002). Unsurprisingly, we find it hard now to think that at the peak of our powers, this contextual setting for everyday activities may be disintegrating under our feet and above our heads.

Modernity in its various manifestations across formations and eras has been about converting, over time, the raw materials and energy found in nature (so-called 'natural capital') into social and economic capital in the name of the 'good society', which was supposed to work for the benefit of all. But this is a process which predates modernity. For Fischer-Kowalski and her colleagues, the history of successive 'socio-ecological regimes' (hunter-gatherers, agrarian, industrial, sustainable) can be understood as the history of 'specific fundamental pattern[s] of interaction between (human) society and natural systems' (2007: 8). They go on to argue that:

> *[i]f we look upon society as reproducing its population, we note that it does so by interacting with natural systems, by organising energetic and material flows from and to its environment, by means of particular technologies and by transforming natural systems through labour and technology in specific ways to make them more useful for*

> *society's purposes. This in turn triggers both intended and unintended changes in the natural environment to which the society reacts.* (2007: 14)

To believe that modernity will remain the cultural frame for imagining the future, it will be necessary to assume that the desired end point of progress is a globalised material modernity for all, and that there are sufficient resources and materials to make this happen without a fundamental restructuring of the economic and political power structures that defend and manage the existing flows of natural and financial resources. This, in essence, is what lies at the centre of the audacious neoliberal vision articulated so clearly by Fukuyama at a time when this project believed it could conquer all (Fukuyama 1992). To validate this vision, it is necessary to assume that existing technologies are basically adequate, and that somehow 'science' will generate the solutions (from genetically modified organisms, to desalination, nuclear fusion, CO_2 capture, and solar power). However, for most of those engaged in one or other branch of sustainability science, this is a project that must collapse in the face of two harsh realities: hard evidence (see Chapter 2) that growing, modern economies and the large, emerging economies have already started to face the consequences of global warming, dwindling raw materials and rapidly degrading eco-system services; and gross and growing inequalities that result in the daily, premature deaths and excruciating suffering across a sixth of the world's population (and growing). Furthermore, these are not unrelated conditions — there is mounting evidence that the increasingly desperate scramble for dwindling resources is exacerbating poverty in many regions, which in turn is the basis for a rising number of failed states and resource wars. Will deepening poverty and eco-system destruction overwhelm modernity, or will these twin crises trigger the renewal of modernity? Is unsustainable modernity irredeemable? Are we witnessing the death of modernity? Or does modernity contain within it the clues for salvaging the promise of emancipation from suffering without destroying the natural means for our collective survival? Or maybe, like James Lovelock (Lovelock 2006), we should give up on modernity and sustainability altogether and focus rather on preparing for collapse? Maybe sustainable development is, in reality, little more than the retarded collapse of modernity? Did modernity ever, in fact, exist or was it part of the myth of our emancipation from nature?

The answers to these questions will hinge on whether the aspiration to progress can be separated from modernity. The latter has conflated the notion of progress with the exclusive fate of humankind so fundamentally that it is difficult to imagine how we can commit to a conception of progress which includes a future for the entire web of human and other life while retaining the paradigmatic trappings of modernism. There are three responses to this conundrum. Environmentalists have essentially given up on modernity, blaming it for the mess we are in — the so-called 'doom and gloom' school of thought. Progress for the human species, in this view, is impossible to imagine because existing trajectories are so path-dependent and so defended by vested interests that the chances of change on the required scale and timeframes are regarded as almost nil. James Lovelock must surely be the chief exponent of this view.

The polar opposite is 'ecological modernisation', which is premised on an unshaken belief in the capacity of modernity—and modernist science in particular—to turn ecological disasters into potential sources of future innovation and growth on terms that do not threaten the existing structures of economic power. In the words of the best reviewer of this trend:

> *Ecological modernisation (EM) is currently the dominant environmental social science, environmental sociology, or environmental policy theory... EM has been developed by environmental social scientists as a critical response to radical environmentalism and environmental movements such as deep ecology. It has shifted the focus from the failures of the state, industry and technological systems to address both environmental problems and success stories of environmental improvements. In other words, EM has provided a tool for those who believe in the continuing and sustainable progress of modernity.* (Korhonen 2008: 1331)

In this paradigm, human progress is made possible by giving a monetary value to eco-system services so that these can be calculated into the full costs and benefits of a particular economic policy. Specifically, a modern society becomes more sustainable if the rate of resource consumption, and therefore environmental degradation, is reduced as the rate of economic output increases.

Bruno Latour captures a third alternative which helps to liberate a conception of progress from the idea of modernity by questioning whether there was such a thing as a 'modern society' in the first place, and if not, what is it then that has suddenly become so environmentally unsustainable? (Latour 2008) Progress, for Latour, is not about a linear progression towards an ever greater condition of emancipation from nature, as envisaged by the modernist dream and implied by the criteria for modernity set up by Stuart Hall *et al.*, cited earlier in this chapter. Instead, for Latour, 'we were never modern' in the first place because progress has always been about *both* emancipation *and* ever demanding attachments to the 'imbroglios' of modern life, which the dream and the science simply denied. For many environmentalists and ecological postmodernists, these 'imbroglios' are so monstrous they prove there is no future for modernity. For Latour, progress is salvageable if we take responsibility for *both* the gains in human well-being *and* the unintended consequences that threaten to overwhelm us: 'If I am right, the breakthrough consists in no longer seeing a contradiction between the spirit of emancipation and their catastrophic outcomes, but to take it as the normal duty of *continuing* to take care of the unwanted consequences all the way, even if this means going always further and further down into the imbroglios' (Latour 2008: 11–12). 'Further and further down', we would argue, in order to rethink and redirect what we mean by 'emancipation'. Just 'taking care' of the imbroglios is not enough, as Chapter 7 on 'resource wars' so clearly demonstrates. This comes remarkably close to what Cilliers means when he talks about giving up 'abstract forms of meaning' and entering the 'agonistics of the network' (Cilliers 1998: 118). From this perspective, sustainability is not about giving up on development as the environmentalists have done, or simply 'greening' development as the ecological modernisers do; rather it is about agonistic engagements across diverse paradigms, disciplines and interests to help redefine what is generally understood by the notion of progress. Progress can be meaningful only if it includes the entire web of all life in ways that deal directly and boldly with the imbroglios that have engulfed modernity as we know it.

Conclusion

If we wanted to identify a twenty-first-century equivalent of C.P. Snow (Snow 1993), Paul Ehrlich (Ehrlich 2002) would be an ideal candidate. Like Snow, he came from the sciences and called for a dialogue between the sciences and humanities in order to be better prepared for the future. Like Snow, he wants the humanities to wake up and do something. But the differences could not be greater. Whereas Snow wanted the humanities to recognise the virtues of the unbound Prometheus of global modernisation driven by science, Erhlich is horrified by the consequences of allowing this Prometheus to remain unbound. Snow, in turn, would have been horrified by the scale of global poverty as the population grew from 2.5 billion in the 1950s to 6 billion by the turn of the millennium, with 2 billion living in poverty. For Erhlich (and others), the unbound Prometheus has become a threat to the sustainability of the environment and therefore the survival of the human species. He calls for a process of 'conscious evolution' inspired by the possibility of sustainability. We have suggested that complexity theory helps to create a language for building this culture of 'conscious evolution' because it charts a midway between certainty and hopelessness. But following Latour, this means starting by taking full responsibility for both what has been achieved and the imbroglios that now threaten our existence.

It may be worth concluding with Vandana Shiva, the tireless Indian academic and activist who wrote an article shortly before the 2002 World Summit on Sustainable Development.

> *How do we turn from the ruins of the culture of death and destruction, to the culture that sustains and celebrates life? We can do it by breaking free of the mental prison of separation and exclusion and see the world in its interconnectedness and non-separability, allowing new alternatives to emerge... We need once more to feel at home on this earth and with each other. We need a new paradigm that allows us to move from the pervasive culture of violence, to a culture of non-violence, creativity, and peace...* (Shiva 2002: 32)

The agonistic engagements are in place as we finally face the reality of our imbroglios — the movement of conscious evolution has begun.

Chapter Two

What is so Unsustainable?

Introduction

In 1987 the World Commission on Environment and Development (WCED), more commonly known as the Brundtland Commission after its leader, Gro Harlem Brundtland, published *Our Common Future* (World Commission on Environment and Development 1987). This report attempted to reconcile the ecological 'limits to growth'[1] articulated by the northern green movement since the early 1970s, with the need for growth to eliminate poverty, as articulated by developing countries in the south, many of whom had recently broken free from colonial control. This commission catapulted the term 'sustainable development' into global developmental discourse because it was a term which was able to capture—and for the optimists reconcile—the tension between the opposing interests of the north and the south. The most frequently quoted definition of sustainable development originated in this report: 'Sustainable development is development that meets the needs of the present without compromising the ability of future generations to meet their own needs.' Although this is a definition which is highly contested (Escobar 1995; Hattingh 2001; Sneddon *et al.* 2006), this extremely influential report provided the strategic foundation for the 1992 Earth Summit in Rio, the World Summit on Sustainable Development (WSSD) which took place in Johannesburg in 2002, and numerous international sectoral policy conferences between 1972 and 2002 (United Nations Development Programme—RSA 2004). These global events put in place the fragile, multilateral, global governance system, which is all we have today to face our collective global 'polycrisis'. Since the release of *Our Common Future*, we have learnt much about the challenges we face: numerous crises that we predicted—but done little to avoid—are starting to be noticed by mainstream centres across many nations in the developed and developing world (United Nations Environment Programme 2006; United Nations 2005; World Resources Institute. 2002; World Wildlife Fund 2008). This has given rise to a new literature on sustainability/sustainable development, and the emergence of a field formally designated as 'sustainability science' (Ayres 2008; for useful overviews, see Costanza *et al.* 1993; Dresner 2002; Hattingh 2001; Jacobson & Kammen 2005; Kates *et al.* 2001; Mebratu 1998; Pezzoli 1997; Sneddon *et al.* 2006). The first synthesis of a southern African perspective on sustainability science has also recently been published (Burns & Weaver 2008).

The documents that changed our view of the world

Seven globally significant mainstream documents and the themes they address will, in one way or another, shape the way our generation sees the world which we need to change. These are as follows:

---

[1] This was the title of a founding text of this movement, namely by Donnella Meadows (1972). *The Limits to Growth.*

1. ***Eco-system degradation***: the United Nations (UN) *Millennium Eco-system Assessment*, compiled by 1,360 scientists from 95 countries and released in 2005 (with virtually no impact beyond the environmental sciences), has confirmed for the first time that 60 per cent of the eco-systems upon which human systems depend for survival are degraded (United Nations 2005).
2. ***Global warming***: the broadly accepted reports of the Intergovernmental Panel on Climate Change (IPCC) confirm that global warming is taking place due to release into the atmosphere of greenhouse gases caused by, among other things, the burning of fossil fuels, and that if average temperatures increase by 2 °C or more this is going to lead to major ecological and socio-economic changes, most of them for the worse, and the world's poor will experience the most destructive consequences (Intergovernmental Panel on Climate Change 2007).
3. ***Oil peak***: the *2008 World Energy Outlook* published by the International Energy Agency declared the 'end of cheap oil' (International Energy Agency 2008). Although there is still some dispute over whether we have hit peak oil production or not, the fact remains that mainstream perspectives now broadly agree with the once vilified 'peak oil' perspective (see www.peakoil.net). Even the major oil companies now agree that oil prices are going to rise and alternatives to oil must be found sooner rather than later. Oil accounts for over 60 per cent of the global economy's energy needs. Our cities and global economy depend on cheap oil and changing this means a fundamental rethink of the assumptions underpinning nearly a century of urban planning dogma.
4. ***Inequality***: according to the UN *Human Development Report for 1998*, 20 per cent of the global population who live in the richest countries account for 86 per cent of total private consumption expenditure, whereas the poorest 20 per cent account for 1.3 per cent (United Nations Development Programme 1998). Only the most callous still ignore the significance of inequality as a driver of many threats to social cohesion and a decent quality of life for all.
5. ***Urban poverty***: according to generally accepted UN reports, the majority (i.e. just over 50 per cent) of the world's population was living in urban areas by 2007 (United Nations 2006). According to the UN-HABITAT report entitled *The Challenge of Slums*, nearly 1 billion of the 6 billion people who live on the planet live in slums or, put differently, one-third of the world's total urban population (rising to over 75 per cent in the least developed countries) live in slums (United Nations Centre for Human Settlements 2003).
6. ***Food insecurity***: the International Assessment of Agricultural Knowledge, Science and Technology for Development (IAASTD) (Watson *et al.* 2008) is the most thorough global assessment of the state of agricultural science and practice that has ever been conducted. According to this report, modern, industrial, chemical-intensive agriculture has caused significant ecological degradation which, in turn, will threaten food security in a world in which access to food is already highly unequal and demand is fast outstripping supply. Significantly, this report confirmed that '23 per cent of all used land is degraded to some degree' (Watson *et al.* 2008: Ch. 1, p. 73).

7. ***Material flows***: according to a 2011 report by the International Resource Panel (http://www.unep.org/resourcepanel), by 2005 the global economy depended on 500 exajoules of energy and 60 billion tonnes of primary resources (biomass, fossil fuels, metals, and industrial and construction minerals), an increase of 36 per cent since 1980 (Fischer-Kowalski & Swilling 2011).

The above trends combine to conjure up a picture of a highly unequal urbanised world, dependent on rapidly degrading eco-system services, with looming threats triggered by climate change, high oil prices and food insecurities. This is what the mainstream literature on unsustainable development is worried about. This is the growing shadow of modernity that has been denied for so long. This marks what is now increasingly referred to as the *Anthropocene*—the era in which humans have become the primary force of historico-geophysical evolution (Crutzen 2002).

Significantly, although these seven documents are in the policy domain they reflect the outcomes of many years of much deeper research on global change by scientists and researchers working across disciplines and diverse contexts on all continents (Steffen *et al.* 2004). Although this process of scientific inquiry leading to policy change is most dramatic with respect to climate science (Weart 2008), it is also true for the life sciences that fed into the outcomes expressed in the Millennium Eco-system Assessment (MEA), the resource economics that has slowly established the significance of rising oil prices and, most recently, the rise of material flow analysis (more on these later). The development of our ability to 'see the planet' has given rise to what Clark *et al.* have appropriately called the 'second Copernican revolution' (Clark *et al.* 2005: 6). The first, of course, goes back to the publication of *De Revolutionibus Orbium Coelestium* by Copernicus in 1530, but only 'proven' a century later by Galileo, who established by observation that Copernicus was correct when he claimed that the Sun rather than Earth was the centre of the universe. This brilliant act of defining the planetary system through observation was a—perhaps *the*—defining moment that paved the way for the Enlightenment and the industrial epoch that followed.

Clark *et al.* date the second Copernican revolution to the meeting in 2001 when delegates from over 100 countries signed the Amsterdam Declaration, which established the 'Earth–System Science Partnership' (Clark *et al.* 2005: 7). Like many similar gatherings since, this was a vast collaborative dialogical effort rather than the outcome of a founding genius. Using the advanced tools of 'Earth–system science' made possible by the ICT revolution and creative modes of social organisation, we now have the ability to see the planet as a vast living organism; as a complex system which we imperfectly understand in ways that are inseparable from our engagement with it as agents of change. The logical outcome of this profound paradigm shift is an increasingly sophisticated appreciation of what Rockstrom *et al.* have called our 'planetary boundaries' which define the 'safe operating space for humanity' (Rockstrom 2009). The significance of the Rockstrom article is that it managed to integrate, for the first time, the quantifications of these 'planetary boundaries' that had already been established by various mono-disciplines. These included some key markers, such as not exceeding 350 parts per million of CO_2 in the atmosphere; extracting 35 million tonnes of nitrogen from the atmosphere per year; an extinction rate of 10 per million species per year; global freshwater use of 4,000 km^3 per year, and a fixed percentage of global land cover converted to cropland (Rockstrom 2009: 473). Without the 'second Copernican revolution', a new science appropriate for a more sustainable world and the associated ethics (see Chapter 10) would be unviable.

Eco-system services

The Millennium Eco-system Assessment (MEA), referred to earlier, was called for in 2000 by the UN Secretary-General, Kofi Annan. The project, initiated in 2001 and funded by the UN Foundation with a US$24 million grant, presented its report in 2005. It is one of the great iconic documents of our time (this, despite the fact that hardly anyone beyond the environmental sciences has ever heard of it). The significance of the MEA lies not in its rather weak policy prescriptions,[2] but in the fact that it is the first comprehensive analysis of the relationship between what it refers to as 'human well-being' and 'eco-system health'. The report was a remarkable global collaboration between 1,360 experts from 95 countries. The research was peer reviewed by specialists drawn from UN agencies, universities and research institutes across the world.

The MEA analysed the following 'eco-system services' upon which socio-economic systems depend:

- ***Provisioning services:*** food (crops, livestock, capture fisheries, aquaculture, wild foods); fibre (timber, cotton, hemp, silk, wood fuel); genetic resources; biochemicals, natural medicines, pharmaceuticals; water.
- ***Regulating services:*** air quality; climate regulation (global, regional and local); water regulation; erosion regulation; water purification and waste treatment; disease regulation; pest regulation; pollination; natural hazard regulation.
- ***Cultural services:*** spiritual and religious values; aesthetic values; recreation and ecotourism.
- ***Supporting services:*** nutrient cycling; soil formation; primary production.

The MEA's four main findings were the following:

1. Over the past 50 years humans have changed eco-systems more rapidly than ever before in human history to meet demands for food, fresh water, timber, fibre and fuel. This has caused substantial and 'largely irreversible loss in the diversity of life on Earth'.
2. Although eco-system change has contributed to gains in human well-being, the costs are degradation of eco-systems, increased risk of non-linear changes, and increased poverty.
3. Degradation will get worse over the next 50 years and is a barrier to achieving the Millennium Development Goals (MDGs).
4. Reversing eco-system degradation is possible, but will require 'significant changes in policies, institutions and practices'.

[2] Unfortunately, and surprisingly, these prescriptions are informed by a naïve form of market economics. It is puzzling why the more mainstream perspectives of institutional economics were not used to frame the economic policy proposals.

Some examples of the conditions referred to in the MEA include:

- 60 per cent (or 16 of the 24) of the world's eco-system services have been degraded or are used unsustainably
- About a quarter of the Earth's land surface is now cultivated
- People now use between 40 per cent and 50 per cent of all available fresh water running off the land — water withdrawals have doubled over the past 40 years
- Since 1980, about 35 per cent of mangroves have been lost
- About 20 per cent of coral reefs were lost in just 20 years, and 20 per cent were degraded
- Nutrient pollution (generated mainly by chemically produced materials) has led to eutrophication of waters and coastal dead zones
- Species extinction rates are now 100 to 1,000 times above the so-called background rate.

Based on these and many other findings, the MEA argues:

> *The consumption of ecosystem services, which is unsustainable in many cases, will continue to grow as a consequence of a likely three- to six-fold increase in global GDP by 2050, even while global population growth is expected to slow and level off in mid-century... An effective set of responses to ensure the sustainable management of ecosystems requires substantial changes in institutions and governance, economic policies and incentives, social and behaviour factors, technology, and knowledge... Costs of unsustainable resource use are rising, but get displaced from one group to another (in particular the poor) and to future generations.* (United Nations 2005)

The significance of the MEA is that it asks us to look beyond a narrow focus on climate change. Although environmental groups, the media and global policy tend to focus on carbon and global warming, this is only one dimension of the polycrisis, which includes eco-system breakdown, depletion of oil and the degradation of key resources for everyday living, such as air quality and water supplies. The enduring impact of the MEA will be the notion that the eco-systems on which we depend for our survival are rapidly degrading and collapsing. The tragedy, of course, is that the first to suffer the consequences will be the world's poor as soils become less productive, water more expensive, life-giving resources such as air and water get polluted, fisheries collapse, climates warm up, deserts spread and forests get felled. Others will be affected, but they can often buy themselves time or more expensive means of survival.

Global warming

The Fourth Assessment Report (Fourth AR) of the Intergovernmental Panel on Climate Change (IPCC) published in 2007[3] confirmed the general trends of the previous assessment

[3] The Fourth Assessment Report was awarded the Nobel Peace Prize in 2007, together with Al Gore for his documentary *An Inconvenient Truth*.

reports, namely that global temperatures are rising, and that these temperature increases are due to an increase in concentrations of greenhouse gases in the atmosphere caused by human activities (Intergovernmental Panel on Climate Change 2007). The International Energy Agency forecasts that if policies remain unchanged, world energy demand is set to increase by 45 per cent by 2030 (International Energy Agency 2008). At the same time, since 1988 the IPCC has warned that nations need to stabilise their concentrations of CO_2 equivalent emissions, requiring significant reductions in the order of 60 per cent or more by 2050. In the Fourth AR, the IPCC argues that dangerous global emissions need to start declining by 2012–2013, and that by 2020 global cuts of 25 to 40 per cent are needed. By 2050, cuts of at least 80 per cent are necessary. The main human activities that have resulted in a 70 per cent increase in greenhouse gas emissions since 1970 are the burning of fossil fuels, deforestation and agricultural production. The projections for the future suggest that even if we act now to build low-carbon economies, temperatures will still rise by 2 °C. If we make only moderate changes along the lines envisaged by the Kyoto Protocol, we could face runaway global warming with devastating consequences (see Figure 2.1). Either way, it may be worth quoting a conservative source on the impact on the poor, namely Sir Nicholas Stern[4] who wrote in his report to the UK government:

> *All countries will be affected. The most vulnerable — the poorest countries and populations — will suffer earliest and most, even though they have contributed least to the causes of climate change.* (Stern 2007)

However, the Fourth AR has significance for this discussion of sustainability for two particular reasons. The first relates to its dire predictions for Africa, the continent least equipped to respond. The second relates to the admission that the solutions go beyond the scope of climate science.

The Fourth AR suggests that the African continent, which has contributed least to global warming, will be drastically affected by climate change. The main findings are that between 75 and 250 million people will suffer the consequences of increased water stress by 2020; by the same date, productive outputs from rain-fed agriculture could drop by 50 per cent with obvious negative consequences for food security; by the end of the twenty-first century, sea-level rise will have negatively affected most of the low-lying coastal cities around the coast of Africa; and by 2080, arid and semi-arid land areas will have increased by between 5 per cent and 8 per cent. There is little evidence that researchers and decision-makers in Africa have registered the full implications of the multiple impacts of global warming for the way in which development policies are designed in Africa.

The Fourth AR has made it clear that climate policy alone will not generate the required solutions. Making a clear link to the structure of the global economy and national economies, the Fourth AR argues that unless economic development policy choices are informed by the need for both mitigation (of GHG emissions) and adaptation (to the consequences of global warming), current trends and associated negative feedback

[4] Sir Nicholas Stern is a former Chief Economist of the World Bank, and was commissioned to write this report on the economics of global warming by Gordon Brown when he was still Chancellor of the Exchequer.

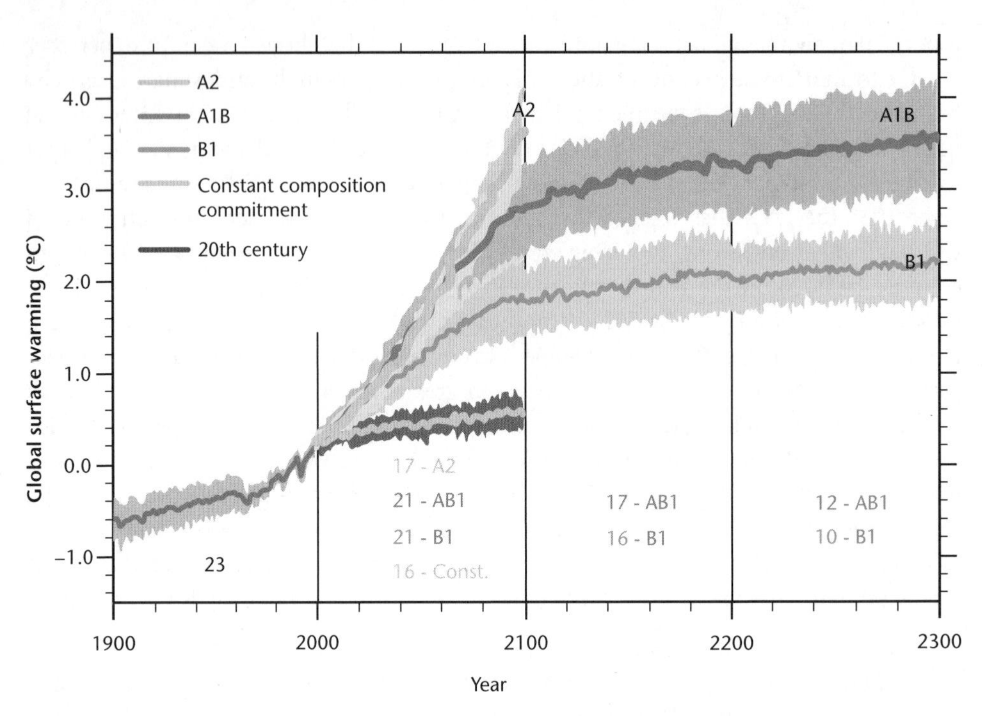

Figure 2.1: The rise in global temperatures as a result of greenhouse gas emissions

(Source: *Climate Change 2007: The Physical Science Basis*. Working Group I Contribution to the Fourth Assessment Report of the Intergovernmental Panel on Climate Change, Figure SPM.5. Cambridge: Cambridge University Press.)

Note: A2 = runaway global warming; AIB = limited changes—warming by 3 °C or more; B1 = warming limited to 2 °C (Copenhagen Agreement 2009).

loops will continue well into the future. In short, authentic sustainable development is seen as the key to significant climate change solutions. In particular, this will mean reviewing the regulation, financing, monitoring and strategic management of key economic sectors, in particular energy, forestry, agriculture, transport, construction and bulk urban infrastructure (water, sanitation, roads, energy and solid waste). Once again, these sectors are dominated by large corporates configured as a set of value chains which are designed, specified, financed and managed by people trained to think in ways that reinforce the logic of these value chains, and their personal material interests are tied to tried-and-tested technologies embedded in these systems. Changing these sectors is, therefore, far easier said than done. Change, when it comes, will be the outcome of some intense power struggles waged by a wide range of socially, economically and spatially constituted interests which will all be differentially affected by the increasingly serious ecological crises that will unfold as they discuss and negotiate. Sedate roundtable stakeholder discussions will play a role, but the real action will lie in other terrains of contestation and competition.

The United Nations Development Programme, an agency mandated to focus on development, decided that its 2007/08 Human Development Report will focus on climate change and the implications for 'human solidarity' (United Nations Development

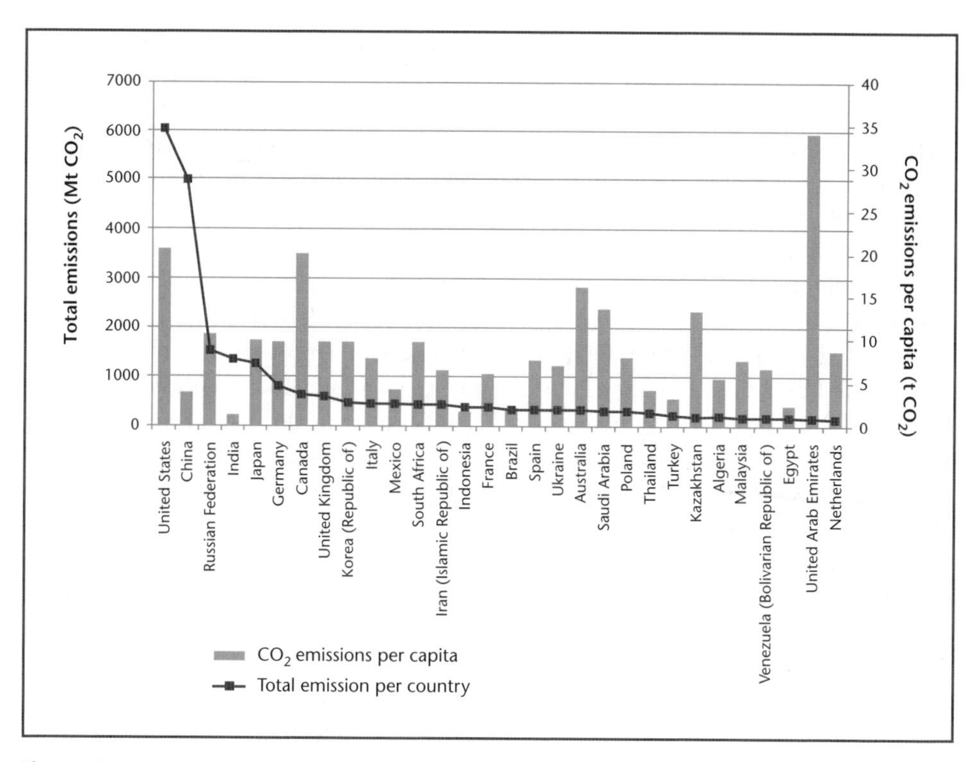

Figure 2.2: Measuring global carbon emissions per country, 2004
(Adapted from United Nations Development Programme 2007)

Programme 2007). Figure 2.2 reveals the state of play with respect to CO_2 emissions per country and per capita for each country. Significantly, three African countries have made it into the top 30, one of which is South Africa which is ranked number 12, despite the fact that it is a country of only 49 million people. Brazil, which is a fast-developing country like South Africa with four times the population, ranks number 16. South Africa's 'resource curse' of abundant coal and cheap electricity is the reason for its high total emissions level. Large developing countries, such as China and India, are in a different position: they insist that emissions must be measured on a per capita basis. This is not surprising. Although China is ranked number 2 in this list, it is now officially number 1 when it comes to emissions per country. However, it is number 25 when it comes to emissions per capita. Similarly, India is ranked number 4 when it comes to emissions per country, but number 30 when it comes to emissions per capita. Large populations are useful when it comes to per capita calculations. South Africa is ranked number 11 when it comes to emissions per capita. In other words, no matter on what basis the global deals are done (i.e. per country or per capita calculations), South Africa will be in the top 12. This has major strategic implications because it means that South Africa is unlikely to avoid carbon taxes in the near future as the world finds ways to pay for carbon sequestration and mitigation. What is also interesting is that developed countries, such as the UK, Japan, Germany, Italy, Australia and even Korea, have managed to grow economically over the period 1990–2004, but with marginal growth in CO_2 emissions — with Germany actually recording minus growth rates. South Africa, on the other hand, is ranked number 15 when it comes to percentage growth in CO_2 emissions for the period, growing by 32 per cent. Countries such as Venezuela, Brazil, China, Spain, Turkey, India and Egypt all grew within the 47 to 110 per cent range over the same period.

The significance of these figures is not simply the contestation over the basis for determining carbon taxes (per country or per capita), but also what they say about the relationship between growth in GDP per capita and CO_2 emissions. It is clear that this is not a fixed relationship. Policy interventions make a difference. Successive German governments have focused on increasing efficiencies to bring down total emissions with positive economic spin-offs and an improved quality of life. Japan, Italy and Korea show similar patterns. However, Brazil's CO_2 emissions per capita are very low, and have grown only moderately over this period, despite high economic growth as Brazil emerged as one of the world's top 10 industrial economies. South Africa has the worst of all worlds: high CO_2 emissions on a per country and per capita basis (similar to Russia), relatively moderate economic growth, together with the threat of carbon taxes that may prevent rapid increases in CO_2 emissions — perceived (incorrectly) by decision-makers as a constraint on much-needed economic growth to deal with poverty challenges. China and India, by contrast, have all the room they need to manoeuvre — all they have to do is to measure themselves on a per capita basis in order to justify massive investments in high carbon infrastructures (which, of course, will give them problems later on).

The Stern Report introduced a new principle into the discourse on sustainability, arguing that it would be cheaper to fix problems now rather than later — the later they are addressed, the greater the scale and complexity of remediation. The report recommended a commitment of 1 per cent of global GDP per annum to finance the changes that are required. Although this is substantial, the report argued that this is preferable to losing between 5 per cent and 20 per cent of global GDP if nothing is done. This may be why global elites and climate scientists talk about an annual expenditure of US$200 billion to finance mitigation. Although everything depends on who controls these funds and what exactly they are used for, we believe that the notion that fixing things now rather than later should become a guiding principle for sustainability.

Oil peak

There is, of course, a joker in the pack of resource cards, and that is the future of oil. Although roundly criticised by the climate change community, who argue we will run out of atmosphere before we run out of oil, the so-called 'peak oil' community are convinced that we either have already — or will soon — hit what they define as peak oil production (Aleklett & Campbell 2003; Campbell 1997; Darley 2005; Deffreys 2001; Goodstein 2005; Heinberg 2003; Kunstler 2005; Roberts 2005; Strahan 2007).

The notion that there is a point in the cycle of oil discovery and production that can be defined as a 'peak' is derived primarily from the US experience, but also other national contexts (for example, the Russian and North Sea oil fields). Oil discovery peaked in the US in the 1930s and production 30 to 40 years later. The oil peak protagonists use similar timeframes to predict global oil peak. In other words, we know when oil discovery peaked globally (in the 1970s/early 1980s), which means the production peak should be some time during the decade after 2005.

What is interesting about peak oil production is that it is only possible to determine whether production has peaked with hindsight. In other words, in the year that production peaks it is highly unlikely that oil companies will tend to predict a decline in the following years. Even if production does decline the following year, this could be seen as a one-to-three-year blip — it has happened before. In 2008 the oil price shot up to US$140 per barrel. Was this a sign of the peak or was it driven by speculators? The debate goes on, but higher oil prices do make it possible to capture more expensive resources, such as residues in old oil wells, tar sands, deep sea deposits and, as the ice melts, oil under the polar ice caps. The sceptics use these kinds of market dynamics to raise questions about the validity of predictions that we have either hit or are about to hit oil peak (Lynch 2003). Nevertheless, oil prices are rising as demand outstrips supply and even oil companies advertise the end of cheap oil and the coming post-oil era. As Figure 2.4 reveals, actual production remained flat between 2005 and 2010, despite significant price hikes. Although peak oil is the cause of much alarm in developed economies, it could spell disaster for emerging economies which have managed to find ways against all odds to play the economic growth game (Association for the Study of Peak Oil and Gas — South Africa 2007; for an application to the South African case, see Wakeford 2007). Figure 2.3 reflects the latest predictions from the Oil Peak Analysis Centre for peak oil production if all existing conditions remain equal.

It is important to emphasise that 'oil peak' does not mean the 'end of oil' as implied by some of the more hysterical voices in the popular media. The most significant consequence of 'peak oil' is the inevitable rise in the price of oil, despite increased investment to expand capacity. Based on a survey of 800 of the world's top oil wells, the International Energy Agency reversed its long-held view that peak oil is a myth by admitting that conventional crude oil production is going to peak sooner rather later (see Figure 2.4). Its 2008 *World Energy Outlook* reluctantly concluded that 'it is becoming increasingly apparent that the era of cheap oil is over' (International Energy Agency 2008: 15). With this simple little sentence, the body that represents the mainstream views of the oil industry told the world that everything we take for granted in everyday life will be fundamentally transformed.

Although cheap oil meets 60 per cent of the world's energy needs, production seems to have levelled off at around 85 million barrels per day. Oil has become ubiquitous in so many different ways. Most people simply associate oil with fuel for motor vehicles. Few realise that most of the polymer that goes into the plastics on which we depend is derived from oil,[5] as

[5] Examples include many middle-class household consumables, as well as packaging, bottles for various liquids, clothing, and an increasing percentage of built structures and motor vehicles.

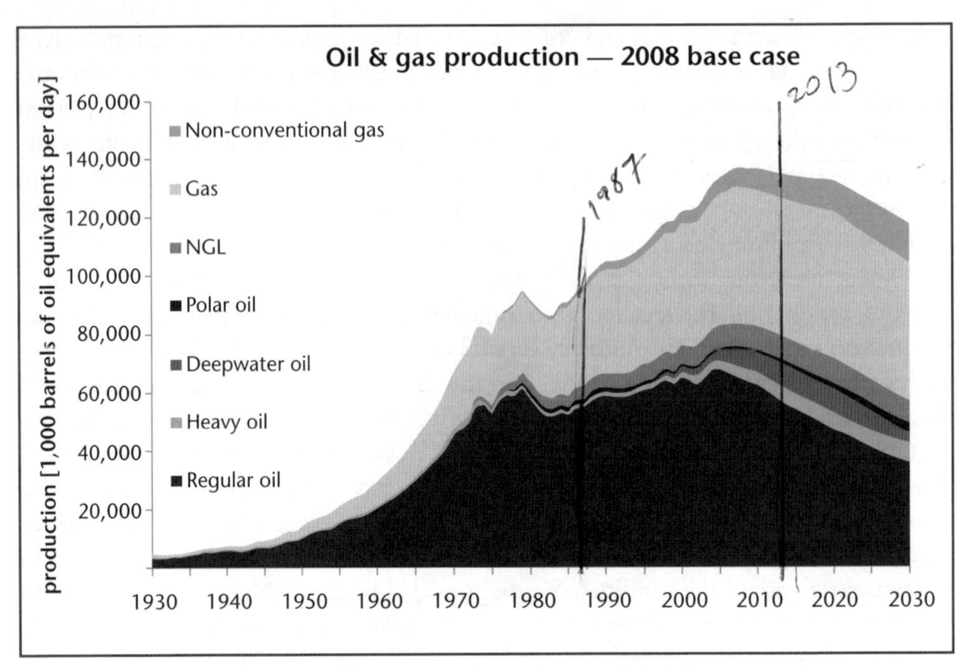

Figure 2.3: Predictions for peak oil and gas production if all existing conditions remain equal
(Source: Reproduced with permission from The Association for the Study of Peak Oil [ASPO])

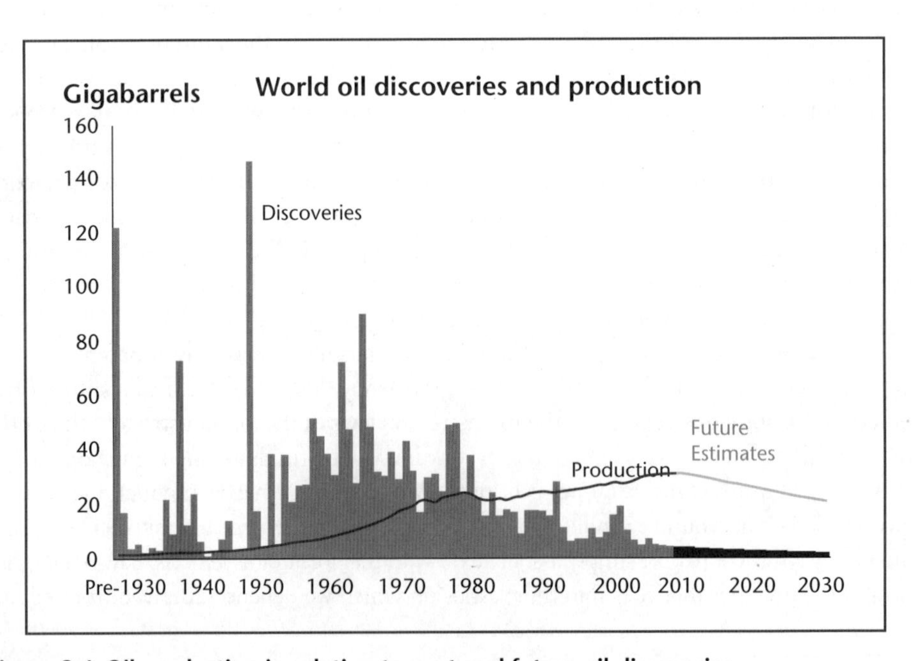

Figure 2.4: Oil production in relation to past and future oil discoveries
(Source: Reproduced with permission from The Association for the Study of Peak Oil [ASPO])

are nearly all the fertilisers and herbicides that are used to grow our food on commercial farms around the world. The energy used to produce the cement for our modern towns and cities[6] is oil dependent, and many countries burn oil to generate electricity. The rapid transportation of people and goods is assumed to be a cornerstone of the globally connected economy—without cheap oil, this would be impossible.

If oil production is peaking currently, it is happening at precisely the same time as the giant oil-dependent economies such as China, India, Brazil and Russia are all escalating their growth rates. Lester Brown calculated that if every Chinese consumed half the amount of oil that the average American consumed, China would need 100 million barrels of oil per day (Brown 2006). This alone would exceed total current global production. The billion or so people on the planet who live a middle-class life (20 per cent of the total population) and consume 58 per cent of all generated energy (United Nations Development Programme 1998) will probably experience the most drastic changes as they learn to depend less on fast, cheap transport and on the consumption of goods from all over the world. The world's working poor could also be drastically affected if the transition is crisis-driven and therefore potentially destructive of labour-intensive economies in the developed and developing world. The billion or so who live below the poverty line will be affected, but probably little will change for most of this group given that they consume only 4 per cent of total generated energy (United Nations Development Programme 1998).

Read together, the MEA report, the reports of the IPCC and the oil peak predictions all provide evidence that the natural resources and eco-system services that have hitherto sustained the phenomenal growth of the global economy are being rapidly depleted or degraded beyond the point of no return. The consumption levels of the 1 billion people who are responsible for 86 per cent of total consumption expenditure are the primary cause of this, even though it is their consumption that drives the global production system upon which so many jobs depend. The consequences will be felt by all, but particularly by the poor who are more dependent on the eco-systems on which they have depended for millennia and who often live in localities (particularly in Africa) with the least capacity to adapt.

Poverty and inequality

Poverty is now, possibly for the first time ever, at the centre of the global development debate. A contributing factor has been the mushrooming of academic and policy literature on poverty over the past two decades. This literature has had a strong quantitative bias, and has been inscribed into the Millennium Development Goals (MDGs) with reference to 'halving poverty by 2015'. Besides the obvious problem of reducing poverty to quantitative measures, poverty is often discussed as if it were unrelated to inequality. This is partly because the literature on inequality is substantially smaller, less

[6] Lime dug out from the Earth's crust is what is used to produce cement in kilns that need to be heated up to 2,000 °C. Oil and coal are the primary resources used for this.

influential and articulated in qualitative terms as a key element of the globalisation of the economy (Peralta 2003; Schaffer 2008). The *1998 Human Development Report* cited earlier stands out as a key exception because it managed to demonstrate its arguments about inequality during the heyday of neoliberal economics. From a sustainability perspective, however, poverty cannot be detached from inequality, especially when it comes to unequal consumption of finite natural resources and eco-system services. If the core principle of ecological economics is accepted, namely that resources are finite, then over-consumption by a few inevitably means less for the majority within a context of degrading natural resources and economic services (Hayward 2006). When the lens is widened from poverty to inequality, what emerges is not simply the quantification of poverty but also the power relations that preserve the global structures of inequality that ensure the continuation of poverty (Benn & Hall 2000; Hurrell & Woods 1995; Nederveen-Pieterse 2002; Peralta 2003; Shaffer 2008). The few that benefit from inequality also have the market power and political might to ensure both the legitimation and institutionalisation of the mechanisms that reproduce global inequality. The fallacy of 'trickle-down economics' has led many to believe the rather absurd notion that for the poor to become less poor, the rich must get significantly richer to fund the gradual increases in consumption by the poor (Kanbur 2009).

Debating how many poor people there are in the world, and whether one uses the US$1 or US$2 per day measurement has become a growth industry. It is fashionable in mainstream development circles to quote the World Bank estimate that there were 200 million fewer poor people in 1998 than there were in 1980. This stems from the World Bank's 2002 report entitled *Globalization, Growth and Poverty* (World Bank 2002). However, in an important (but rarely cited) World Bank research paper, two World Bank researchers (Chen & Revallion 2004) re-examined the available data for the 1981–2001 period and concluded that if the US$1 per day measurement is used, there were 390 million *fewer* very poor people in 2001 than in 1981. However, if you used the more realistic US$2 per day measurement, there were 286 million *more* poor people in 2001 than there had been in 1981. By 2001, just over 1 billion people were very poor (using the $1/day standard) and 2.7 billion were poor (using the US$2/day standard). Based on the work by Chen and Revallion, this means that in percentage terms, in 1981 33 per cent of the world's population of 4.5 billion were *very* poor ($1/day), whereas 53 per cent were poor ($2/day). By 2001, the percentage of very poor ($1/day) had dropped to 16 per cent of the global population of 6 billion, and the size of the poor ($2/day) had dropped marginally to 45 per cent of the global population. However, as already pointed out, this masks increases in the absolute number of people living below $2/day. It also masks significant regional disparities. In 1981, 57.7 per cent of the very poor and 84.8 per cent of the poor were in East Asia (mainly China), dropping remarkably to just under 15 per cent and, less remarkably, to 47.4 per cent respectively in 2001. This is because 400 million (plus) Chinese moved out of this definition of material poverty during the two decades leading up to 2001. Chen and Revallion therefore find: 'For the developing world *outside* China, the number of poor living under $1/day increased, from 840 million to 890 million over 1981–2001' (Chen & Revallion 2004: 16). In other words, excluding China which is a model of state-planned and driven development, during the 20 years that macroeconomic policies were dominated by neoliberal economic prescriptions for market-driven development, the number of poor people increased.

Significantly, sub-Saharan Africa is the only region in which the percentage of the very poor and poor increased consistently during the 1981–2001 period: 41.6 per cent of the sub-Saharan African population were very poor ($1/day) in 1981, rising to 46.9 per cent in 2001; whereas 73.3 per cent were poor ($2/day) in 1981 rising to a staggering (when compared to East Asia) 76.6 per cent in 2001 (Chen & Revallion 2004).

One consequence of the analytical separation of poverty from inequality is the implication that there is a non-causal relation between growing inequalities and increasing poverty. The profound significance and impact of the *1998 Human Development Report* cited at the outset of this chapter is that this connection was forcefully and clearly established (United Nations Development Programme 1998). According to this report, by 1998 total public and private consumption expenditures were six times greater than they had been in 1950 and 16 times greater than in 1900. During the last quarter of the twentieth century, consumption per capita in developed industrialised countries increased by about 2.3 per cent annually. In Africa, consumption for the same period declined by 20 per cent. Even the spectacular annual increases in consumption over the same period in East Asia (6.1 per cent) and South Asia (2 per cent) did not reduce consumption inequalities between these regions and the developed world. The socio-economic consequences of non- or under-development in many parts of the developing world are well known: three-fifths lack basic sanitation; a third have no access to clean water; a quarter have inadequate housing; a fifth have no access to modern health services; a fifth do not get enough protein; and worldwide, 2 billion people are anaemic. What is not so well known is that these indicators of poverty are directly related to

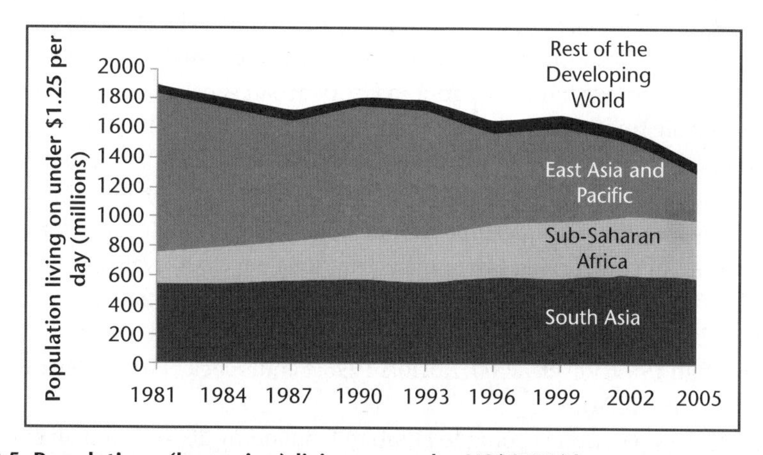

Figure 2.5: Populations (by region) living on under US$1.25/day

(Source: Graph drawn from figures in S. Chen & M. Revallion 2010. The developing world is poorer than we thought, but no less successful in the fight against poverty. *The Quarterly Journal of Economics*, November 2010: 1577–1625.)

increasingly unequal access to the world's primary natural resources (most of which, as argued earlier, are reaching their ecological limits). As the *1998 Human Development Report* demonstrated, the richest fifth of the world's population consumes 45 per cent of all meat and fish, the poorest fifth, 5 per cent; the richest fifth, 58 per cent of total energy, the poorest fifth, less than 4 per cent; the richest fifth, 74 per cent of all telephone lines, the poorest fifth, 1.5 per cent; the richest fifth, 84 per cent of all paper, the poorest fifth, 1.1 per cent; and unsurprisingly, the richest fifth, buy 84 per cent of the world's vehicles (made mainly from metals and polymers made from oil), the poorest fifth, less than 1 per cent (United Nations Development Programme 1998: 2).

Without denying that rising consumption is essential for human development (defined as the enlargement of capabilities and opportunities), the *1998 Human Development Report* articulated its landmark clarion call for sustainability as follows:

> *Today's consumption is undermining the environmental resource base. It is exacerbating inequalities. And the dynamics of the consumption-poverty-inequality-environment nexus are accelerating. If the trends continue without change — not redistributing from high-income to low-income consumers, not shifting from polluting to cleaner goods and production technologies, not promoting goods that empower poor producers, not shifting priority from consumption for conspicuous display to meeting basic needs — today's problems of consumption and human development will worsen.* (United Nations Development Programme 1998: 1)

Our urban futures

There seems to be a high degree of consensus that the world's population will grow to 8 billion by 2030, and at least 9 billion by 2050 (United Nations 2006). Given that there were only 2.5 billion people on the planet as recently as 1950, it is not surprising that the Malthusian ghost has come back to haunt many discussions about the Earth's 'carrying capacity'. More importantly, however, is that we have crossed the 6 billion mark at precisely the moment we have become a majority urban species. If we are interested in a more sustainable future, both development theory and sustainability perspectives will need to accept that this will be an urban future (see Pieterse 2008). This, at least, is what underlies the emergence of the rapidly growing so-called 'sustainable cities' literature, which deals with difficult questions about cities being the locales for contesting sustainable ways of configuring urban infrastructures, services, everyday life and urban cultures (Beatley & Newman 2009; Beatley *et al.* 2009; see Beatley 2000; Evans *et al.* 2005; Evans 2002; Girardet 2004; Hardoy *et al.* 1996; Low *et al.* 2000; Portney 2003; Pugh 1996; Pugh 1999; Ravetz 2000; Roelofs 1996; Sandercock 2003; Satterthwaite 2001; Wheeler & Beatley 2004).

If the world's population is going to climb to 8 billion by 2030, and if all else remains equal the majority of the next 2 billion people are most likely to be living in Asian and African cities (United Nations 2006). The reason for this is that population growth and urbanisation rates have effectively levelled off everywhere except in these two regions.

The bulk of this expansion will be in secondary and tertiary cities, however, not in the existing sprawling mega-cities such as Cairo, Calcutta, Mumbai, Shanghai, São Paulo, Seoul, Rio de Janeiro, Dhaka, Karachi, Buenos Aires and Manila (National Research Council 2003). It has been projected that by 2015 nearly 60 per cent of the total world population will be living in cities of less than a million people. Around 25 per cent will live in cities of 1–5 million people, and about 5 per cent will live in cities of 6–10 million people. This means that by 2015 only 10 per cent of the total urban population will live in cities of 10 million or more, and the number of these so-called 'mega-cities' is likely to remain static at between 25 and 30. It has been estimated that by 2015 there will be 511 so-called 'million cities' (populations of around 1 million)—up from 276 in 1990 (National Research Council 2003). Significantly, populations have been declining in the following mega-cities: Mexico City, São Paulo, Buenos Aires, Calcutta and Seoul. This has major implications for sustainability (explored further in Chapter 5). The obvious overriding implication is that the old mega-cities are the nodes of smokestack industrialisation, whereas the emerging smaller urban systems could, potentially, be more easily reconfigured to be more sustainable from a resource consumption and equity perspective.

Herein lies the significance of the UN-HABITAT Report entitled *The Challenge of Slums* referred to earlier (United Nations Human Settlements Programme 2003). This report estimated that in 2001, 1 billion of the world's 6 billion people lived in urban slums. In other words, whereas Chen and Revallion tell us about the quantitative dimensions of poverty, this report documents the socio-spatial dimensions of poverty. Poor families build, via thousands of community-based groups and family labour, more homes for the world's poorest than any other state, public or non-profit sector (De Cruz & Satterthwaite 2005).

As Davis suggests in his well-known book, *Planet of Slums* (Davis 2005), progress today looks as if the human species is systematically reversing the promise of modernity by flocking in millions into slums, in which daily life is—in those famous words of Thomas Hobbes—'nasty, short and brutish'. For Davis, the promise of urban modernity has been exposed as mere hypocrisy by the persistent failure of developmental agencies, states and (especially) NGOs to find real and lasting solutions to this global 'problem'. Although this might be an overstatement and in some ways an over-simplification (see review by Satterthwaite 2006), there is some truth in the notion that we can no longer represent the rate and level of urbanisation as indicators of progress. Escaping rural poverty in many developing countries often means getting locked into urban poverty, with no escape, sometimes for generations. Cities themselves have become the spatial context for twenty-first-century struggles to (re-)define progress and, in particular, what it means for poor households and communities to experience material improvements in their living conditions despite their locations within entrenched systems of politically sanctioned global inequalities.

Food insecurity, soils and the future of agriculture

The International Assessment of Agricultural Knowledge, Science and Technology for Development (IAASTD) was released after a final plenary meeting in Johannesburg in

2008 (Watson *et al.* 2008). It was co-sponsored by an impressive alliance of multilateral institutions, namely the Food and Agriculture Organisation (FAO), Global Environment Facility (GEF), United Nations Environment Programme (UNEP), United Nations Education, Scientific and Cultural Organisation (UNESCO), UNDP, the World Bank and the World Health Organisation (WHO). It initially included representatives from the genetically modified (GM) seed industry, but these companies withdrew when the scientists and researchers came to conclusions that were not in line with their way of thinking. Although IAASTD paints a complex picture, there are basically four important issues at stake: increasing demand for a wider range of products as the global middle class expands, increasing food prices, worsening malnutrition in developing countries (if China is excluded), and degrading soils and related eco-system services. It is worth quoting a key finding:

> *Quantitative projections indicate a tightening of world food markets, with increasing resource scarcity, adversely affecting poor consumers. Real world prices of most cereals and meats are projected to increase in the coming decades, dramatically reversing trends from the past several decades. Price increases are driven by both demand and supply factors. Population growth and strengthening of economic growth in sub-Saharan Africa, together with already high growth in Asia and moderate growth in Latin America drive increased growth in demand for food. Rapid growth in meat and milk demand is projected to put pressure on prices for maize and other coarse grains and meals. Bioenergy demand is projected to compete with land and water resources. Growing scarcities of water and land are projected to increasingly constrain food production growth, causing adverse impacts on food security and human well-being goals. Higher prices can benefit surplus agricultural producers, but can reduce access to food for a larger number of consumers, including farmers who do not produce a net surplus for the market. As a result, progress in reducing malnutrition is projected to be slow.* (Watson *et al.* 2008: Ch. 5, p. 3)

According to the report (Watson *et al.* 2008: Ch. 3, p. 3), over 800 million people, predominantly in developing countries, are malnourished (and this is projected to increase) while 1.6 billion, mainly in developed countries, are overweight and suffer the consequences of over-eating (diabetes and heart disease), with major implications for health care expenses in these countries.

As Figure 2.6 reveals, rising oil prices are a key driver of rising food prices, because modern agriculture depends heavily on chemical inputs derived from oil and supplied by a handful of global chemical companies. This is true for almost all agriculture in developed countries, and for 40 per cent of the 437 million farms in developing countries, which support the livelihoods of 1.5 billion rural dwellers (Madeley 2002: 21). Chemical fertilisers, together with irrigation, hybrid seeds and micro-credit, formed the mainstay of the so-called Green Revolution that transformed agricultural practices on a global scale from the 1960s onwards (with India leading the way, but starting off mainly in the USA in the 1930s).

It would, however, be incorrect to focus purely on oil prices and growing consumer demand to explain the crisis of rising food costs and the related decline in food security.

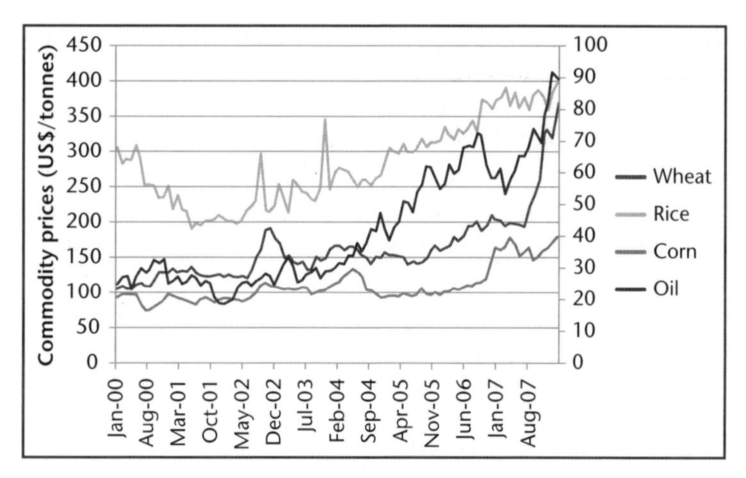

Figure 2.6: Commodity prices (US$/tonnes), January 2000–September 2007 (oil on right scale)

(Source: Calculated from FAO and IMF data. Von Braun, J. 2007 *The World Food Situation: New Driving Forces and Required Actions.* Food policy report, International Food Policy Research Institute, Washington DC.)

As argued in more detail in Chapter 6, what lacks adequate discussion is the impact that rapidly degrading soils are having on supply. It is obvious that if soils are degrading, this must contribute to rising prices because supply is negatively affected. Despite this logic, this connection is hardly ever mentioned in the recent research and policy literature on the food crisis. The IAASTD refers to the problem when it points out that '23% of all used land is degraded to some degree' (Watson *et al.* 2008: Ch. 1, p. 73) and it goes on to suggest that 'declining soil fertility' (Watson *et al.* 2008: Ch. 3, p. 4) is the major challenge for agricultural science. The report states that:

> *Agricultural use of natural resources (soils, freshwater, air, carbon-derived energy) has, in some cases, caused significant and widespread degradation of land, fresh water, ocean and atmospheric resources. Estimates suggest that* resource impairment negatively influences 2.6 billion people. (Watson *et al.* 2008: Ch. 3, p. 3 — emphasis added)

The notion that 'resource impairment' (unsustainable resource use) affects 2.6 billion people is a truly remarkable statement of the immense dimensions of the crisis that the agricultural socio-ecological system is actually facing, and puts into perspective the role that degrading ecological resources play in driving up food prices.

Material flows

In order to function, the global economy depends on a flow of materials that are extracted from the Earth, processed via production and consumption processes to meet human needs, and then disbursed as wastes generated by the extraction, production and consumption processes. The most important materials extracted for use are biomass, fossil fuels, ores, industrial minerals and construction minerals. These material flows, which make GDP growth possible and are referred to as the 'metabolic rate' of the global economy, are measured in tonnes per capita or per unit of GDP (tonnes per US$1 billion of GDP). Material flow analysis (MFA) is the methodology that has emerged to calculate these material flows. The advantage of MFA is that it makes it possible to quantify upstream material flows, that is, the total amounts extracted, the total amounts used, and the total amount extracted but not used (Behrens *et al.* 2007; Bringezu *et al.* 2003; Bringezu *et al.* 2004; Haberl *et al.* 2004; Krausman *et al.* 2008; Krausmann *et al.* 2009; National Research Council of the National Academies 2003).

Estimates of the global quantity of used extracted materials range from 55 billion in 2002 (Behrens *et al.* 2007) (see Figure 2.7) to nearly 60 billion tonnes in 2005 (Fischer-Kowalski & Swilling 2011). This represents an increase of 36 per cent compared to the year 1980 (Behrens *et al.* 2007: 446). For the International Resource Panel this represents an increase by a factor of 8, from approximately 7 billion tonnes in 1900 to nearly 60 billion tonnes in 2005. Population growth and rising income levels have been

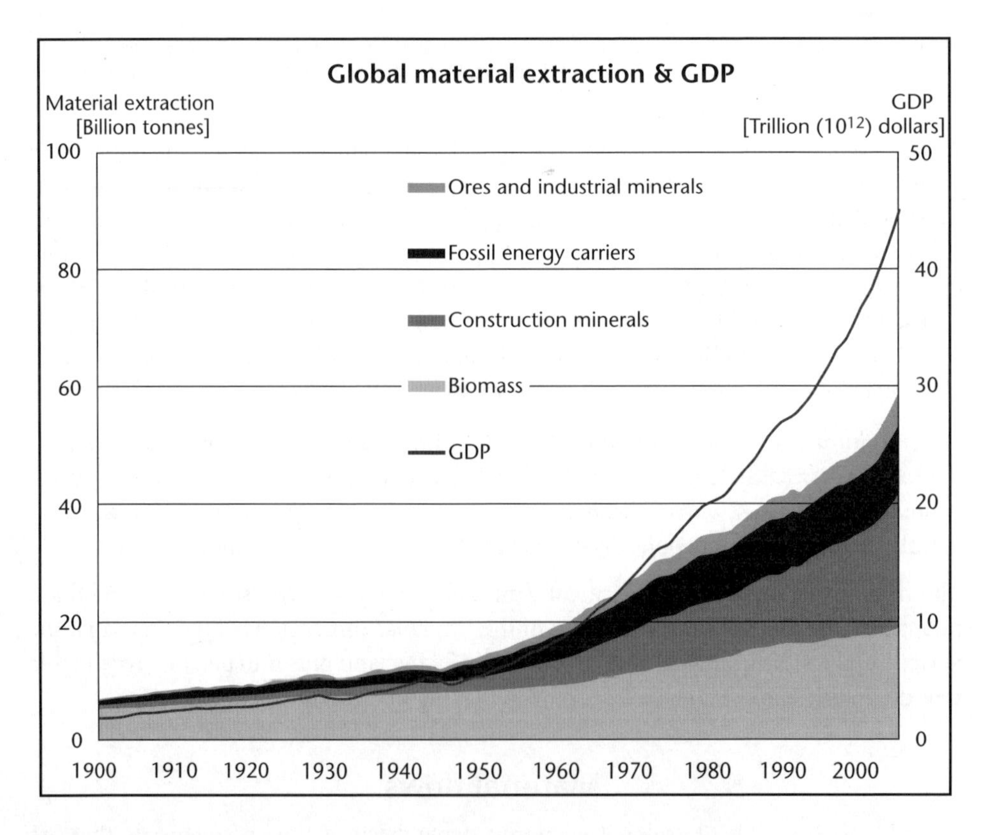

Figure 2.7: Global material extraction in billion tonnes, 1900–2005

(Source: United Nations Environment Programme (2011). Decoupling natural resource use and environmental impacts from economic growth. A Report of the Working Group on Decoupling to the International Resource Panel. Fischer-Kowalski, M., Swilling, M., Von Weizsäcker, E.U., Ren, Y., Moriguchi, Y., Crane, W., Krausmann, F., Eisenmenger, N., Giljum, S., Hennicke, P., Romero Lankao, P., Siriban Manalang, A.)

the primary drivers of rising resource use. However, as Figure 2.7 shows, resource use has not escalated at the same rate as GDP growth, which increased 23-fold over the same period. Overall, for the century starting in 1900, biomass extraction has tended to decline as a percentage of total material extraction from nearly 75 per cent to only a third of that by 2005. Construction minerals increased by a factor of 34 (partly as a result of the second urbanisation wave), and ores/industrial minerals increased by a factor of 27 for the same period.

If we accept that there are physical limits to the absolute quantity of materials that can be extracted for human use, it follows that we need to know what these limits are. Although the International Resource Panel is slowly moving towards a quantitative assessment of these limits (see http://www.unep.org/resourcepanel/), rising resource prices may well be an indication that resource depletion may explain why demand is starting to outstrip supply. Figures 2.8 and 2.9 show that during the twentieth century resource prices tended to fall, and that 2002 might have marked the start of a new era of steadily rising resource prices. The argument in the International Resource Panel Report that rising resource prices may well be related to the economic consequences of resource depletion (Fischer-Kowalski & Swilling 2011: 13) is supported by a similar argument from GMO, the established US-based investment consulting firm (Grantham 2011). Both refer to declining quality of extracted materials and the associated rising costs of accessing low-grade materials and/or previously inaccessible materials (Fischer-Kowalski & Swilling 2011: 23).

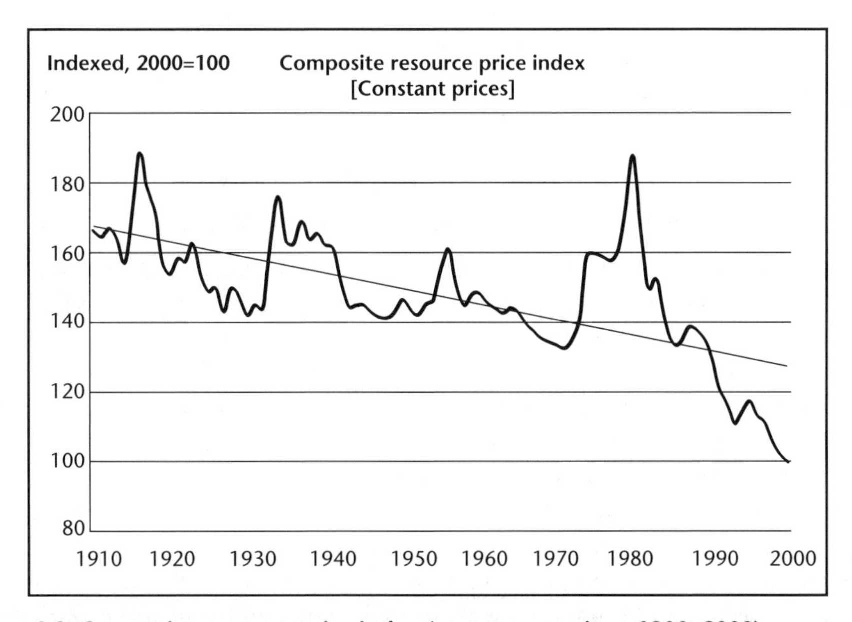

Figure 2.8: Composite resource price index (at constant prices, 1900–2000)

(Source: Wagner, L., Sullivan, D. & Sznopek, J. (2002) *Economic Drivers of Mineral Supply*. U.S. Geological Survey Open-File Report 02–335. Washington DC.)

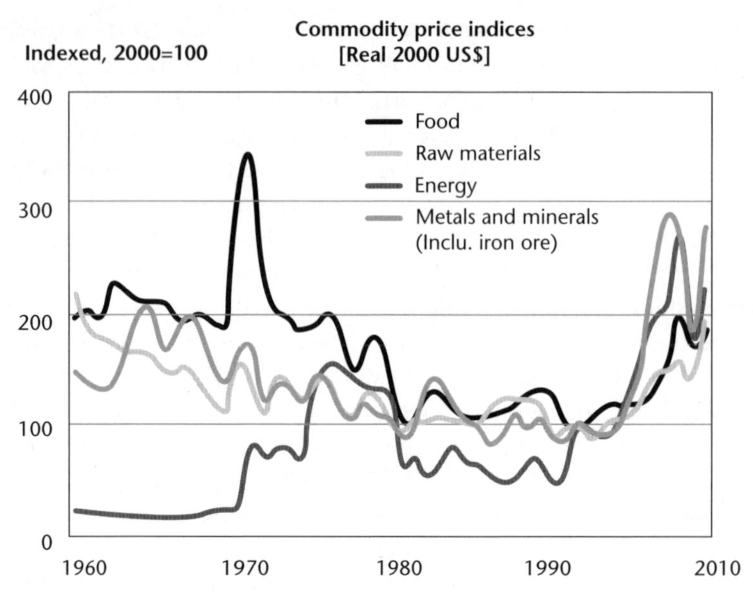

Figure 2.9: Commodity price indices, 1960–2010 (real 2000 US$)

(Source: World Bank commodity prices, cited in United Nations Environment Programme (2011). Decoupling natural resource use and environmental impacts from economic growth. A Report of the Working Group on Decoupling to the International Resource Panel. Fischer-Kowalski, M., Swilling, M., Von Weizsäcker, E.U., Ren, Y., Moriguchi, Y., Crane, W., Krausmann, F., Eisenmenger, N., Giljum, S., Hennicke, P., Romero Lankao, P., Siriban Manalang, A.)

MFA not only makes it possible to analyse the global, national and local economies from a material flow perspective, it is also indispensable when it comes to conceptualising future options and the possibility of a transition to a more sustainable global economy (the focus of Chapters 3 and 4).

Sustainability, inequality and the limits of ecological modernisation

After all is said and done, the challenge of sustainable development in the current global conjuncture is about eradicating poverty, and doing this in a way that rebuilds the eco-systems and natural resources on which we depend for our collective survival.

An argument that is pursued in this book is that poverty eradication through a more equitable distribution of the world's resources can only be achieved if ways are found to restructure the global economy. To do this, we will need to consider ways of achieving what the Latin American theorist of sustainability, Gallopin, has called 'non-material economic growth' (Gallopin 2003). He makes useful distinctions between *development* (improvements in well-being plus material economic growth), *maldevelopment* (material economic growth with no improvements in well-being), *underdevelopment* (no material economic growth and no improvements in well-being), and *sustainable development* (improvements in well-being plus non-material economic growth) (Gallopin 2003: 26). Gallopin argues as follows:

> *In the very long term, there are two basic types of truly sustainable development situations: increasing quality of life with non-material growth (but no net material growth) and zero-growth economies (no economic growth at all). Sustainable development need not imply the cessation of economic growth: a zero growth material economy with a positively growing non-material economy is the logical implication of sustainable development. While demographic growth and material economic growth must eventually stabilize, cultural, psychological, and spiritual growth is not constrained by physical limits.* (Gallopin 2003: 27)

Gallopin represents this conception in Figure 2.10:

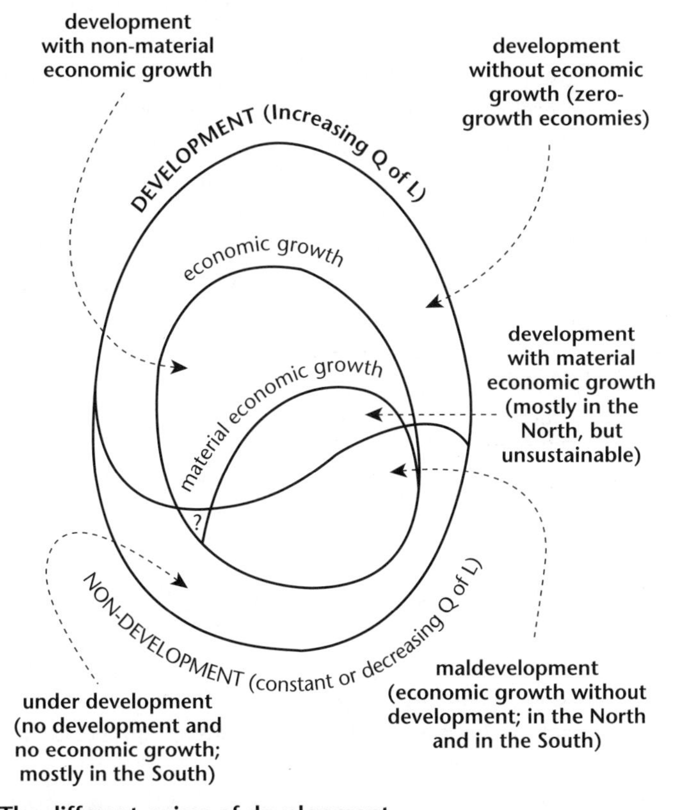

Figure 2.10: The different guises of development

(Source: Gallopin, G. 2003. *A Systems Approach to Sustainability and Sustainable Development.* Project NET/00/063. Santiago: Economic Commission for Latin America.)

The logic of Gallopin's framework is that development strategies for developing countries should be split into two modes (which could be consecutive phases in certain circumstances). The first mode would entail moving from *maldevelopment/ underdevelopment* to *development* whereby improvements in well-being for the majority are achieved via inclusive material economic growth. This is what mainstream development economics is all about, and it is the central focus of *The Growth Report* that brings together the perspectives of the most influential economists in the world today (Commission on Growth and Development 2008). *The Growth Report*, however, virtually ignores ecological sustainability.

The second development mode would entail a shift into *sustainable development* whereby improvements in well-being are achieved via *non-material economic growth*. When references are made to 'leap-frogging', this usually means either shortening the transition considerably from the first to the second mode, or as proposed by Wolfgang Sachs *et al.*, skipping the first phase altogether (Sachs 2002). Leap-frogging, however, will depend entirely on whether the capacity for innovation exists within a particular developing country and whether, in turn, an appropriate set of institutional arrangements are in place to incentivise and harness innovations that demonstrate economically viable 'leap-frog' technologies.

For many in the developed world, the sustainability crisis is synonymous with global warming. However, an exclusive focus on global warming runs the danger of reinforcing the notion that global warming is just a hitch along the path of progress which will be resolved by some kind of grand techno-fix (legitimised by a narrow conception of 'mitigation'). Global warming is, in reality, not just an unfortunate side-effect of the global industrial system, it is an intrinsic part of how this system is constituted, fuelled and financed. As argued by Sachs *et al.* in their influential paper published in the lead-up to the World Summit on Sustainable Development in Johannesburg in 2002, unless we are prepared to deal with the root causes in the way our economic system is configured, solutions to global warming and eco-system breakdown will elude us (Sachs 2002). This means recognising that the most powerful corporations in the world profit from value chains which contribute directly to the worst aspects of global warming: mass private transit, oil production, cement-based building construction, energy production and distribution, large-scale commercial agriculture and deforestation. Very few of the mainstream global reports blame the core structure of this capitalist economic system and the over-riding logic of capital accumulation for the mess we are in and the implications for billions of people who will suffer the consequences. The 2008 financial crisis might raise some awareness about the linkages, but it is too early to tell. It is time, however, for the world's corporate elites to account for the products they produce, and the impacts of the sources of raw materials and processes of transforming these materials into final products.

Global business groups have, of course, begun to address the challenge of sustainable development (Crane & Matten 2004; Hamman 2006; Hawken *et al.* 1999; Marsden 2000; United Nations Research Institute for Social Development 2002). Business perspectives on sustainable development tend to reflect an underlying conceptual framework which has been usefully depicted as 'ecological modernisation' (for an overview, see Korhonen 2008). Not surprisingly, efficiency lies at the centre of this perspective because 'doing more with less' makes business sense. This is also why Life-cycle Analysis—as the primary tool for engineering for efficiency—has risen so rapidly into a very sophisticated tool of analysis and system change. However, ecological modernisation is an approach which may dominate the 'cleaner production' movement but 'there is a danger that the term may serve to legitimize the continuing instrumental domination and destruction of the environment and the promotion of less democratic forms of government, foregrounding modernity's industrial and technocratic discourses over its more recent, resistant and critical ecological components' (Christoff, cited in Korhonen 2008). More radical perspectives are depicted by business-linked sustainable development proponents as 'unrealistic' and a 'threat to the financial health of the global economy'. This has a historical parallel in the response to William Wilberforce in the early 1800s when it was argued by the defenders of slavery that abolition would wreck global trade and modern industrial production. In fact, global capitalism did rather well after slavery was abolished. Sometimes doing the right thing (even if there is no apparent 'business case' for it) turns out to be the best thing for business to do.

The language of 'ecological modernisation' is often expressed in terms of the 'triple bottom line'—an approach associated with an influential text by John Elkington (1998). The core argument was that in addition to *financial value*, corporations must also factor in the need to generate *social value* and *environmental value*, the latter defined as 'adding value by actively promoting sustainable development and reducing public "bads" like pollution, waste, global warming'. In this conception, environmental value is placed alongside and equated to *economic value/financial value* and *social value*. This approach is widespread and is found in virtually every statement on sustainable development in business circles, the mainstream multi-lateral institutions (such as the UN, UNEP, World Bank and EU), and many large environmental NGOs. The triple bottom line approach essentially sees sustainable development as a point at which the three spheres (economic, social and environmental—often depicted as interlocking circles) overlap. Following Hattingh (Hattingh 2001), the problem with this approach is that it locks analysis into a language of trade-offs, with each sphere retaining its own respective logic (an economy driven by markets, society glued together by welfarism, and the environment protected by conservationism). A complex systems perspective offers an alternative that depicts these spheres as embedded within each other. Following the logic of institutional economics, the economy is embedded within the social–cultural system, and following ecological economics, both are embedded within the wider system of eco-system services and natural resources. The result is a way of thinking about sustainability as the organising principle for an expanded 'complex systems' conception of public value which encompasses all three spheres. An elegant definition of sustainability that captures this process-oriented systems perspective emerged from the US-based National Science Foundation Workshop on Urban Sustainability that took place in 1998[7]:

> *In light of ... countervailing definitions based on conflicting economic and political agendas, we propose a definition of sustainability that focuses on sustainable lives and livelihoods rather than the question of sustaining development. By 'sustainable*

[7] Although this workshop took place in 1998, the report on the workshop was only published in 2000.

> *livelihoods' we refer to processes of social and ecological reproduction situated within diverse spatial contexts. We understand processes of social and ecological reproduction to be non-linear, indeterminate, contextually specific, and attainable through multiple pathways.* (Centre for Urban Policy Research 2000)

It is now technically possible for entire communities to meet their material needs by reusing all their solid and liquid wastes, using renewable energy instead of burning fossil fuels to meet up to 50 per cent of their energy requirements (Monbiot 2006), renewing rather than degrading soils for food production (Badgley & Perfecto 2007), cleaning rather than polluting the air, preserving instead of cutting down forests and natural vegetation, under- and not over-exploiting water supplies, and conserving biodiversity instead of killing off other living species (in particular marine species). If it is technically possible, what's left is to make the necessary policy and financial decisions that will change living and behavioural patterns. However, it would be naïve to ignore the fact that this will cut across the way most production and consumption systems are currently configured and, therefore, also entrenched vested interests and investment structures. This, in turn, means that sustainability will more than likely be opposed by some of the most powerful economic stakeholders obsessed with short-term financial gains.

Contrary to what most development economists think, the depleted resource base is such that we can no longer first eradicate poverty and then 'clean up the environment'. This was the third of the four main findings of the MEA and, if true, it calls into question the core foundations of development economics. Nor is there much sense in the neoliberal resource economics argument that tries to suggest that the poor benefit from unsustainable resource use by the rich because this is what drives global growth, and that as scarcities kick in, the market will trigger demand for more sustainable production and consumption. The alternative perspective sees sustainable resource use as a precondition for poverty eradication, not simply because of scarcities but also because sustainable resource use can be a driver of innovation and new value chains with implications for future (dematerialised) growth. This will mean dealing with inequality, which is the root cause of poverty and, in particular, the economic and political power structures that reproduce these inequalities. Over-consumers will have to cut back and be satisfied with sufficient to meet their needs, and the savings this generates will be needed to eradicate poverty. The call for 'sufficiency' seems to capture what this means — or to use a slogan used by the South African Government's Department of Water Affairs and Forestry, 'some for all forever'. This is very different to the current global consumerist culture which can be depicted as 'all for some for now'.

Conclusion

It has been more than 20 years since the publication of the Brundtland Commission Report that established the most widely accepted definition of sustainable development. In light of the trends reviewed in this chapter it may be worth concluding by revisiting the following lesser known but far more contentious paragraph from this report:

> *The concept of sustainable development does imply limits—not absolute limits but limitations imposed by the present state of technology and social organisation on environmental resources and by the ability of the biosphere to absorb the effects of human activities. But technology and social organisation can be both managed and improved to make way for a new era of economic growth. The Commission believes that widespread poverty is no longer inevitable... Meeting essential needs requires not only a new era of economic growth for nations in which the majority are poor, but an assurance that those poor get their fair share of the resources required to sustain that growth... Sustainable global development requires that those who are more affluent adopt life-styles within the planet's ecological means—in their use of energy, for example... Yet, in the end, sustainable development is not a fixed state of harmony, but rather a process of change in which the exploitation of resources, the direction of investments, the orientation of technological development, and institutional change are made consistent with future as well as present needs.* (World Commission on Environment and Development 1987: 8)

This highly ambiguous paragraph has been read and reinterpreted in many ways over the past two decades to justify all sorts of activities, both good and bad for the planet and society. The outright rejection of the notion that there are 'absolute limits' to the stock of natural resources available for human consumption and use made it possible to justify (resource-extractive) economic growth in the developing world as the panacea for poverty eradication. The two key riders to this statement were that there are limits to 'the ability of the biosphere to absorb the effects of human activities', and that the poor receive their 'fair share of resources'.

The review of global trends presented in this chapter suggests that the poor did not receive their 'fair share' of global growth over the past two decades, that biophysical limits to the absorption of the 'effects of human activities' were breached on several counts, and that the 'more affluent' expanded rather than reduced their ecological footprints to the detriment of the poor.

In light of the massive expansion of our scientific knowledge about our natural resources and eco-systems, it may be necessary in future to accept what the Brundtland Report rejected, namely that there are indeed 'absolute limits' that should not be breached. This would mean endorsing, for example, the IPCC recommendation that average CO_2 emissions per capita should be 2.2 tonnes rather than the current 4.5 tonnes; or the suggestion by the International Resource Panel that the average consumption of extracted materials should be 6 tonnes per capita rather than the current 8 tonnes. Furthermore, it is not just about the biophysical limits to *absorption* of the effects of human activities that matter, but also limits to the quantities of remaining strategic non-renewable resources (such as oil and metals) and limits to how far eco-systems such as fisheries, water cycles, soils and atmospheres can be exploited and modified.

Equally, we may need to question the notion that economic growth is by definition in the interests of the poor. The last two decades were dominated by economic theories which assumed that free markets, unregulated capital flows and cheap resources would result in more equitable outcomes for developing country economies (especially the resource-rich ones). As these theories lose their positions of policy influence and evidence mounts that the resulting policies reinforced the ransacking of global resources for the benefit of the global over-consumers, we have an opportunity to rethink economic development theory and strategies in light of the twin challenges of persistent poverty and the need to accept that there are in fact 'absolute limits' to the natural resources we can consume and use for human benefit.

Chapter Three

Crisis, Transitions and Sustainability

Introduction

The financial crash of October 2008 marked a turning point in the contemporary history of governance and the role of the state in society. In 2009 the global economy contracted for the first time since World War II. If China and India are excluded, the real GDP of developing countries declined by 2.2 per cent in 2009. The 13 October 2008 edition of *Newsweek* marked this moment by devoting the front cover to a burning dollar bill under the headline: 'The future of capitalism: The end of the age of Reagan and Thatcher, and what will follow'. Between its covers one of the most aggressive prophets of the neoliberal revolution, Francis Fukuyama, declared the end of the road for neoliberalism when he wrote: 'Many commentators have noted that the Wall Street meltdown marks the end of the Reagan era. In this they are doubtless right' (Fukuyama 2008). This breathtakingly simple confession of ideological retreat removes a powerful pole of global certainty and clears the way for a much more open eclectic discussion about our future development options, especially for those located in the global South. It opens a refreshing space for the search for connections between the epochal and industrial transitions that will shape future policy options.

The significance of the crash of October 2008 is that the new openings for rethinking governance coincide with the emergence of a scientific consensus that the future of human development—indeed, possibly even the modernist project—is being undermined by the rapid depletion of the natural resources and eco-system services on which societies depend for their survival and ongoing prosperity. It must be more than just a coincidence that in the months leading up to October 2008 global leaders gathered twice to deliberate two closely related crises: rocketing oil prices, which went over US$140 per barrel, and unprecedented increases in food prices which resulted in marches by starving people across the developing world. These crises relate directly to two underlying primary resources—oil and soil.

Are these separate crises unfolding in parallel, or are they related in ways that our current paradigms cannot fully comprehend? If they are related, what are the implications for the way we understand the solutions? More specifically, is the emergence of a more appropriate conception of governance for the post-neoliberal era related to the volatile epochal dynamics of socio-ecological transition? Is development during the era beyond traditional Keynesianism, contemporary neoliberalism and mainstream developmental statism simply about more durable long-term growth as proposed by some of the world's leading economists who sit on the Growth Commission? (Commission on Growth and Development 2008). Or must we completely rethink what growth and development will mean in light of the polycrisis described in Chapter 2? If so, what does this mean for the way we reconfigure the state in the post-neoliberal era as we deal with our ecological crises?

Conceptualising transitions

As the 'gales of creative destruction'[1] sweep across the economic, institutional and ecological landscapes, we need to pay attention to the historical role that innovations have played in past transitions at two different scales. The first is at the more familiar scale of the transitions that have taken place during the course of the industrial era over the past 230 years. These industrial transitions have to be contextualised, however, within the much wider historical scope of socio-ecological transitions that started with the transition to the agricultural epoch at the end of the last Ice Age some 13,000 years ago. Transitions at both these scales have been driven by crisis, with the evolution of new modes of existence instigated by innovations that partially or provisionally resolved the crisis and, in so doing, destroyed the basis of pre-existing modes of existence, technologies and hierarchies of power. Our contention here, however, is that the potential transition to a more sustainable future may well be driven by simultaneous transitions at the shorter-term industrial scale and the longer-term socio-ecological scale. It is necessary to focus on both the industrial and epochal scales in order to comprehend the significance of transitions brought on by the dynamics of the polycrisis.

The conceptual synthesis developed in this chapter and Chapter 4 aims to 'build up from below' a more general conception which connects with, but is also distinct from, the three main conceptions of transition in the sustainability literature, namely the social innovations approach that has emerged from the management literature, the Multi-Level Perspective (MLP) developed by the Dutch school, and the resilience approach associated with the Resilience Alliance (for a review see Westley *et al.* 2011).

As discussed in the Introduction, the MLP takes into account the context-specific interactions between transitional dynamics at the *landscape, regime* and *niche* levels (Grin *et al.* 2010). What we refer to as the *industrial socio-ecological regime* that began 250 years ago with the transition from the agricultural to the industrial socio-ecological regime (following Fischer-Kowalski & Haberl 2007), the MLP would depict as the *socio-technical landscape* that defines the geographical, historical, environmental, technological and physical context of modern society. All else happens within the landscape, and as such it shapes and is shaped by human action. What we refer to as the *five industrial transitions* that have taken place since the *start* of the industrial era (Perez 2002), the MLP would define as *socio-technical regimes.* Both concepts refer more or less to the same succession of particular configurations of technologies, markets, rules and productive routines that have emerged and declined since the 1770s. Using the language of the MLP, our argument is that the resolution of the crisis at the *regime* level, triggered by the financial crash of 2007/08, will be limited and shaped by the crisis of resource limits at the *landscape* level. However, following the logic of the MLP, much will depend on whether sufficiently mature *niche*-level innovations have been replicated in sufficient numbers across the globe to become the focus of accelerated financial investment. Without this, the path-dependent dogmas of existing unsustainable regimes will not

[1] This famous phrase was coined by Joseph Schumpeter (1939).

be replaced and so the crises at the landscape level will become more severe and harder to resolve later. The key here will be state interventions that enhance the kinds of niche innovations that are responsive to the root causes of the problems at the landscape level and which exacerbate the economic crisis at regime level. Unfortunately, the rather bland conception of governance offered by the Dutch school (see Parts II and III of Grin *et al.* 2010) does not help very much to understand the political modalities of these interventions, especially when it comes to the notoriously mercurial dynamics of states in developing and former communist countries (quite a few of which are by no means democracies). Nor are the social innovation and resilience approaches much help in this regard. What is needed is a new theory of the developmental state that emphasises capabilities for innovation and transition, which is the subject of Chapter 4. First we need to consolidate our understanding of transitions.

Industrial transitions since 1771

Echoing the sensibilities of an earlier generation of economists (Russia's Kondratieff 1935; Austria's Schumpeter 1939), Perez identified five 'transitions' that she associates with specific technological innovations that emerged at particular historic moments since the dawn of the industrial era in the 1770s (Perez 2002). These transitions were roughly 50-year cycles that began with technological innovations (steam, then steel, followed by oil, and finally information) which were funded initially by high-risk investors, enticed by the profits that revolutionary new processes could generate. After disrupting the previous structure of industrial organisation over two or even three decades, these innovations were themselves displaced after enjoying fairly long periods of technological dominance during a 'deployment period' which could last for two to three decades (Perez 2002; Perez 2007).

The five 'transitions' are reflected in Table 3.1. The first transition — the effective start of the Industrial Revolution in Britain — dates back to 1771 when 'canal mania' underpinned the development of machines and the emergence of the mechanised cotton industry as the template for industrialisation. The second transition — the Age of Steam and Railways — started in 1829 when the steam engine, fuelled by coal, made it possible to build transportation systems (railways) and factories powered by fossil fuels extracted from the Earth using new mining technologies. This established the dependence of manufacturing on fossil fuel extraction and mining. The third industrial transition — the Age of Steel, Electrification and Heavy Engineering — started in 1875 as so-called 'heavy industry' emerged and consolidated itself on a global scale, all made possible by cheap steel, electrification, steel ships and the start of mass consumption. Heavy engineering made possible the socio-technical revolutions that resulted in the birth of mass electrification, water reticulation and transportation — the so-called '*lights, water, motion*' paradigm that made it possible to invent the modern city. The fourth transition — the Age of Oil, Automobiles and Mass Production — started in 1908 in the USA with the production of the Model-T Ford which signalled the start of 'Fordism', that is, the mass production of a limited range (by today's standards) of consumer goods,

Table 3.1: Five technological revolutions in 250 years: Main industries and infrastructures

Technological revolution	*New technologies and new or redefined industries*	*New or redefined infrastructures*
FIRST: From 1771 The *Industrial Revolution* Britain	Mechanised cotton industry Wrought iron Machinery	Canals and waterways Turnpike roads Water power (highly improved water wheels)
SECOND: From 1829 *Age of Steam and Railways* In Britain and spreading to Continent and USA	Steam engines and machinery (made of iron, fuelled by coal) Iron and coal mining (now playing a central role in growth)* Railway construction Rolling stock production Steam power for many industries (including textiles)	Railways (use of steam engine) Universal postal service Telegraph (mainly nationally along railway lines) Great ports, great deports and worldwide sailing ships City gas
THIRD: From 1875 *Age of Steel, Electricity and Heavy Engineering* USA and Germany overtaking Britain	Cheap steel (especially Bessemer) Full development of steam engine for steel ships Heavy chemistry and civil engineering Electrical equipment industry Copper and cables Canned and bottled food Paper and packaging	Worldwide shipping in rapid steel steamships (use of Suez Canal) Worldwide railways (use of cheap steel rails and bolts in standard sizes) Great bridges and tunnels Worldwide Telegraph Telephone (mainly nationally) Electrical networks (for illumination and industrial use)
FOURTH: From 1908 *Age of Oil, Automobiles and Mass Production* In USA and spreading to Europe	Mass-produced automobiles Cheap oil and oil fuels Petrochemicals (synthetics) Internal combustion engine for automobiles, transport, tractors, airplanes, war tanks and electricity Home electrical appliances Radio and television Refrigerated and frozen foods	Networks of roads, highways, ports and airports Networks of oil ducts Universal electricity (industry and homes) Worldwide analog telecommunications (telephone, telex and cablegram) wire and wireless National broadcasting networks
FIFTH: From 1971 *Age of Information and Telecommunications* In USA, spreading to Europe and Asia	The information revolution Cheap microelectronics Computers, software Telecommunications Control instruments Computer-aided biotechnology and new materials	World digital telecommunications (cable, fibre optics, radio and satellite) Internet/Electronic mail and other e-services Multiple source, flexible use, electricity networks High-speed physical transport links (by land, air and water) Global 'narrow-casting' networks

*Note: These traditional industries acquire a new role and a new dynamism when serving as the material and the fuel of the world of railways and machinery.

(Source: Perez, C. 2002. *Technological Revolutions and Financial Capital: The Dynamics of Bubbles and Golden Ages.* Cheltenham, U.K.: Elgar.)

transportation and communication systems. The fifth transition—the Age of Information and Telecommunications—started in 1971 in the USA and spread across the world transforming production, consumption, distribution, finance and communication.

Significantly, each of these transitions depended on the exploitation of a key natural resource/eco-system service: soils, cotton and wood from the colonies for the first transition; coal and iron ore for the Age of Steam; coal, iron ore, copper and agricultural produce for the Age of Steel and the start of mass production; oil plus all the other resources—especially food—for the fourth; and a conglomeration of oil, metals, minerals, biomass, agricultural produce and microscopic, digitalised resources (from fibre optics to DNA), secured via digitally networked trading relationships, for the Information Revolution. Up until the 1950s, industrial society was primarily dependent on biomass, but since the commencement of the twin processes of informationalisation and globalisation from the 1970s onwards (with roots in the 1960s), fossil fuels, metals and minerals have rapidly escalated, eventually displacing biomass as the primary resource base of industrial society (Krausmann *et al.* 2009). As UNCTAD economist Charles Gore has observed, each of these transitions is characterised by

> *the introduction of a few leading sectors which provide cheap inputs to a wide range of economic activities; the installation of large-scale transport, communications and energy infrastructures; induced investment and innovation in economic activities, which are related to the leading sectors and the new infrastructures through forward and backward linkage effects; and the creation of new organisational and managerial practices.* (Gore 2010: 719)

Each of the five periods resulted in unique configurations of primary resource inputs, infrastructures and associated innovations in order to consolidate the hegemony of the lead sectors that had become the focus of financial investment.

Perez argues that

> *... each technological revolution irrupts in the space shaped by the previous one and must confront old practices, criteria, habits, ideas and routines, deeply embedded in the minds and lives of the people involved as well as the general institutional framework, established to accommodate the old paradigm. This context, almost by definition, is inadequate for the new.* (Perez 2002)

Significantly, Perez demonstrates that each transition goes through distinct periods, starting off with an 'irruption' phase during which the innovations are generated, followed by a phase of 'frenzy' as investors rush for a stake in the businesses spawned by the innovations. After this crowding-in of investments in search of capital gains triggers a bubble and associated financial crisis (devaluation), the state steps in to reorganise institutions to absorb the new technologies, leading to a phase of 'synergy' during which there is a generalised global dispersion of production systems across the economy as a new 'golden age' (of steady growth and long-term profitability from dividends) sets in. The transition ends with a 'mature' phase during which the new technologies reach saturation point and production systems are stabilised with diminishing returns on investment.

Each transition has generally spread out across 50-year periods, with internal crises driving the shift from one phase of the transition to the next. However, it is normally financial capital working with innovative entrepreneurs that drive the first two phases—what Perez calls the 'installation period'—as the former search for profits in new high-return ventures. This then generates techno-financial crises as over-investments accumulate in 'bubbles' of technological innovation that have not yet generated the much hoped for returns; nor have the institutional structures of society changed sufficiently to foster the expansion of the innovations into generalised systems of production and consumption. As Figures 3.1 and 3.2 illustrate, the result is what Perez calls a 'turning point' marked by a major financial crisis which, in turn, triggers new institutional responses that drive the progression from the 'frenzy' to the 'synergy' phase, with the latter marking the start of the 'deployment period' during which more low-risk, long-term 'productive capital' becomes the main player. If state intervention during and after the crisis cannot restrain financial capital in order to clear the way for the more sedate and formalised investment modalities of productive capital, the chances are that the benefits of the new technologies would be limited to elites rather than dispersed across the whole of the economy and society. This shift in power from financial capital and its thirst for quick capital gains to productive capital's appetite for steady profits over the long term is the defining feature of a successful 'turning point'.

Significantly, the financial crisis in each case triggers massive state interventions aimed at managing the crisis, but also to fundamentally restructure the institutional and moral orders of society to prepare the way for the mass deployment of the new technologies and production systems. As power shifts from financial capital that drove the installation phase, to productive capital that drives the deployment stage, innovations get embedded in newly structured regimes of accumulation and governance, which are often seen as the 'golden ages' that follow the crisis period as solutions are introduced to 'resolve' the problems that are seen to be causes of the crisis. Without new and central roles for the state, such a transition would be impossible. Although in each case the ideological language of interventionism was different ('Keynesianism' after 1920 and 'free-market economics' in the 1980s), the goals and modalities of the interventions were similar.

The 1929–1933 global economic crisis, for example, prepared the way for the consolidation of the Age of Oil, Automobiles and Mass Production ('Fordism'), with the USA emerging after the World War II as the powerhouse of this transition. State interventions that became known as the New Deal were justified by the economic theories of John Maynard Keynes. Without intervention, this transition would have been inconceivable. After recovering from the turning point and their loss of power to a much more stable accumulation regime, financial capital eventually moves off in search of returns that the next generation of innovators start to develop, which anticipates the next irruption-driven transition. Along the way there are detours and reversals, in particular as globally uneven development opens up investment opportunities in societies that lag behind those that entered the transitions first.

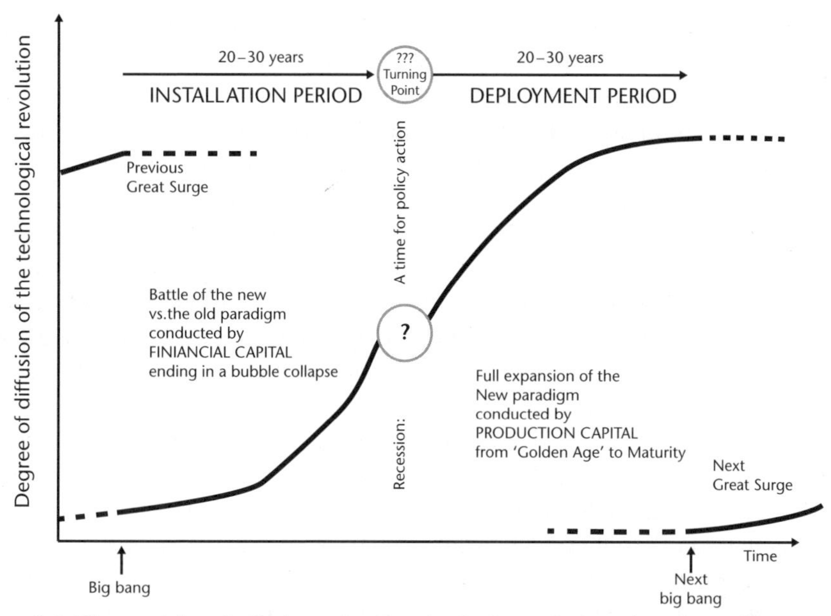

Figure 3.1: The social assimilation of technological revolutions breaks each great surge of development in half

(Source: Perez, C. 2006. Respecialisation and the deployment of the ICT paradigm: An essay on the present challenges of globalisation. In Compano *et al.* (eds) *The Future of the Information Society in Europe*, Technical report EUR22353EN, IPTS, Joint Research Centre, Directorate General, European Commission, pp. 27–56.)

GREAT SURGE		INSTALLATION PERIOD 'Gilded Age' Bubbles	TURNING POINT Recessions	DEPLOYMENT PERIOD 'Golden Ages'
1st	1771 The Industrial Revolution (Britain)	Canal mania	1793–97	Great British leap
2nd	1829 Age of Steam and Railways (Britain)	Railway mania	1848–50	The Victorian Boom
3rd	1875 Age of Steel and Heavy Engineering (Britain/USA/Germany)	London funded global market infrastructure build-up (Argentina, Australia,USA)	1890–95	Belle Époque (Europe) 'Progressive Era' (USA)
4th	1908 Age of Oil, Autos and Mass Production (USA)	The roaring twenties Autos, housing, radio, aviation, electricity	Europe 1929–33 USA 1929–43	Post-war Golden age
5th	1971 The ICT Revolution (USA)	Emerging markets dotcom and Internet mania financial casino	2007 -???	Sustainable global knowledge-society 'golden age'?

We are here

Figure 3.2: The historical record: Bubble prosperities, recessions and golden ages

(Source: Perez, C. 2011. The advance of technology and major bubble collapses. In Linklater, A. (ed.), *On Capitalism*. Engelsberg Seminar, Axon Johnson Foundation, Stockholm, pp. 103–114.)

The obvious question, therefore, is whether the analytical framework developed by Perez helps us to understand the global economic crisis that began with the sub-prime financial crisis in the USA in 2007. What makes this a tricky question is that the Information Age has, in fact, experienced what Perez has called a 'double bubble'—the so-called 'dot com' bubble of 1997–2000, followed by the financial bubble of 2004–2007. Perez has argued that these 'two bubbles of the turn of the century are two stages of the same phenomenon' (Perez 2009: 780). She argues against the neoliberal argument that explains financial crises in terms of irrationalities such as state intervention, and against the Keynesian argument that debt markets have an in-built tendency towards financial instability, which can only be mitigated by increased state spending (Krugman 2008). Instead, she argues that the most significant crises are triggered by the financial opportunities created by new technologies which result in 'major technology bubbles (MTBs)' that eventually burst. This is what the Internet mania of 1997–2000 was all about. However, instead of a deep economic recession that would have necessitated extensive state intervention to prepare the way for productive capital to take over from financial capital after the bubble burst in 2000/01, the post-crisis recession was mitigated by the rapid financialisation of the global economy that the IT revolution had made possible, and the China factor which brought down the cost of mass consumer goods. Indeed, the preference for liquid assets and quick operations within the paper economy generated skyrocketing capital gains between 1996 and 2000, while profits in the real economy remained flat (Perez 2009: 787).

After the 'dot com' crash, instead of interventions to restrain financial capital, the opposite happened as various interventions by the Federal Reserve and neoliberal governments around the world effectively allowed the paper economy to mushroom into a gigantic unregulated global casino. The resulting bubble was not, according to Perez, another MTB, but rather a Ponzi-type 'easy liquidity bubble' (ELB) driven by massive concentrations of investments in paper assets (or what Warren Buffett famously called 'financial weapons of mass destruction') which eventually lost their value in 2007–2008 (Perez 2009).

Ironically, it was the last remaining Communist state—the Chinese government—that bought massive quantities of US Treasury bonds, which created the mega-flows of cash that powered the ELB. Instead of benefiting Chinese workers, this excessive cash was literally forced out into households across the developed economies (with their properties acting as security) to fuel a consumer boom that kept the formal growth rates of these economies up, wages flat and created the global markets for cheap Chinese goods (Gowan 2009). Political leaders struck a global bargain: by stimulating global demand, the Chinese government could avoid the politically risky prospect of increasing the wages of Chinese workers to create their own domestic consumer market, and cheap consumer goods effectively increased the real value of salaries and wages in the developed world, thus shoring up the neoliberal political coalitions in power across many parts of the developed and developing world. Households across the world (but in particular in low-interest-rate regimes such as North America and Europe) were given access to low-cost debt, secured by property values that the financial institutions assumed would consistently improve in value over time (Gowan 2009). Whereas this debt-financed consumer bubble drove global growth, the Chinese manufacturing boom cushioned the global economy from the consequences of the 'dot com' bust of 2000/01. But when the ELB finally burst in 2007–2008, the underlying crisis of the real economy was laid bare for all to see and decisive intervention was inevitable. Keynesianism returned to the corridors of economic power, albeit to bail out banks and car companies in ways that must have made Keynes turn in his grave.

In short, whereas for Perez the 1997–2000 MTB was driven by lucrative investment opportunities created by the new information and communication technologies, the 2004–2007 ELB was driven by excessive liquidity caused by financial deregulation, 24/7 global trading that computerisation made possible, accelerated globalisation and the Chinese addiction to US Treasury bonds as an alternative to investments in the quality of life of Chinese workers. Both the MTB and the ELB, however, form part of a single turning point which has triggered extensive state interventions to manage the damage and restructure financial flows and institutions (Perez 2009). What remains unclear, however, is whether bodies such as the G20 and the so-called 'major economy' governments have what it takes to discipline financial capital, which is now embedded in a highly complex global institutional architecture, largely beyond the reach of national governments. For some this means that financialisation has become endemic and is here to stay, and for many Marxists it is *the* defining characteristic of contemporary capitalism (Altvater 2009; Bond 1999; Dore 2008; Epstein 2005). Perez, however, was optimistic when she wrote in 2009:

> *What came after the internet bubble was not the restructuring of the real economy that tends to occur in the aftermath but a casino revival that only fulfilled part of that task. There can be, however, little doubt that this second major bust and its consequences are likely to follow the script and facilitate the necessary institutional recomposition to unleash the deployment period of the current surge.* (Perez 2009: 800)

While pointing out that a spate of mergers and acquisitions brought on by the crisis has put in place the conditions for productive capital to take the lead, she admits that this time round it will not be easy to discipline financial capital. The Stiglitz Report has described, in practical terms, what it will take to restructure the global financial system in order to prepare the way for the re-emergence of the 'real economy' as the centre of global economic gravity (Stiglitz 2010b). Stiglitz himself, however, repeatedly complains in his regular columns that the banks and financial regulators are not doing enough. At the time of writing (early 2011), there was little evidence of fundamental restructuring of the global financial system. Instead, the evidence suggests an interregnum between the old, which has not died and the new waiting to be born: we have the rivalry between China and the USA about the value of the Chinese currency; the ongoing financial instabilities in the EU, exacerbated by the Greek sovereign debt crisis; the *de facto* bankruptcy of the USA; the relatively unfettered flow of speculative finance through global markets; the hoarding of cash as investors wait for short-term capital gains opportunities to return, instead of looking for long-term productive investments in the real economy; and national governments who, having experienced massive devaluations in the past, continue to build up currency reserves to counteract financial shocks, thus keeping much-needed investment capital away from productive investment. It is clear that what is needed is decisive and visionary political leadership which recognises the historical significance of this conjuncture. Krugman, the Nobel Prize-winning Keynesian economist, is correct when he argues that increased spending and fiscal deficits are essential to counteract recessionary conditions (Krugman 2008), but his analysis is limited to the ELB component of the crisis. Unless the MTB component of the crisis is also recognised, prescriptions for intervention will not emphasise the need to shift the balance of power from financial to productive capital (or what is referred to in the popular financial press as the 'real economy') to harness the potential created by the information technology revolution. But even if this broader perspective is adopted, it remains insufficient, because it ignores the implications of the much wider challenge of epochal transition to a more sustainable socio-ecological regime. Unless the ecological and resource limits to the industrial epoch as a whole are recognised, purely economic interventions will continue to have a limited effect.

When Perez considers the deployment period that is bound to follow the October 2008 crash, she envisages a global governance system and a return of a developmental state that will promote 'an alternative mode of globalization, fully compatible with the [ICT] paradigm and capable of unleashing a worldwide steady expansion of production, markets and well-being. It would need to be *production-centred and led; pro-growth and pro-development*; with dynamic, *locally differentiated markets*, enhancing national and other identities and reaching towards optimum worldwide welfare' (Perez 2007).

This optimistic vision for the *deployment period of the Information Age* is, in short, a vision for global (re-)industrialisation which drives a new period of global growth that is more equitable and contributes to unprecedented well-being (see also Sen 2009). It is all good news from a developing country context, not only because a 'production-centred' view of the world recognises that the state is needed to subordinate the profits of specific corporations to the national objectives of industrialisation, but also because many developing countries are well positioned to become the nodal centres of this new world of globalised industrial production.

The only thing more surprising than the optimism of this vision is the failure of Perez—and nearly every other development economist with an optimistic view of the future—to realise that this kind of vision rests on the assumption that the natural resources and eco-system services are in place to support this massive expansion of production (Soderbaum 2009). Interestingly, though, Stiglitz, Perez and Krugman started in 2010/11 to refer to resource depletion as key limits to Keynesian economic recovery programmes which focus only on expanding consumer demand to drive new growth (Krugman 2010b; Parker 2011; Perez 2010). Both Stiglitz and Krugman wrote significant recent interventions that accepted the IPCC's predictions and advocated global cooperation to implement a new carbon tax regime (Krugman 2010a; Stiglitz 2009). UNCTAD economist Charles Gore has called for a new development paradigm which recognises the consequences of climate change and rising resource prices, but remains sceptical due to the power of vested interests (Gore 2010: 734). However, echoing Latour, Gore worries that the next great leap forward will mean even greater attachments to the imbroglios that will cut this all off at the knees if we continue to think and live in denial.

Socio-ecological transitions since 13,000 BCE

The transition to a globalised ICT-based system of production which benefits the poor majority will depend on the dynamics of a much wider socio-ecological transition that is also simultaneously underway. Fischer-Kowalski and her colleagues at the Institute for Social Ecology (2007) redefine economic history by referring to the evolution of successive 'socio-ecological regimes': the hunter-gatherer regime, followed by the agrarian regime that began 13,000 years go, and then finally the industrial regime that began 250 years ago. They argue that these successive socio-ecological regimes can be understood in terms of the history of 'specific fundamental pattern[s] of interaction between (human) society and natural systems' (Fischer-Kowalski & Haberl 2007). They go on to argue that

> *[i]f we look upon society as reproducing its population, we note that it does so by interacting with natural systems, by organising energetic and material flows from and to its environment, by means of particular technologies and by transforming natural systems through labour and technology in specific ways to make them more useful for society's purposes. This in turn triggers intended and unintended changes in the natural environment to which societies react.* (Fischer-Kowalski & Haberl 2007: 14)

In order to understand socio-ecological transitions, Fischer-Kowalski *et al.* utilise 'material flow analysis' (MFA) — a method which has become increasingly popular among European ecological economists (Bartelmus 2003; Behrens *et al* 2007; Fischer-Kowalski & Haberl 2007; Giljum *et al.* 2007; Haberl *et al.* 2004; Krausmann *et al.* 2009).[2] Instead of focusing on long lists of indicators of 'environmental impact' as in mainstream environmental science, MFA makes it possible to analyse material and energy flows *into and through* socio-ecological systems in ways that make possible an integrated understanding of specific regimes (see Figure 3.3). The end result is a 'metabolic rate' expressed as a simple metric: tonnes of materials consumed per capita per annum, and energy used per capita measured in gigajoules (GJs). It follows, therefore, that to become more sustainable means using less energy and materials while simultaneously pursuing conventional development targets, such as human well-being and ensuring more equitable access to all resources. As Fischer-Kowalski *et al.* put it, MFA is useful for 'analysing and understanding the metabolic exchange relations between human societies and their natural environments, the feedbacks that transform both social and natural systems and the biophysical limitations of the systems involved.' (Fischer-Kowalski & Haberl 2007: 16)

---

[2] For a history of materials flow analysis, see Fischer-Kowalski 1998; Fischer-Kowalski 1999.

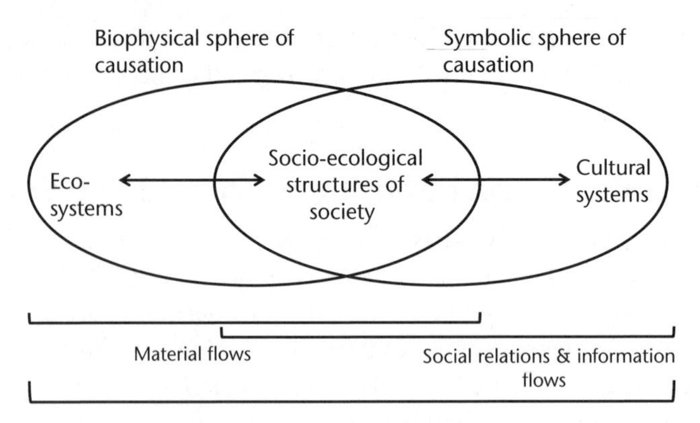

Figure 3.3: Socio-ecological systems as the overlap of a natural and a cultural sphere of causation

(Source: Haberl, H., Fischer-Kowalski, M., Krausman, F., Weisz, H. & Winiwarter, V. 2004. Progress Towards Sustainability? What the Conceptual Framework of Material and Energy Flow Accounting (MEFA) can offer. *Land Use Policy*, 21: 199–213.)

Fischer-Kowalski *et al.* identify three major socio-ecological regimes:

- The hunter-gatherer socio-ecological regime that existed for at least 100,000 years prior to the agricultural revolution that took place after the last Ice Age some 13,000 years ago
- The agrarian socio-ecological regime that existed until the Industrial Revolution nearly 250 years ago (and persists into the present in quite a few predominantly agricultural societies dependent on agriculture)
- The current industrial socio-ecological regime that began in the 1770s and dominates the global economy.

They also envisage a 'sustainable socio-ecological regime' that will be brought about by the next major socio-ecological transition as the conditions of existence of the industrial era disintegrate (Fischer-Kowalski & Haberl 2007: 12–16). Of course, as will be discussed later, it is not just the unviability of the *ancien régime* that determines the transition to a new one, but also conscious and purposive innovations to create pathways to a new order. Innovations do not simply happen because they 'need to', but because quite a specific set of conditions are in place to promote and foster them. Hence the significance of the developmental state as promoter of innovations that is discussed in Chapter 4, and the spatial significance of cities as the geographical contexts for innovations as discussed in Chapter 5.

Using a 'stocks and flows' methodology, Fischer-Kowalski *et al.* depict the successive socio-ecological regimes in terms of their respective metabolic modes of energy and material consumption (Fischer-Kowalski & Haberl 2007: 15–16). Hunter-gatherers depended on passive solar energy use embodied in biomass and animal meat. They were unable to accumulate stocks of material goods other than basic weaponry, and their intervention in natural processes to create more energy was limited to the use of fire for controlling localised micro-territories and for protection against wild animals. Because their innovations were limited, their population numbers remained small and so they existed in a manner that was generally compatible with the sustainability of the eco-systems on which they depended. However, there is considerable evidence that as hunter-gatherers migrated out of the African continent, they eliminated significant numbers of the larger animals they encountered on other continents, which had evolved without the threat of humans carrying deadly weapons, especially in Australia and North America (for an account of this process, see Diamond 1997).

As agriculture spread from its origins in the so-called Fertile Crescent along an East-West axis across the Eurasian land mass, new systems of production and consumption were forged that survive to this day. The key innovations that made this possible were the discovery (mainly in what is now Iraq) of cultivatable seeds, the domestication of animals (cows, horses and pigs) and the construction of agricultural implements (Diamond 1997). For Fischer-Kowalski *et al.*, the agrarian socio-ecological regime was characterised by active solar energy use because agriculturalists transformed natural systems via biotechnologies and mechanical devices. They also accumulated material stocks as they constructed social systems, buildings and weapons to monopolise and control territories with high value resources. However, agrarian socio-ecological societies were always at risk, depending on how successfully they managed 'to maintain the delicate balance between population growth, agricultural technology, labour-force needed to maintain the productivity of the agro-ecosystems, and the maintenance of soil fertility' (Fischer-Kowalski & Haberl 2007: 15). Many of these agrarian societies — especially in Africa — either moved into a new territory after degrading the soils, removing the trees and exhausting water supplies, or they simply withered away over time, or they were invaded by groups with access to more remote eco-systems.

The industrial socio-ecological regime is based on fossil fuels and is currently the dominant economic system globally, even though half the global population is still dependent on agrarian socio-ecological systems (although 40 per cent of the farmers that operate within these agrarian systems are, in turn, dependent on chemical inputs produced by the industrial system). The key innovations relate to the chemical and mechanical devices to extract and harness these fuels. Massive stocks of material resources were accumulated, in particular in urban infrastructures and global transportation systems. Fossil fuels are a limited resource, however, and their consumption has generated greenhouse gases which are rapidly transforming all of the most important eco-systems on which human societies depend. The urban-centred industrial socio-ecological regime will gradually be forced into a transition as fossil fuel supplies dwindle and the environmental effects of consumption undermine the current industrial conditions of existence.

The MFA is a useful framework because it makes a comparison between the agrarian and industrial regimes possible, and potentially sets up criteria for defining a sustainable socio-ecological regime. As Fischer-Kowalski and her colleagues demonstrate, the agrarian socio-ecological regime had a population of less than 40 cap/km^2; energy use of 50–70 GJ/cap/yr; biomass met 95–100 per cent of domestic energy consumption needs; and material use was 2–5 t/cap/yr. Under the industrial socio-ecological regime, by the year 2000 population density was 100–300 cap/km^2; energy use was 150–400 GJ/cap/yr; biomass met 10–30 per cent of domestic energy consumption needs; and material use was 15–25 t/cap/yr. The unsustainability of the industrial regime arises from the fact that the total requirements of the system, as it expands to include another billion potential middle-class consumers, will outstrip the available natural resources and eco-system services, if existing production and consumption systems remain intact. With respect to the UK, during the period of the industrial era (1750–2000) population density increased from 30 to 247 cap/km^2 (making it one of the most densely populated societies in the world); energy use increased from 63 to 189 GJ/cap/yr; biomass met only 12 per cent of domestic energy consumption needs compared to 94 per cent in 1750; and material use increased from 1.7 t/cap/yr to 28.7 t/cap/yr. It follows from this that the transition to a more sustainable socio-ecological regime will mean substantial reductions in these resource consumption levels and related adjustments in well-being and ways of living.

Understanding the crisis

If we now bring together the work reviewed here by Perez on the one hand, and Fischer-Kowalski *et al.* on the other, what emerges is a potentially integrated conceptual framework for understanding global change. For Perez, the MTB of the late 1990s and the subsequent ELB brought on the current global economic crisis, which marks the turning point between the 'installation' and the 'deployment' phases of the Information Age (or what is, in her terms, the *fifth industrial transition*). The extensive state interventions that have taken place across all continents to 'resolve' the crisis are both about damage control and, to some extent, the reorganisation of the global financial system in order to (hopefully) redirect investment back into the 'real economy'. This is, at least, the case that the Nobel Prize-winning, neo-Keynesian economists have been punting for nearly a decade (Krugman 2008; Stiglitz 2010a). Others are far more sceptical about what enlightened states can do by manipulating monetary and fiscal flows to re-establish demand. They call for more radical counter-movements (Patel 2009), a return to publicly accountable banking institutions (Gowan 2009) or an acceptance that, finally, we have hit oil peak as *the* key natural limit to economic growth, which means that any future attempts to stimulate growth will trigger cost spikes that will undermine growth (Heinberg 2009).

Finance capital in the traditional Western economies might be strong enough to prevent or manipulate attempts to redirect financial flows into the 'real economy'. Alternatively, productive capital might be too much of a captive of finance capital, hence unable to exercise its historic post-crisis mission as funder of the deployment phase. What is significant, though, is that China's publicly owned banks (the largest five of which are now within the top 10 largest banks in the world) have vigorously channelled increased investments into infrastructure and production since the onset of the 2007/08 economic crisis. Similar trends are evident across the developing world and parts of Europe, while the cases of Japan and the USA suggest that breaking from the addictions of debt-financed consumerism is much easier said than done.

Just as significant, however, is that information technologies are breaking free from their historic concentration in the financial and services sectors in search of new markets in the 'social' and 'real' economies, emerging beyond the networked nodes of global finance. For Castells, the deployment phase of the Information Age will result in new modes of 'communication power' within the social economy that could revolutionise what citizenship means, reconfigure the traditional relationships between state power and society, redefine what democracy means in practice, restructure the way key public services like health are delivered, and create the opportunity for a new generation of social enterprises to break out of their marginalised niche positions within the global economy (Castells 2009). Perez, in turn, has noted in her most recent work on what the deployment phase could look like that there is evidence of a rising level of investments in innovations that merge the expanded role of information technologies with 'green investments', with a special emphasis on 'smart grids', 'telecommuting', renewable energy systems, digitising manufacturing and a new generation of production-linked financial instruments (Perez 2010). The 2010 World Congress on Information Technology in Amsterdam was devoted to discussions about how the IT industry could provide technology platforms to facilitate more sustainable use of resources in the energy, water, waste, mobility, security, health, governance and urban planning sectors. A central theme was 'smart grids', which are essentially IT systems for digitising the integrated management of urban infrastructures (including health systems) to optimise efficiencies—systems that were included in nearly all of the major recovery investment budgets (especially the USA, China and South Korea).[3] In the introduction to the programme for the congress, the World Information Technology and Services Alliance said: 'ICT developments are fuelling industrial energy savings, smart-grid technologies and new virtual and teleworking opportunities that will form a digital road to recovery' (World Information Technology and Services Alliance 2010).

This raises an obvious question: how do we explain this 'greening' of the 'digital road to recovery'? Our argument is that this can best be explained by seeing the deployment phase of the Information Age as coinciding in some way with the beginnings of an epochal transition from the industrial to a (as yet undefined) sustainable socio-ecological regime. In other words, the limits to the successful unfolding of the deployment phase of the Information Age are not simply the deregulated catastrophes of market fundamentalism, but also the real threats of interlocking resource constraints and eco-system breakdown. Oil is a good case in point.

Heinberg noted that oil output did not rise very much as oil prices increased from 2006 onwards, spiking in July 2008. Why? In short, because supply could not keep up with demand due to the geophysical limits to the amount of oil available (the first real

[3] Mark Swilling was invited to talk at this conference.

signs that confirm the 'oil peak' thesis). So what is the connection between this and the sub-prime crisis that triggered the financial crisis in the USA and, in turn, the global crisis? Citing research by economist James Hamilton from the University of California, Heinberg shows that by 2007/08 the price of houses far from urban centres had started to fall as demand dried up for homes which necessitated long, increasingly costly commutes by private vehicle (Heinberg 2009). In an economy fuelled by a consumer boom, funded by debt secured by property values that were assumed to be permanently buoyant, this was the beginning of the end. And the first domino to fall that triggered this effect was rising oil prices, which is a key symptom of the oil peak phenomenon. For Heinberg, if future growth means another rise in the demand for oil, and if the oil peak thesis is empirically correct, then growth as we know it (that is, growth through consumption spending) will always be tripped up by rising oil prices.

As resource depletion rates express themselves in various (often unpredictable) price trajectories and new investment opportunities, investment patterns emerge that connect the deployment phase of the Information Age with responses to the resource limits of the industrial epoch. The key is whether these investments take place in technological shifts which could result in the successful decoupling of economic growth rates from the rates of resource consumption. This has conceptual and strategic implications for our understanding of the role of the state in this transition (discussed further in Chapter 4) and the spatial co-ordinates of these transitionary dynamics (discussed further in Chapters 5 and 9).

Overlapping industrial and epochal transitions

While there is a gathering mainstream consensus that a business-as-usual approach will mean unsustainable accelerated rates of resource extraction and global warming, there is very little consensus about how to build low-carbon, resource-productive and zero-waste economies that are also more equitable. When Gordon Brown was the UK Prime Minister, he articulated a popular ideological perspective that emerged across various global think-tanks during the 2008/09 period, which depicted the investments that speed up the transition to a more sustainable low-carbon economy as the best kind to resolve the global economic crisis. In his words:

> *There can be little doubt that the economy of the 21st century will be low-carbon. What has become clear is that the push toward decarbonisation will be one of the major drivers of global and national economic growth over the next decade. And the economies that embrace the green revolution earliest will reap the greatest economic rewards... Just as the revolution in information and communication technologies provided a major motor of growth over the past 30 years, the transformation to low-carbon technologies will do so over the next. It is unsurprising, therefore, that over the past year governments across the world have made green investment a major part of their economic stimulus packages. They have recognised the vital role that spending on* energy efficiency and infrastructure *can have on demand and employment in the short term, while also laying the foundations for future growth.* (Brown 2009: 26–27 — emphasis added)

In this optimistic technocratic vision, the transition to a more sustainable socio-ecological regime will be coterminous with the greening of the deployment phase of the Information Age. We are not citing Brown here as an authority, but rather as a good example of the connections made at the popular ideological level between industrial and epochal transition. This is what needs explaining — it is not just a coincidence. It can be explained by overlapping the two scales of transition — that is the transitions within the industrial era conceptualised by Perez, and the transitions from one socio-ecological regime to another conceptualised by Fischer-Kowalski *et al.* Our aim here is not to test whether Brown is correct or not, but to contextualise the significance of the ideological perspective his statement represents.

If the October 2008 financial crisis marks a turning point that could pave the way for an alternative 'back-to-basics' industrialisation of the 'real economy', then this implies increasing material and fossil-fuel consumption for all countries to levels that currently exist only in developed industrial economies. As argued in Chapter 2, the scientific consensus about the nature of the underlying resource base is that this will be impossible. The key material resources (such as fossil fuels, metals, construction minerals and biomass) will simply be insufficient, the atmospheric and terrestrial impacts will be devastating, the existing degradation of eco-system services will accelerate, and the pollution created by waste streams will create increasingly toxic living environments. All of this will intensify resource wars in areas in which remaining resources are concentrated (see Chapter 7). However, if we can imagine a process of global re-industrialisation driven by sustainability-oriented innovations that 'dematerialise' the economies, then a simultaneous industrial and epochal transition becomes theoretically possible. This kind of decoupling may be easier than some economists may think. The historical evidence does suggest that the relationship between GDP per capita growth and resource consumption is not structurally fixed, but is a matter of policy choice (Bringezu *et al.* 2004; Fischer-Kowalski & Swilling 2011). If this is true, it suggests that there is considerable room for manoeuvre through appropriate investments in sustainability-oriented innovations. This, however, is not inevitable and needs to take into account the logic of growth, prices and technological evolution.

To make the links between technological change and economic growth, we turn to the work of UNCTAD economist Charles Gore. He has located the socio-technical cycles described by Perez within the Kondratieff-like 'global development cycle' that took off in the 1950s and ended with the global economic contraction of 2009. As Table 3.2 suggests, this global development cycle is divided into two phases (see rows 1, 3 and 7 of Table 3.2): the first phase comprises a 'spring' of accelerated growth (1950s/60s) followed by a 'summer' of growth deceleration that ended in a stagflation crisis (1970s); while the second phase consists of an 'autumn' of another growth acceleration (1980s/90s) followed by a 'winter' of contraction that began with the technology and liquidity bubbles of the 2000s. The full Kondratieff-type cycle — from 'spring' to 'winter' — is what Gore calls the *long-term development cycle*. Not only is this *growth-crisis-growth-crisis* pattern (typically lasting 50–60 years) common to all five previous Kondratieff cycles, Gore notes that the 'spring/summer' period of all Kondratieff cycles tends to be dominated by heavy investments in new long-term communications, transport and energy infrastructures that are influenced by the technologies of the time. Indeed, the mid-cycle crisis has a lot to do with the over-investments in these infrastructures relative to the level and pace of GDP growth. The next Kondratieff cycle — or, rather, the next *long-term development cycle* that should follow the current crisis — will not be different, hence the importance of tracking current and future investment flows into communications, mobility and energy infrastructures. The kinds of infrastructures that are being designed and built will provide the clues to the nature of the next long-term development cycle. As will be argued in Chapter 5, the most significant are those that will redefine what *urbanism* will come to mean as we head into the next *long-term development cycle*. Significantly, this will be the first Kondratieff cycle that starts with the majority of the population living in cities.

The post-World War II, long-term development cycle (1950s–2007/10) does not correspond to the *socio-technological* cycles that Perez describes (referred to earlier and summarised in rows 4, 5 and 6 of Table 3.2). Although Perez tried to link investments in technology to economic growth cycles, in her later work she gave up this effort. Gore, however, has completed the picture by showing that the 'irruption' of the Information Age (or what is the fifth industrial transition for Perez) in the 1970s/80s, and the subsequent 'frenzy' of the 1990s that led to the 'dotcom' crisis of 2000/01, is what fuelled the second (autumn/winter) phase of the post-World War II development cycle. For both Gore and Perez, if the information technology revolution is to proceed through to the 'synergy' and 'maturity' phases (row 4, last 2 columns), the old socio-institutional order and the out-dated technologies on which it depends will need to be dislodged and replaced. However, for Gore 'it is not only information and communication technologies which could potentially "carry" the next Kondratieff cycle but also new renewable sources of energy and the deployment of a low-carbon economy' (Gore 2010: 725). In other words, the problem is not just institutional blockages for the information revolution, but deeper resource constraints which could undermine the initiation of the next long-term development cycle. What is required, Gore argues, are investments in energy technologies that have been neglected because of the 'political economy of oil' and the associated marginalisation of renewable sources of energy (Gore 2010: 725).

Gore's argument provides the connection we need between the Information Age (that is, the fifth industrial transition) as described by Perez, and the transition to a more sustainable socio-ecological regime anticipated by Fischer-Kowalski *et al.* If Gore's application of Kondratieff cycles to the current crisis is correct, it means that the spring/summer period of the next *long-term development cycle* will need to be powered by the *deployment phase of the Information Age* (the fifth industrial transition in Perez's terms) and the socio-technical imperatives of the transition to a *sustainable socio-ecological order* (what Fischer-Kowalski *et al.* would call the third socio-ecological transition). But this is by no means inevitable. In her recent work Perez is concerned about the logics of the global financial system that could block the shift in power from finance to productive capital that is needed (Perez 2009). Similarly, the interests of the current, global, mineral-energy complex are already blocking the transition to a

Table 3.2: The global development cycle, 1950s–2030s (adapted from Gore 2010)

		1950s	*1960s*	*1970s*	*1980s*	*1990s*	*2000s*	*2010s?*	*2020s?*	*2030s?*
1	Phase of Kondratieff cycle (Gore)	← spring →		summer	← autumn →		winter	← spring →		summer
2	Price cycle	rising price inflation			falling price inflation			rising price inflation (driven in part now by rising resource prices)		
3	Growth cycle	growth acceleration		growth deceleration	growth acceleration		growth deceleration	growth acceleration		growth deceleration
4	Phase of industrial transition (Perez)	← synergy →		maturity	← irruption →		frenzy	← synergy →		maturity
5	4th industrial transition	deployment period of the age of oil & coal, post-WWII institutional matrix			persistence of post-WWII US hegemony plus globalisation, new industrialisers			decline		
6	5th industrial transition			invention, then irruption and frenzy period of the Information Age				deployment phase of the Information Age, new post-oil, post-US hegemony institutional matrix		
7	Nature of financial crisis			stagflation crisis			deflationary depression			
8	Pattern of economic development	equalising			unequalising			equalising?		
9	Resource flows (Swilling & Fischer-Kowalski)	mainly biomass, 10–20 bt/yr	doubling of non-biomass materials, 20–30 bt/yr	non-biomass materials become dominant, increase to 50 bt/yr				two-thirds non-biomass, 60 bt/yr, relative decoupling	absolute resource reduction?	absolute resource reduction?
10	Socio-ecological regime (Fischer-Kowalski)	industrial socio-ecological regime						sustainable socio-ecological regime?		

(Source: Gore, C. 2010. Global recession of 2009 in a long-term development prospective. *Journal of International Development*, 22: 714–738.)

low-carbon economy—this being a critical element of the overall transition to a sustainable socio-ecological regime. It therefore follows that the next long-term development cycle will emerge only when ways are found to reconcile the rapidly evolving information technologies that are moving into a deployment phase in the social and real economies, the sustainable use of resources to counteract rising resource prices and underlying resource depletion (including limits to atmospheric absorption of carbon), and a set of institutional and financial arrangements appropriate for tackling the challenges of an unfair resource-constrained world.

This sums up the challenge facing the world today. We can pretend there are no natural limits to growth and hope that the next long-term development cycle will emerge from purely economic decision-making. In all likelihood, this cycle will be truncated as larger ecological crises overwhelm the global economy harming, in particular, the economies of the new drivers of the global economy (China, India, Russia, Brazil, Indonesia, Turkey, Mexico, South Korea, South Africa, and others). Or we can recognise that rising resource prices reflect a deeper reality which needs to be addressed if we want the next long-term development cycle to result in a world of 9 billion people who have sufficient to live decent lives without destroying the natural systems on which the entire web of life depends. If the next long-term development cycle achieves this, then the sustainable socio-ecological transition envisaged by Fischer-Kowalski *et al.* and the deployment period of the Information Age envisaged by Perez (and Castells) will have been reconciled in ways that fundamentally transform our understanding of progress, prosperity and everyday living. This, in economic terms, is what Lester Brown means when he calls for a programme to 'save civilization' (Brown 2011).

Decoupling and the next development cycle[4]

It will be argued that the notion of decoupling has emerged to address the key condition that needs to be met to ensure that the next long-term development cycle is sustainable. For writers who are critical of ecological modernisation, decoupling is of little use because, in their view, it reinforces the myth that infinite economic growth in a finite world is reconcilable with environmental sustainability (Jackson 2009; Nass & Hoyer 2009). While they are not opposed to investments in resource productivity and reduced environmental impacts, they insist that without 'de-growth' in developed economies, rising incomes and related increases in consumption will always cancel out the gains made by sustainability-oriented improvements and efficiencies (the so-called 'rebound effect' or 'Jevons Paradox'—see Berkhout *et al.* 2000).[5] However, the necessity for economic growth up to a certain level in developing countries is accepted by proponents of this view. The question, of course, is what kind of growth?

4 This section draws on work that Mark Swilling did for the International Resource Panel (see Fischer-Kowalski & Swilling 2011).

5 In any case, one needs to wonder whether the rebound effect in a world of rising resource prices really is as much of a threat as it is made out to be. Jevons wrote at a time when new technologies were making it possible to access vast quantities of cheap coal.

While the critics of decoupling are undoubtedly correct to focus attention on consumption-driven economic growth models, even if the growth model changed as a result of a radical change in the balance of political forces, it will still be necessary to figure out ways to construct high-density settlements dependent on public transport and renewable energy; to build factories which are energy efficient and zero waste; to establish food-production systems which rejuvenate rather than destroy the soils and to construct water and sanitation systems which are efficient and non-polluting. Quite correctly, doing all this merely to increase consumption among those who have too much makes little sense; but it does make sense to do this for the billion or so new consumers who aspire to consume like the average European. The alternative is just far too unimaginative—de-growth for over-consumers, and conventional growth for the under-consumers. Surely this is not the sum total of the sustainability agenda? Surely sustainability-oriented innovations which create new markets, jobs and value chains as the new drivers of the next long-term development cycle are a far more imaginative alternative to this rather stale growth-decoupling dualism? And would this not redefine what growth usually means? Even for the developed world, as many are now suggesting, retaining GDP growth as the measure of progress makes very little sense.

As described in Chapter 2, the International Resource Panel's *Decoupling Report* has shown that by 2005 the global economy needed 60 billion tonnes of materials and 500 exajoules of energy (Fischer-Kowalski & Swilling 2011). At the same time, the report argues that the era of declining resource prices has ended. Although it is suggested that rising resource prices are related to resource depletion (by way of references to declining qualities of extracted materials), the report keeps the door open to the possibility that the upward trend might not last. This is significant, because the economics literature has identified four major economic drivers of rising resource prices: low interest rates (which means there is an incentive to keep materials in the ground thus reducing supply as demand increases which, of course, pushes up prices); speculation as investors move out of financial instruments and into commodities; the China factor (massive demand for materials to feed China's supersonic industrialisation drive); and risk due to geo-political uncertainties and instabilities in areas that are key resource producers (Akram 2009; Frankel and Rose 2009; Kellard & Wohar 2006). Grantham, however, questions these economic drivers and demonstrates convincingly that resource depletion is emerging as a key driver of rising resource prices (Grantham 2011).

What we do not yet know is whether rising prices have triggered new investments in increased output, because it takes on average 11.8 years for initial investments to materialise as productive outputs.[6] However, even if this is happening on the required scale to make a difference, the chances are that this may increase output, but at prices that are unlikely to be low enough (due to declining quality and less accessible points of extraction) to reverse the upward trend in resource prices. At best there might be short periods in which resource prices level out as additional capacity comes on stream and/ or when saturation levels are reached in high-demand nodes such as China, but it is

[6] Personal communication, Prof. Tom Graedel, Yale University, May 2011.

unlikely that the result will be a reversal in the general upward trend over the coming decades. Indeed, if the 'peak everything' thesis is correct, no matter how much is invested the absolute quantities of physical output will not rise (Heinberg 2010). This has major implications for global inequality as higher input costs undermine employment creation in developing countries and rising food prices push more people into poverty. It follows, therefore, that if the next long-term development cycle has any chance of delivering a fairer world, rapid improvement in resource productivity will be an absolute necessity.

To refine our understanding of decoupling we draw on the distinction between resource and impact decoupling proposed by the *Decoupling Report* (Fischer-Kowalski & Swilling 2011). *Resource decoupling* refers to decoupling the rate of use of (primary) resources from economic growth rates, which is equivalent to 'dematerialisation'. It implies using less material, energy, water and land resources for the same economic output. If there is resource decoupling, there is an increase in resource productivity or, in other words, an increase in the efficiency with which resources are used. Resource productivity can usually be measured unequivocally: it can be expressed for a national economy or for an economic sector or even for a certain economic process or production chain, by dividing added value by resource use (for example, GDP/domestic material consumption or human well-being/domestic material consumption). If this quotient increases with time, resource productivity is rising. Another way to demonstrate resource decoupling is by comparing the gradient of economic output across time with the gradient of resource input: if the latter is smaller, there is resource decoupling (see Figure 3.4).

Impact decoupling, by contrast, is what the environmental movement has traditionally been concerned about and refers to the relation between economic output and (various) negative environmental impacts. There are environmental impacts associated with the extraction of resources required (such as groundwater pollution due to mining or

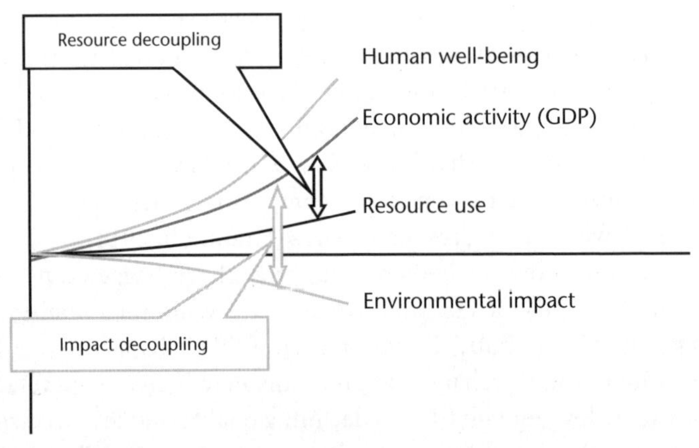

Figure 3.4: Stylised depiction of resource and impact decoupling

(Source: United Nations Environment Programme (2011) Decoupling natural resource use and environmental impacts from economic growth, A Report of the Working Group on Decoupling to the International Resource Panel. Fischer-Kowalski, M., Swilling, M., Von Weizsäcker, E.U., Ren, Y., Moriguchi, Y., Crane, W., Krausmann, F., Eisenmenger, N., Giljum, S., Hennicke, P., Romero Lankao, P., Siriban Manalang, A.)

agriculture), environmental impacts from production (such as land degradation, wastes and emissions), environmental impacts associated with the use phase of commodities (for example, mobility resulting in CO_2 emissions), and there are post-consumption environmental impacts (again, wastes and emissions). Methodologically, these impacts can be estimated by life-cycle analysis (LCA) in combination with various input-output techniques. If environmental impacts become dissociated from added value in economic terms, there is impact decoupling. On aggregate system levels, such as a national economy or an economic sector, it is methodologically very demanding to measure impact decoupling, because there are numerous environmental impacts to be considered, their trends may be quite different or not even monitored across time, and system boundaries as well as weighting procedures are contested.

The notion of resource decoupling can be deployed to evaluate the ecological limits of future growth and development strategies (Fischer-Kowalski & Swilling 2011). Figure 3.5 depicts the scenarios that the International Resource Panel has developed to frame its future work. The baseline refers to the metabolic rate of 55 billion tonnes used in 2000. This equates to a global average of 8 tonnes per capita at 6.5 billion people. If the global economy continues to be managed on a business-as-usual basis, then by 2050, with a global population of 9 billion people, a staggering 140 billion tonnes of used extracted material will be required, which will equate to 16 tonnes per capita (which is equal to what is consumed by the average European). However, if an absolute decline in resource use in the developed world is introduced to reduce material consumption to 8 tonnes per capita in all developed economies, and developing economies implement development strategies to bring them up to this level of consumption, 75 billion tonnes per annum will be required. If, however, a more radical decision is taken to freeze the metabolic rate at 55 billion tonnes, this would result in each person being entitled to consume 6 tonnes of materials per capita by 2050. This will require far-reaching global restructuring and is in line with what is required to achieve the IPCC goal of 2.2 tonnes of CO_2 per capita. Both the 75 billion tonne and the 55 billion tonne scenarios imply radical changes in consumption in nearly all developed economies, where consumption levels in low-density countries such as the USA, Australia and Canada can be around 30 tonnes per capita. They also imply that developing countries should give up on the assumption that the end-points of their development strategies can — or should — be consumption levels of 30 tonnes per capita for their countries. Instead, sufficiency for all over-consumers may well be a precondition for poverty eradication in poorer countries.

In short, the notion of decoupling sets the terms of reference for what a sustainable long-term development cycle would need to achieve: 2.2 tonnes of CO_2 per capita per annum as proposed by the IPCC, and 6 tonnes of resources per capita per annum as suggested by the International Resource Panel scenarios.

If global re-industrialisation during the next long-term development cycle means taking resource consumption up to 140 billion tonnes, then this is a development strategy that is clearly doomed to failure. It is difficult to imagine that there is a respectable scientist today who will be able to generate convincing evidence that the natural resource base and climate system can support these levels of extraction and use. However, if current global

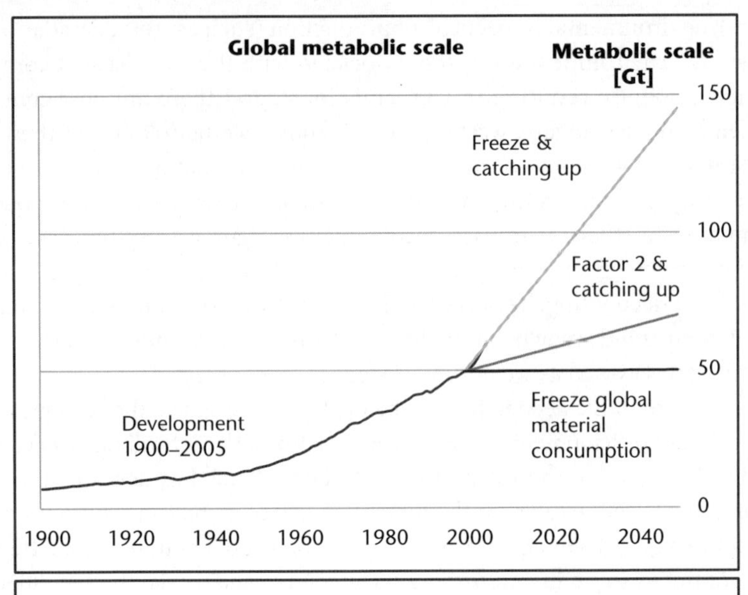

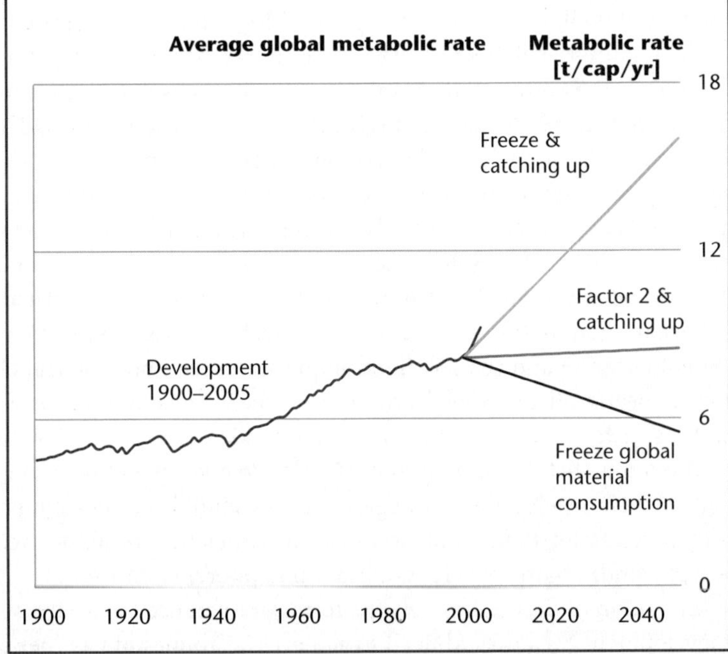

Figure 3.5: Resource use according to three different scenarios up to 2040

(Source: United Nations Environment Programme 2011.)

Note: 'Freeze and catching up' refers to zero growth in resource use in developed economies, while developing economies catch up. 'Factor 2 and catching up' refers to reductions in resource use in developed economies by a factor of 2, while developing economies catch up to this level. The 'freeze' option refers to zero growth in total resource use at global scale.

commitments to ending poverty are taken literally, if the failure to reach agreement at climate talks persists, and if the present system of production and consumption is maintained, it means that there is a wide-spread assumption that it *is* possible to ignore planetary boundaries and drive economic growth by increasing extraction and use from the current 55 billion tonnes to 140 billion tonnes.

Without in any way reinforcing the popular ideological nonsense that the introduction of information technology does by itself somehow lead to reductions in resource use, it is impossible to ignore the fact that ICTs are the *means* for producing new innovations that could massively reduce the energy and material content of production and consumption systems.[7] As the mouthpiece of the ICT industry put it: 'With the digitally connected world, life on Earth has changed and the pace of innovation has significantly accelerated. We no longer live with networking and digital media, but in networks and in media' (World Information Technology and Services Alliance 2010). Innovations are announced daily with respect to new approaches to motor vehicles and transportation, design of buildings and communities, and the re-planning of cities, energy generation, sustainable food production, water supplies, waste management, city management, public health, financial incentives to stimulate green behaviour, new low-energy materials, energy-efficient and zero-waste production systems, re-establishing eco-systems, and the introduction of 're-manufacturing' into industrial manufacturing (for cases and insights from a range of perspectives, see Bang 2005; Beatley *et al.* 2009; Beatley 2000; Benyus 1997; Bioregional 2002; Birkeland 2002; Goodall 2008; Hawken *et al.* 1999; Jackson & Svensson 2002; Jenks & Dempsey 2005; Marinova *et al.* 2006; McDonough & Braunart 2002; Newman & Kenworthy 2007; Organisation for Economic Co-Operation and Development 2002; Revi *et al.* 2006; Satterthwaite 2001; Smith *et al.* 2010; Smith *et al.* 2010; Smith *et al.* 2007; Von Weizsacker *et al.* 2009; Von Weizsacker *et al.* 1997; Wheeler & Beatley 2004; Worldwatch Institute 2007).

More and more peer-reviewed journals dedicated to sustainability are emerging, and established ones devote more and more space to case studies and overviews, while conferences are convened to share learning on virtually every aspect of contemporary economic life. Hundreds of thousands of websites actively promote sustainability innovations (both authentic and hoax) and massive university-based (usually inter- or trans-disciplinary) research programmes have been mounted (quite often as the primary strategic focus for research of the entire university). Paul Hawken's book, *Blessed Unrest,* chronicles the unfolding of a highly networked, global, grassroots movement of community-based and non-governmental organisations which are simply getting on with implementing changes from below, which together have a huge impact (Hawken 2007). At the same time, financial institutions and venture capital funds are rapidly escalating their investments in these new technologies as they see in them potential returns, which

[7] The ICT community and elements of the environmental movement have recently found one another. The result is a new movement that has now even given itself a name, namely the Bright Green movement. This movement likes to distinguish itself from Dark Green perspectives which are depicted as the doom-and-gloom brigade (Robertson 2007).

are characteristically over-optimistic (Goodall 2008). Large-scale public funds are being created to subsidise the transition (such as the Global Environmental Facility and the Clinton Global Fund) and the Stern Report called for the allocation of 1 per cent of global GDP for expenditure on both mitigation and adaptation measures (most of which will be about investments in innovations, both good and bad) (Stern 2007).

How do we make sense of the massive increase in investments in sustainability innovations (managed in and through ICT networks) given the framework of analysis developed in this chapter? One option is that the material and energy constraints of the industrial epoch have been—or will be—recognised, and that the sustainability innovations arc part of preparing the way for the deployment period of the Information Age that will see ICTs become embedded in a revived globalised industrial system that will be far less material- and energy-intensive than industrial production and consumption processes have been in the past (for two of the best recent empirically rich arguments along these lines, see Smith *et al.* 2010; and Von Weizsäcker *et al.* 2009). This is certainly the vision that inspires the IT industry (World Information Technology and Services Alliance 2010). It is also the perspective that lies at the centre of the notion of a Green New Deal—an idea adopted at G20 meetings and which both President Barack Obama and former UK Prime Minister Gordon Brown referred to often during 2009, and which spread globally during 2010 after formal adoption by the United Nations. Although some of the key Green New Deal documents envisage massive investments in 'green industries' to create jobs to resolve the global economic crisis and reduce poverty (Barbier 2009; Worldwatch Institute 2008), mainstream perspectives articulated by political and business leaders have tended to reflect an ecological modernisation approach (Korhonen 2008).

The strategic question this all raises is whether the 'greening' of the deployment period of the fifth industrial transition (that is, the Information Age) will be sufficiently fundamental to facilitate the *third socio-ecological transition* from the industrial epoch to the *sustainable epoch*. If so, we can indeed be optimistic about the possibility of a *sustainable long-term development cycle* emerging from the current crisis. But there is another possibility. Maybe what we are seeing is the beginnings of a *sixth industrial transition* which paves the way for the *third socio-ecological transition* a decade or two from now. In other words, the third socio-ecological transition does not materialise during the first growth period (that is, the spring/summer) of the next long-term development cycle, but rather materialises in the second growth period (that is, the autumn/winter) of the cycle following a mid-point crisis. This crisis would be brought about in part by the ordinary dynamics of economic growth (Kondratieff logic); in part because of the emergence of a new 'green-tech bubble' due to over-investment in technologies that have yet to be deployed as the basis for structuring society (Perez logic); but also in part by catastrophic ecological crises that will be devastating large swathes of the globe (implications of the Fischer-Kowalski *et al.* logic).

These are two different scenarios for what the next long-term development cycle as envisaged by Charles Gore could look like. In the first, relative decoupling means that decarbonisation and dematerialisation are tamed and incorporated into an ecological modernisation agenda, formulated largely by elites gathered at World Economic Forum meetings and related networks. In the second, delayed action to deal with the consequences of rising levels of CO_2 and resource depletion results in shocks and breakdowns which reinforce a more radical transformation driven by grassroots social movements, millions of niche innovations in more sustainable living at the community level, and accelerated investments in the new technologies of decarbonisation, resource productivity and ecological restoration.

The options and variations in between are all possible. How actors respond will be determined by how the consequences of the economic and ecological crises get mixed together and interpreted. Rising food prices and/or the consequences of austerity packages might well trigger grassroots action from below in the spirit of Hawken's *Blessed Unrest*, the contemporary Arab uprising or the street protests of European youth. It could just as easily provide a context for global fragmentation into warring political blocs that ramp up what are currently local resource wars into global resource wars as states respond defensively (with precedents set already in the way they responded in a more limited way to the 2007/08 financial meltdown). These are poles on a continuum with options in between that will be determined by political choices and a wide range of possible strategic coalitions.

These two options can be imagined by extending the logic of the cycles represented in Figure 3.2. In reality, however, there may be a messy mixture of both. Much will depend on the intensity and severity of the socio-ecological crises (in particular resource scarcities and CO_2 impacts) over the short- to medium-term and how these translate into prices, in particular food prices. It is, therefore, most likely that the first phase of the next long-term development cycle will involve a modicum of dematerialisation, resource productivity and decoupling in response to ecological challenges, but this will fall short of the more radical changes envisaged by the *Decoupling Report* (Fischer-Kowalski & Swilling 2011) and key researchers (Bringezu & Bleischwitz 2009; Brown 2011; Brown 2008; Smith *et al.* 2010; Von Weizsäcker *et al.* 2009). Long-term investments in particular technologies and value chains will be made that will result in quite pervasive resistance to disruptive innovations. Our guess, therefore, is that the initial growth phase of the next long-term development cycle will be short-lived (10–15 years) while a set of really disruptive innovations emerge as part of a very different logic that we have (following Fischer-Kowalski *et al.*) referred to as both the third epochal transition (this time from the industrial to the sustainable socio-ecological regime) that will run conterminously with the logic (following Perez) of the sixth industrial transition driven by so-called 'green-tech' investments. This will reach its own 'turning point', but this turning point will not only be driven by over-investments in a new techno-bubble as finance capital once again overreaches itself, but will also be over-determined by the consequences of severe ecological breakdowns and resource depletions, which will have been only partially addressed. If the dominance of financial capital remains intact and productive capital subordinated to a minor role, the end result will be the worst of both worlds: the continuation of a global casino economy with limited benefits for the global poor and mounting ecological crises. The continuous failure of climate change negotiations seems to confirm that global agreements to resolve the ecological challenges are not about to happen soon, and the continued bickering at G20 meetings about the structure of the global financial system suggests that big decisions along the lines suggested by Stiglitz (Stiglitz 2010b) are not being made. As Gordon Brown put it in *Newsweek* in May 2011:

> *[I]f the world continues on its current path, the historians of the future will say that the great financial collapse of three years ago was simply the trailer for a succession of avoidable crises that eroded popular consent for globalisation itself... Those who believe that the world has learned from the mistakes that led to the crash are mistaken.* (Brown 2011)

Conclusion

The increasingly intense interactions within innovation niches that are attracting increasing quantities of public and private investment will not only be benign processes of mutual learning by new networks, but will in time become the basis of new ideological movements and power struggles that will play themselves out in sectoral niches, spaces (new nodal locations of innovation), capital markets and political arenas (new interest-based alignments at local, national and global levels). The patterns that will shape the future are already evident in one form or another, but remain opaque because current paradigms do not have the cognitive capacities to make sense of what is really going on. Things are moving very fast and key concepts lose meaning too quickly for shared understandings to evolve. This is why we have attempted in this chapter to articulate a historical sensibility for comprehending these complex shifts and transitions. We have used this historical sensibility to conceptualise the interplay between the 'greening' of the Information Age (or what at the ideological level is called the Green Economy) and the initiation of a new 'irruptive phase' that could trigger the sixth industrial transition (the makings of a 'green-tech' MTB) that coincides with an epochal shift from the industrial to sustainable epoch. This would constitute the key elements of the next long-term development cycle.

What is less clear now is when this will take place, how it will overlap with the deployment phase of the Information Age, and what role ecological crises will play in accelerating the transition. One alternative is to argue (following Heinberg and Jackson) that significant economic growth in future is unlikely (due mainly to oil peak) and even undesirable from a sustainability perspective (Heinberg 2010; Jackson 2009; see also Nass & Hoyer 2009). For ecological modernisers, growth should not be jettisoned, but rather needs to be reconciled with limits, resulting in what the Netherlands Environmental Assessment Agency calls 'growing within limits' (Netherlands Environmental Assessment Agency 2009).

The alternative, of course, is to redefine what we mean by growth. Instead of deriving this purely from the conclusion that the present global economy is unsustainable, our synthesis of the long-wave historical work by Perez, Fischer-Kowalski *et al.* and Gore makes it possible to trace some of the historical trajectories which could emerge from the current polycrisis, culminating, possibly, in the evolution of a sustainable long-term development cycle.

PART II

Rethinking Development

Chapter Four

Greening the Developmental State

Introduction

The global financial system needs to be restructured to constrain the worst aspects of financialisation and redirect investments into the 'real economy'. Developmental states across the developing world are sharpening their interventionist instruments to target all the holy cows of the neoliberal era: import/export tariffs, deficit spending, expanded fiscal expenditure, controls over capital flows, taxes on repatriated profits, subsidies of key industries and technologies, ensuring that resource rents have developmental benefits, and land expropriations. As G8 and G20 leaders continue to extol the virtues of free markets at international meetings, regulators and businesses are quietly introducing a wide range of protectionist measures. 'Getting the prices wrong' from a pure market perspective may well be the hallmark of the new generation of developmental states. The crash of October 2008 and the (partial) demise of neoliberalism that this represents has effectively cleared the way for some of the more progressive developmental states that are keen to promote a more inclusive form of capitalism than the kind championed by neoliberalism. Some socialist alternatives (whatever this may mean in practice) might also emerge, but probably on the margins of the global economy. China's leaders will feel that their state-managed capitalist model has been vindicated.

The question is whether there is an opportunity for developmental states to take advantage of the transitions underway to mount a progressive pro-poor developmental agenda, or whether they will simply reinforce the inequalities of the current structure of global capitalism. Will these states learn how to do this in ways that sustain rather than destroy the natural resources and eco-systems on which all development strategies depend? And what will it take for these states to realise that unless they address the second question, they are unlikely to find adequate answers to the first?

We attempt to answer these questions by suggesting that we may need to synthesise recent trends in development economics (expressed most clearly in the work of Peter Evans) with insights from ecological economics in order to generate a conception of the developmental state that is appropriate for the era of transition into which we may have moved. Indeed, if the dual industrial and epochal transitions discussed in Chapter 3 have any chance of mutually reinforcing each other, this will be because a theory of the state has emerged which is appropriate to this particular context.

Unfortunately, not all states have the capacity to respond appropriately to rapid economic and ecological change. This is particularly true for many states in the developing world, far too many of which are in Africa. This reality is dealt with more directly in Chapter 7 when we discuss the challenge of resource wars and failed states. So, although this chapter discusses the 'developmental state' in general terms, it is not about the 'state in the developing world,' which is a subject that is far too complex to deal with here. Instead we have in mind a wide range of state formations in mainly developing countries (some more democratic than others) that have some sort of capacity for strategic analysis and policy-driven intervention. Discussions of the developmental state are not usually about the North American, Western European, Scandinavian, Japanese and Australian/New Zealand states. Even the Chinese state is often excluded because of its unique character. However, much of the theoretical argument about the need to synthesise ecological and development economics in order to reinvent our conception of the state has relevance for those contexts in which there is a coherent state formation with the capacity for significant interventions to reshape the political ecologies of their respective countries.

Defining the developmental state

The discussion about governance, growth and development that has taken place mainly in Europe and North America in recent decades has been dominated essentially by the various alternatives within the Keynesian welfarist paradigm and versions of contemporary neoliberalism. Whereas the former was about full employment, high taxes, expanded welfare services and extensive state intervention, the latter was about markets, privatisation, deregulation, low taxes (in theory) and a minimalist role for the state (again, in theory). However, since the reconstruction of Japan and Taiwan after World War II, followed by the rapid development of the Asian Tigers since the 1960s, there has been a third governance trend which was neither Keynesian welfarism nor neoliberalism; instead it can be referred to as the developmental state. Developmental states have tended to be interventionist, productivist, ideologically opportunist, protectionist, obsessed with industrialisation (to 'modernise'), resource intensive and quite often authoritarian — this is why Korean economist Ha-Joon Chang calls them the 'Bad Samaritans' (Chang 2007). The decline of neoliberalism in many developing countries, the apparent 'success' of state-managed capitalism in China and Singapore, the resurgence of nationalist economic policy-making (from Russia to Malaysia), the survival of large chunks of social democracy in Europe (even under right-wing governments in Germany, France and more recently, The Netherlands), the emergence of some stable developmental states in Africa over the past decade, the rise of the neo-Keynesians in Latin America (now no longer demonised by the so-called 'post-Washington Consensus'), and the re-appearance of socialist options (up until recently in Venezuela, or the 'Kerala Option' mounted by successive Communist Party governments in India's state of Kerala since the 1950s) has triggered a rapid expansion of literature on the developmental state, which is now concerned with a much wider agenda than simply state-driven industrialisation (Amsden 1989; Amsden 1995; Bagchi 2000; Beaseley-Murray *et al.* 2009; Chang 2002; Chibber 2002; Chibber 2003; Evans 1995; Evans 2006; Haggard & Kaufman 1992; Hobson & Ramesh 2002; Isaac & Franke 2000; Jayasuriya 2001; Johnson 1982; Khan 2004; Leftwich 1995; Leftwich 2000; Mkandawire 2001; Onis 1991; Ritzen *et al.* 2000; Swilling 1999; Swilling 2008).

Chalmers Johnson is the pioneer of the concept of the 'capitalist developmental state', elaborated in his monumental study of Japanese industrialisation (Johnson 1982). The distinguishing feature of the developmental states is that their 'political purposes and institutional structures ... [are] developmentally driven, while their developmental objectives ... [are] politically driven' (Leftwich 2000).

The legitimation of these developmental states derived primarily from their ability to promote sustained growth and development via aggressive industrialisation. Their sustained growth and development is derived from their unique ability to:

- Extract and deploy capital productively
- Generate and implement national and sectoral plans
- Effect dynamic egalitarian and productivity-enhancing development programmes in land, education and training, small enterprise, infrastructure and housing sectors
- Manipulate private access to scarce resources through, amongst others, financial sector re-engineering, subsidies, taxes, concessions and high levels of lending
- Cultivate close and productive relationships with business, wherein state leadership is more important than its followership
- Manage interest groups through state corporatism (authoritarian top-down imposition of the state's agenda versus social corporatism)
- Coordinate the efforts of individual businesses by encouraging the emergence and growth of private economic institutions
- Target specific industrial projects and sectors
- Resist political pressure from popular forces and, at times, also brutally suppressing them
- Mediate and/or insulate their domestic economies from (extensive) foreign capital penetration
- Most importantly, sustain and implement a project of productivity improvement, technological upgrading and increased market share that break them out of their existing path-dependent economic trajectory (Khan 2008).

The institutionalisation of the developmental state has received much attention over the past decade. Leftwich summarises what many regard as the key institutional characteristics of the developmental state as follows (Leftwich 1995):

- A 'determined developmental elite' committed to the modernisation project
- 'Relative autonomy' from major capitalist economic interests who are always keen to capture the state
- 'A powerful, competent and insulated economic bureaucracy' that enjoys the highest possible political support but operates without too much political interference
- A 'weak and subordinated civil society' which means there are no rival centres of alternative policy formation
- The 'effective management of non-state economic interests' via formal structured compacts, incentives and penalties
- Accessible and usable institutions of 'repression, legitimacy and performance'.

Significantly, the developmental state is often regarded as appropriate to Asia and certain Latin American countries, but not to Africa. What emerges in the literature on Africa is the so-called 'impossibility thesis': what has obviously worked for other 'late industrialisers' is simply a non-starter in Africa. While it is now admitted that the state has played a central role in the development of Asian countries, the replication of the Asian experience is regarded by most Western and African scholars as impossible for Africa. The reasons given in the literature often include the following factors: dependencies of various kinds, including those created by debt; the lack of ideology; the 'softness' of the African state and its proneness to 'capture' by special interest groups; the lack of technical and analytical capacity; the changed international environment, which does not permit protection of industrial policies; and a long record of so-called 'weak governance' and poor performance, including a growing list of 'failed states' (Mkandawire 2001). There is some truth in this scepticism, but this way of seeing the problem becomes a self-fulfilling prophecy rather than the basis on which to build a more positive, optimistic image of a developmental African state.

For at least the past two decades, developing countries have been bullied into accepting the key ingredients of neoliberalism's global developmental ideology — what many have called the 'Washington Consensus'. Despite the fact that neoliberalism emerged in the late 1970s/early 1980s as a programme of Western governments to reinvigorate growth (following the stagflation crisis of the 1970s) by dismantling the Keynesian state within these specific contexts, it was regarded as a developmental panacea for every country, no matter the massive contextual differences (and even if in practice the opposite happened, as in the case of many Asian NICs). This ideology legitimated the ascendance of finance capital as the strategic co-ordinator of global capitalism and 'development' in the global South within the wider context of the globalisation of the information and communication age. The key ingredients were as follows: free trade, capital market liberalisation, flexible exchange rates, market-determined interest rates, deregulated markets, privatisation of state-owned assets, fiscal restraint, balanced budgets, tax reform, secure property rights, decentralisation and the protection of intellectual property rights. The underlying logic was that the large welfare states in the developed world and the bureaucratised developmental states in the developing world were responsible for slow and unequal growth. Neoliberal development economists argued that if restrictions on capital flows were removed and an export-friendly global market created, investments would flow to poor countries where capital intensity was lower, labour was cheap and export-led growth would stimulate development in environments in which savings and tax bases were too small to finance accelerated growth.

Significantly, as evidence mounts of the failure of these policies, neoliberal economists are homing in on capital market liberalisation as a key problem (Held 2004). Stiglitz argued that the East Asian and Latin American crises in the late 1990s show that 'premature capital market liberalisation can result in economic volatility, increasing poverty, and the destruction of middle classes' (quoted in Held 2004: 5). A 2003 study by IMF economists concurred when they concluded that 'there is no strong, robust and uniform support for the theoretical argument that financial globalisation *per se* delivers a higher rate of economic growth' (quoted in Held 2004: 5). By the end of the first decade of the twenty-first century two Nobel Prize winners for economics, Joseph Stiglitz and Paul Krugman, were arguing that there was now a consensus in economic policy circles that liberalisation of markets for short-term capital can be detrimental and should be approached with caution; that poverty should be measured using education and health, as well as income; and that excessive corporate power (market and political) is a problem (Krugman 2008; Stiglitz 2010). They both blamed neoliberal market economics and excessive deregulation of the financial sector for the economic crisis and both insisted on some form of re-regulation of the banking sector.

Warnings about the negative consequences of neoliberal 'one-size-fits-all' solutions comes after nearly three decades of 'advice' to southern governments to the contrary. Warnings from alternative voices from the South were arrogantly ignored. Dissidents were suppressed, often with brute force. The damage was done, now it's time to pick up the pieces.

Rethinking the developmental state

The key question is not whether there should be greater state intervention in the coming decade, but what kind of interventionism and to what end. If we want to 'green' the next long-term development cycle, much will depend on how the state's role and purpose is redefined. The great legacy of the history of the twentieth-century developmental state is the notion that an appropriately configured and governed state has a key role to play in the development process. But in order to conceptualise this role for a very different future, we must realise that the twenty-first century developmental state sits at the intersection of the simultaneous transitions taking place at the industrial and socio-ecological scales discussed in Chapter 3. Our core question is therefore: *can we conceptualise the kind of developmental state that will be needed to facilitate the next long-term development cycle (that both deploys the Information Revolution and ensures a transition to a post-industrial sustainable epoch)?* Our answer is that such a state must have one overriding priority, namely investments in *innovations for sustainability.* By this we mean new ways of building up an appropriate mix of knowledge, capabilities and investments to drive new value chains which simultaneously create productive employment and reduce dependence on increasingly expensive primary materials and fossil fuels. To understand this proposition, we need to rethink the developmental state by synthesising the recent trends in institutional and ecological economics. Although these two fruitful fields stem from vastly different conceptual points of departure, they are converging in ways that make it possible to conceptualise what could be called the *innovational developmental state,* namely a developmental state which invests in sustainability-oriented innovations as an explicit way to drive job-creating growth.

Since at least 2002, but accelerated since the start of the USA-centred financial meltdown in 2007, the notion of a 'developmental state' has returned as the central actor in the global discussion about development policy and strategy. In 2002 the governing party in South Africa, the African National Congress (ANC), formally adopted the 'developmental state' as its official ideology, distancing itself (at least rhetorically) in the process from neoliberal economic prescriptions. For the first time in many years, it is possible for mainstream players in developing countries to be critical of neoliberal orthodoxy and question the logic of what some have called 'capital fundamentalism' (Evans 2005: 91; King & Levine 1994). But is the twenty-first-century development state the same thing as the twentieth-century developmental state? Must it be configured to focus exclusively on modernisation through industrialisation as the Korean-born, Cambridge-based economist Ha-Joon Chang, and others argue (Chang 2007)? Many in the South certainly think so, with the South African government—as Pillay points out—a good example of this kind of political perspective (Pillay 2007). However, as this chapter argues, we need to recognise that the twenty-first-century developmental state is emerging as something quite different from its twentieth-century forbears.

The emergence of diverse experiences of economic development and industrialisation has contributed to a realisation that context matters—or as Dani Rodrik, a leading Harvard economist, put it in the title of a recent article: 'World too complex for one-size-fits-all-models' (Rodrik 2007). This realisation has contributed new strands of thought within development economics, which together pose a serious alternative to the assumptions and prescriptions of neoliberalism (Evans 1995; Evans 2005; Evans 2006). Key threads of an emerging consensus include the critique of neoliberal economic history by Ha-Joon Chang (Chang 2002; Chang 2007); the sustained critique of traditional economistic definitions of development by Indian economist and Nobel Prize winner Amartya Sen (Sen 1999); the new institutionalism of Harvard's Dani Rodrik[1] (Rodrik 2000; Rodrik *et al.* 2004); the neo-Keynesian economics of former World Bank Chief Economist, Joseph Stiglitz (Hoff & Stiglitz 2001; Stiglitz 2010) and his fellow Nobel Prize winner, Paul Krugman (Krugman 2008); Vivek Chibber's seminal study of the Indian phenomenon that reveals the central significance of the developmental state (Chibber 2002; Chibber 2003), and Ching Kwan Lee's remarkable exposé of the real world of the China 'miracle' which demonstrates the institutional drivers that created the space for massive inflows of capitalist investment (Lee 2007).

As already suggested, the core project of the twentieth-century developmental state was to accelerate massively the traditional Western pathway to industrialisation, namely the transition from an agricultural to a manufacturing-based economy via managed industrialisation (Chang 2002; Evans 1995; Leftwich 2000). The underlying economic theory that drove this was, of course, the notion that economic value derived from the build-up of stocks of material and physical plant, infrastructure and consumable goods. One advantage of manufacturing-based industrialisation is that it fits well with nation-building projects that inevitably get symbolically represented in grand modernist city-building initiatives. This provides the legitimation that states required to prioritise industrialisation over profits and to limit excessive inequality via taxation and targeted interventions. The key to success was massive investments in education and human

1 Dani Rodrik is one of the so-called 'Harvard Economists' who have advised the South African presidency.

capital within an urban spatial hierarchy, which was planned to absorb large investment in economic infrastructure to cope with high rates of urbanisation and to create an operating framework for heavy industry. The other key, of course, was systematic exploitation of natural resources and eco-system services with used extracted resources rising from 40 billion tonnes per annum to 55 billion tonnes during the last 20 years of the twentieth century characterised by massive global growth (Behrens *et al.* 2007; Krausmann *et al.* 2009).

Trends in development economics

Following Evans (Evans 2005; Evans 2006), three strands of mainstream thinking in development economics can, when read together, create the basis for rethinking the future of the developmental state. These are endogenous growth theory (Aghion & Howitt 1998), institutional economics (Rodrik *et al.* 2004) and Amartya Sen's theory of capabilities for development (Sen 1999).

By the end of the twentieth century, endogenous growth theory had become a mainstream trend in economic theory (Aghion & Howitt 1998). The core logic of this theory is that the diminishing returns on increasing capital intensity are counteracted by the transformative effects of investments in human capital and technological innovation. Simply put, as profits from an increasingly capital-intensive system of production decline, productivity improvements and innovations create new systems of production which generate higher investment returns. The result, as reflected in the work by Perez and many others, is an interlocking set of innovation-driven economic cycles that drive the overall process of economic development. These innovations, however, cannot be regarded as externally derived intrusions into natural growth cycles as defined by neoclassical economic theory, but are 'endogenous' to the overall pattern of capitalist economic development and dependent on the existence of a range of incentives to stimulate innovation, many of which are a function of policy decisions, institutional arrangements and cultural capacities. Unsurprisingly, for innovations to translate into more profitable systems of production, a stock of skilled labour and an appropriate institutional context is required. In other words, if one accepts that innovations are the key to interlocking continuous and discontinuous growth cycles, one has to accept that these, in turn, depend on the existence of an institutional context that ensures the existence of an appropriate set of incentives for innovation and a supply of appropriately skilled labour. It does not 'just happen' in accordance with some hidden economic law.

In theory, endogenous growth is good news for developing countries because ideas and technologies may seem cheaper to acquire and develop than large stocks of physical capital and infrastructure. But in reality, technologies and brands are captured via intellectual property regimes that ensure returns via royalties and licence fees to developed countries, and skilled labour tends to concentrate (or migrate to) localities in developed countries that have desirable working and living conditions and can pay the higher salaries for their skills. In short, ideas and technologies are contested terrain.

Ha-Joon Chang captures the dynamic of what he calls the technological 'arms race' when he writes:

> *As water flows from high to low, knowledge has always flowed from where there is more to where there is less. Those countries that are better at absorbing knowledge inflow have been more successful in catching up with the more economically advanced nations. On the other side of the fence, those advanced nations that are good at controlling the outflow of core technologies have retained their technological leadership for longer. The technological 'arms race', between backward countries trying to acquire advanced foreign knowledge and the advanced countries trying to prevent its outflow has always been at the heart of the game of economic development.* (Chang 2007: 127)

The combined impact of the ICT revolution and finance-led globalisation made it pretty obvious that value (or more precisely profits) were no longer derived primarily from the stock of physical assets and machinery. Long, complex, globally articulated value chains emerged that linked manufacturing in low-wage, low-margin localities (mainly in developing countries) to high-cost, high-margin centres (mainly in developed countries) that retained control of the intellectual property rights, brands, distribution outlets and trade logistics. Following Castells, without the mushrooming of computer-processing power, this would not have been possible (Castells 1997). The most profitable links in these new value chains derived their high returns from intangible assets, protected by intellectual property regimes, proprietary operating routines/systems, knowledge-intensive, brand-marketing machines, and centrally controlled distribution and tracking systems.[2] But while all this was unfolding, development policies in developing countries remained wedded to the traditional assumption that capital flows in search of higher returns would make it possible for developing countries to 'catch up'. To put it crudely, development — and therefore 'catch up' — was equated to numbers of manufacturing jobs (that is, industrialisation) while simultaneously the ICT revolution made possible a new 'space of flows' that redefined development and the nature of capital accumulation (Castells 1997). Hyper-financialisation of this new 'space of flows' then made it possible to reinvest these global surpluses in the speculative investment instruments that became the rocket fuel of the techno-financial bubble that eventually burst in 2007/08.

Endogenous growth theory is significant for our argument because it places innovation and not capital at the centre of the growth and development story. Innovations become embedded in production systems when economic agents respond to incentives to innovate, generated by the policy and institutional environment. As Evans then argues:

---

2 For a remarkable case study, see Crestanello, P. & Tattara, G. 2006. *Connections and Competences in the Governance of the Value Chain: How Industrial Countries Keep their Competitive Power.* Oslo and Tallinin: The Other Canon Foundation and Tallinin University of Technology. *Working Papers in Technology Governance and Economic Dynamics,* 7. This study reveals how the globally branded northern Italian clothing and shoemakers (e.g. Benetton) relocated their production facilities to Romania to utilise cheaper labour, which made it possible to increase company profits massively by retaining control of branding, marketing, design and the continued production of the most high-value products. The actual profit margins on manufacturing in Romania were in fact very low, but knowledge and skills transfer to Romanians did start to stimulate new — albeit constrained — economic opportunities on the margins.

'The fact that institutions not only mould the incentives to generate new ideas but can be seen themselves as essentially constituted by "ideas" completes the logic that binds the new growth theory to the institutional turn.' (Evans 2005: 94)

Based on comparative analysis of historical trends, the 'institutional turn' in economics further consolidated the critique of capital fundamentalism. In a remarkably bald statement in a mainstream textbook defining 'modern economics', Hoff and Stiglitz wrote: 'Development is no longer seen primarily as a process of capital accumulation, but rather as a process of organisational change' (Hoff & Stiglitz 2001). Development economists write as if institutions are a new discovery. Strangely, they do not recognise that which many outside of their exclusive club (particularly in the urban research and public management fields) have been saying for over two decades, that it is the quality of public institutions that really matters when it comes to achieving positive development outcomes, and not simply the logic of capital flows (see Benington & Moore 2010; Moore 1995; Pieterse 2008; Swilling 1999). Nevertheless, now that the economists are saying this, the policy community is taking note.

Institutional economists are primarily interested in the institutional functionality and integrity of the entire macroeconomic system of a particular country and how this interrelates with the institutional and market dynamics of the global economy (Chang 2002; Commission on Growth and Development 2008; Evans 2005; Hoff & Stiglitz 2001; Rodrik *et al.* 2004). Their argument is that the formal and informal rules of the game are determined by the institutional arrangements that pertain in a given country. These institutional arrangements, in turn, are the product of particular leadership styles, organisational cultures, social histories, moral norms (with respect, in particular, to corruption), informational flows, transactive relations, and constellations of ideas about institutional life. Given this, the formal and informal rules that govern economic life and market behaviour become, over time, deeply entrenched in the cultures of everyday life. These path-dependent behaviours can be conducive or dysfunctional for growth and development in ways that are context specific. If most are dysfunctional, there is very little that can be done to change this in the short term, often with dire economic consequences (for example, Zimbabwe since the 1990s). Significantly, institutional economists in their academic roles recognise the futility of setting up predetermined images of the 'right institutional mix' that can get imposed across all contexts (although in reality they often cannot resist the temptation to do so when playing out their roles as policy advisers). In a revealing passage, the *2008 Report of the Commission on Growth and Development* (which reflects the new post-Washington Consensus mainstream orthodoxy in the global development finance institutions) argues that '[i]n recent decades governments were advised to "stabilize, privatize and liberalize"... But we believe this prescription defines the role of government too narrowly' (Commission on Growth and Development 2008: 5). This report proposes a classic institutionalist alternative when it argues:

> *[M]ature markets rely on deep institutional underpinnings, institutions that define property rights, enforce contracts, convey prices and bridge informational gaps between buyers and sellers.*

> *Developing countries often lack these market and regulatory institutions. Indeed, an important part of development is precisely the creation of these institutionalized capabilities... However,* we do not know in detail how these institutions can be engineered, *and policy makers cannot always know how these institutions can be engineered, and policy makers cannot always know how a market will function without them.* (Commission on Growth and Development 2008: 4 — emphasis added)

This kind of thinking, which has been around in the World Bank since at least 2000 (Ritzen *et al.* 2000), is not only a recognition of the economic efficacy of governance, public management and politics. It also reflects the recognition, by economists, of the voices of countless development practitioners across the developing world, who have been saying for decades that development is not simply a function of capital investments and accumulation. Symbolic and cultural values of human capabilities for development are the narratives used in practise to articulate what the microeconomics of development in the global South is all about (Max-Neef 1991; Mbembe 2004; for an overview see Pieterse 2008: 108–130; Shaffer 2008; Simone 2004; Swartz 2010). Development practitioners have known all along that development programmes work only when the experience and ideas that are deeply embedded in dense networks of lived social relations are directly tapped and mobilised to animate development processes.[3] In both cases, however, it is the quality and configuration of institutions and networks that make it possible — or impossible — to transform embedded and/or commoditised ideas into the drivers of economic growth and development.

Although growth theory and institutional economists spearheaded a critique of capital fundamentalism, these approaches still tend to measure development in terms of GDP per capita growth. Amartya Sen makes the final break from capital fundamentalism (although not from the liberal methodological individualism that underpins it) by replacing this reductionist measurement of development with the elegantly simple notion that development is about building human capabilities for activating developmental processes. This is the theoretical construct that inspires the logic of the globally accepted Human Development Index and the writing of the UNDP's annual *Human Development Report*. Sen goes further, however, by arguing that expanding human capabilities to achieve the developmental goals of a particular group or nation is directly dependent on the creation of democratic spaces for public discourse about what these goals should be, how best to achieve them and what roles different collectivities can play in the various development processes.[4] In other words, unlike the Asian developmental state that bureaucratically determined these goals 'from above', a democratic developmental state aims to facilitate the creation of a 'deliberative democracy' with institutionalised spaces to realise Sen's vision of *development as freedom* (Evans 2006; Sen 1999).

---

[3] This is not to suggest that these ideas can be extracted from these networks — they are accessed only by working with the networks themselves.

[4] Sen does not, in fact, have a notion of collectivities as social actors — this is an interpretive elaboration of his logic. See Evans, P. 2002. Collective capabilities, culture and Amartya Sen's 'development as freedom'. *Studies Comparative International Development.* 37(2): 54–60.

Herein lies the importance of Sen's notion of a deliberative democracy as the institutional context for expanding the 'capabilities for development'—a logic Peter Evans captured recently with the notion of a 'capability enhancing developmental state'.[5] Once capability enhancement is seen as both the means and ends of development, then a reductionist exclusive focus on capital flows becomes, quite simply, ridiculous.

Given the theoretical convergence around technological innovations, institutional functionality, and capabilities discussed above, it is possible to argue that the twenty-first-century developmental state has three basic tasks (Chang 2002; Evans 2006). Firstly, if institutions are key to an environment that fosters innovation, networks and new value chains as the driver of growth, the leadership capabilities to build effective institutions—and networks of institutions—across all sectors becomes the main challenge. This will mean striking a very delicate balance between regulation of shared norms/values and self-managed implementation. The growing consensus amongst those who think about 'corporate citizenship', for example, is that this is clearly about what corporates agree is appropriate behaviour, but state intervention is needed because this will not happen voluntarily (Hamman 2006). For Gelb, the state must demonstrate it has the capacity to 'discipline' particular business interests to fit into the wider strategic direction (Gelb 2006). Similarly for those interested in a strong civil society—this does not imply a weak state. Public leadership, therefore, is critical: capable, uncorrupted political leaders, who are accountable (without depending on media spin or shadowy thuggery), are a necessary condition for building the institutions that foster the key ingredients of a 'capability enhancing', innovation-driven, developmental trajectory, namely trust, reciprocity, mutuality and creativity.

Secondly, no one disputes that knowledge and innovation matter, but these are (using complexity language) emergent properties which stem from dense networks of people, working together across institutional boundaries, unconstrained by outdated (usually hierarchically organised) norms or an atmosphere of fear and conformity. The private sector will always under-invest in human capital, innovation and networks because the direct returns to the investor are impossible to predict. Without state-led investment in these sectors, via universities, NGOs and developmental partnerships/compacts, knowledge-based, innovation-led, economic development will be impossible.[6] These are the alliances and processes that drove the Baltic economic miracle with a population similar in size to that of southern Africa, and the Kerala model that succeeded in raising the HDI of a portion of India to Scandinavian levels without high levels of economic growth. This is what drives local economic development from the simplest agri-centres to the most complex global cities.

[5] Talk presented at the Conference on Democracy and Developmental State in the 21st Century, Rosa Luxemburg Foundation: Johannesburg, 25–27 May 2008.

[6] For an excellent overview of the innovation literature, which has really only developed into a distinct research field since the early 1990s, see Lundvall, B.A. 2007. *National Innovation System: Analytical Focusing Device and Policy Learning Tool.* Ostersund, Sweden: Swedish Institute for Growth Policy Studies. Working Paper R2007: 004.

Thirdly, embeddedness for the twenty-first-century developmental state might mean partnering more with networks of civil society formations, trade unions and small, entrepreneurial associations, than with the business elites that are now fully consolidated in most places, especially in middle-developing economies such as South Africa, Brazil, India, South Korea, Philippines, Turkey, Mexico, Czech Republic, Malaysia and Nigeria. A weak national bourgeoisie is a good reason for the state to get involved in welding together local business elites. But in situations where the national bourgeoisie is fully consolidated, the state has more freedom to integrate a wider set of class alliances. Herein lies the significance of Sen's notion of 'development as freedom'. If money on its own could resolve poverty, poverty eradication would not be so difficult to achieve. Effective solutions are context-specific, which means partnering with the requisite know how, and this is rooted within civil society — especially when it comes to the sprawling, informal cities that dominate the rapidly expanding developing economies. What the trade unions, community-based organisations, NGOs, entrepreneur associations, faith institutions, science and research organisations, and the cultural arts community really need is a state which knows how to engage, listen and co-produce public goods and spaces. This entails a multiplicity of smallish interventions, rather than a few, massive, physical infrastructure investments which satisfy the need for capital deepening (or increasing investment in fixed assets), but do little to redefine the institutional context for the circulation of the benefits beyond the elites that make the investment decisions.

Since 2002 the ANC government in South Africa has shifted away from a neoliberal discourse favouring, rather, a developmental state discourse. Fortunately, the 1990s 'one-size-fits-all' neoliberal state approach to development is dying. This does not mean that the Asian model is the only alternative that is appropriate for the current context. Experimentation is the order of the day across the globe. From Chavez's Venezuela to the China boom, from the Baltic miracle and India's unique Kerala model, to Cuban self-reliance and calls for an 'African way' (Mkandawire 2001), now is the time for innovation and creativity. Public investments in skills, education and human capital were the key to success in each case. An open society that is free to debate development goals is a necessity if it is accepted that creativity is a key energiser and driver of development. Authentic empowerment of the poor is a precondition for success and a strong trade union movement is indispensable. And when developmental local governments think in these terms, remarkably creative initiatives are often the end result. Unfortunately, with some exceptions, it is the more limited, pro-business, industrialising Asian model that seems to be prevalent in the current South African policy discourse on the developmental state (Pillay 2007), despite the fact that South Africa has a consolidated national business class, a well-organised civil society, serious ecological challenges and a proven track record of limited social returns on capital deepening (Frankel *et al.* 2006).

Most recent writing on the South African developmental state remains sceptical of the proposition that replicating the 'Asian tiger' model to build an 'African lion' is possible under current global conditions of capital accumulation, characterised by de-industrialisation everywhere except China and (to some extent) India. Gelb seems doubtful that government can broker a state–labour–business 'pact' aimed at reconciling

Black Economic Empowerment[7] and export-led growth, and he is sceptical about the state's capacity to discipline business investment strategies to favour nationalist priorities (Gelb 2006). Southall questions whether the requisite state capacity exists to effectively implement a fully fledged developmental state approach (Southall 2006); and Pillay argues that the tendency to favour the Asian model, coupled with the way the progressive trade unions mistakenly supported Jacob Zuma for the Presidency, has resulted in the failure to consider seriously alternatives to the Asian model (Pillay 2007). We have questioned whether the state-led infrastructure investment programme will deliver the economic growth rates required to reduce the 40 per cent unemployment levels, due to the impact of oil price hikes, electricity blackouts caused by inappropriate energy policies, a debt-saturated middle class, and investments in carbon-intensive infrastructure projects and resource-intensive industries, which have grown at the expense of nearly all the other sectors (Swilling 2007; Swilling 2008).

These debates highlight the enormous difficulties that progressive political forces in the developing world face when it comes to building the kinds of developmental alliances required to realise substantive public value within a globalised economy, which works to the advantage of the developed economies. The advantage of the current conjuncture, however, is that these countries can benefit from the breakup of self-righteous, neoliberal orthodoxy, which has given way to the recognition of context, experimentation and home-grown solutions. The convergence of growth theory, institutional economics and Sen's theory of capability building bring to centre stage the strategic significance of innovation at many different levels: technological innovations to create new productive value chains; institutional innovations to create new spaces for skills development, social learning and cultural evolution; and relational innovations that open up spaces for deliberative democratic engagements that build capabilities for development. The question we pose is: innovations for what? Chang ends his remarkable book with one answer:

> *Having accepted that increasing capabilities is important, where exactly should a country invest in order to increase them? Industry — or, more precisely, manufacturing industry — is my answer.* (Chang 2007: 213)

We agree, but in line with the overall argument of this chapter, this is subject to three provisos. The first is whether the crash of October 2008 will, in fact, be a 'turning point' that results in the subordination of finance capital to productive capital so that we can shift into the next long-term development cycle (with the deployment phase of the information era as the economic operating system). The second is whether investments in innovation, knowledge and human capabilities, rather than simply physical assets, are recognised as the key priorities of the developmental state. And thirdly, development economists will need to accept that this will happen only if this process of (re-)industrialisation is part of a wider transition to a sustainable socio-ecological regime. Without substantive equitable decoupling and massive increases in resource

[7] This refers to the South African government's policy of promoting black ownership of South African companies and capital assets.

productivity, (re-)industrialisation on a global basis that contributes to general welfare for all will turn out to be a false promise if the ecological constraints are ignored. We need to turn to ecological economics for guidance as to how to think about global re-industrialisation in a way that avoids the kind of ransacking of global resources that previous growth spurts have engendered. Indeed, this may even entail completely rethinking our conception of growth

Trends in ecological economics

The search for a synthesis of economics and ecology has a short history, going back to MacNeill (1991),[8] Boulding (1991) and Costanza *et al.* (Costanza *et al.* 1993) who published their work in the wake of the *Brundtland Commission Report* in 1987. These texts, together with the seminal texts of the late 1990s by Daly (Daly 1996), Ekins (Ekins 2000), Von Weizsäcker *et al.* (Von Weizsäcker *et al.* 1997) and Douthwaite (Douthwaite 1999), established the foundations for the current flowering of ecological economics as an influential (ideologically diverse) discipline sustained by a wide range of peer-reviewed journals, with *Ecological Economics* being the best known (for recent contributions to the discussion from different perspectives, see Ehrlich 2008; Gallopin 2003; Greenwood & Holt 2008; Korhonen 2008; Nass & Hoyer 2009; Reed 2001; Shiva 2005; Smith *et al.* 2010; Sneddon *et al.* 2006; Sterner 2003; Swilling 2010).

Paul Ehrlich's famous formula locked one strand of ecological economics into the language of 'impact' and trade-offs — an approach which has been depicted as 'ecological modernization' (Korhonen 2008). Ehrlich's formula was as follows: I = P x A x T where I = environmental impact, A = affluence, and T = technology (Ehrlich & Ehrlich 1990). The implication was that to reduce impact you must reduce population growth and/or levels of affluence and/or change technologies to use less resources to create the same or an improved level of output. A focus on population growth meant talking about reducing birth rates and, by implication, ignoring that it is the billion over-consumers who are the real problem. Focusing on affluence is often interpreted as meaning over-consumers must sacrifice their lifestyles; and that the billion potential 'new consumers' in the developing world must give up the middle-class dream. Both are politically risky for most governments, despite the fact that over-consumption remains the biggest threat to the planet. So, not surprisingly, the focus has shifted to technological innovation. Inspired by key texts such as *Natural Capitalism* (Hawken *et al.* 1999) and *Factor Four* (Von Weizsäcker *et al.* 1997), and now by the Bright Green movement (Robertson 2007), the core argument is that much more can be done with much less. The key is huge investments in technological innovations aimed at massively increasing the productivity of key natural resource inputs. The carrot, as the slogan goes, is 'green is black', which means there are profits to be made in investments in 'green technologies'. This is what makes it possible to argue that growth and sustainability no longer need to be seen as polar opposites (Lovins *et al.* 2002; Lovins 2005; Smith *et al.* 2010). Unsurprisingly, this

[8] MacNeill was one of the co-authors of the *Brundtland Commission Report.*

approach aims to appeal to businesses to take the lead. The state is depicted as only a policy enabler (by, for example, ceasing to subsidise unsustainable, resource-intensive production and consumption systems), the research capacity for innovation is assumed to be evenly spread across the developed and developing world, and the effects of ever-tightening intellectual property regimes, which restrict access to technologies by developing countries, are often ignored.

Costanza *et al.* go a little further than a faith in technology and markets and have, instead, tried to theorise a systems model of a sustainable economy (Costanza *et al.* 1997a). Indeed, some who think in these terms are sceptical of how much can be achieved via innovation-driven resource productivity within the existing capitalist economic framework (see, for example, Ayres *et al.* 1996; Nass & Hoyer 2009). Costanza's model (Figure 4.1) tries to achieve what most ecological economists have in

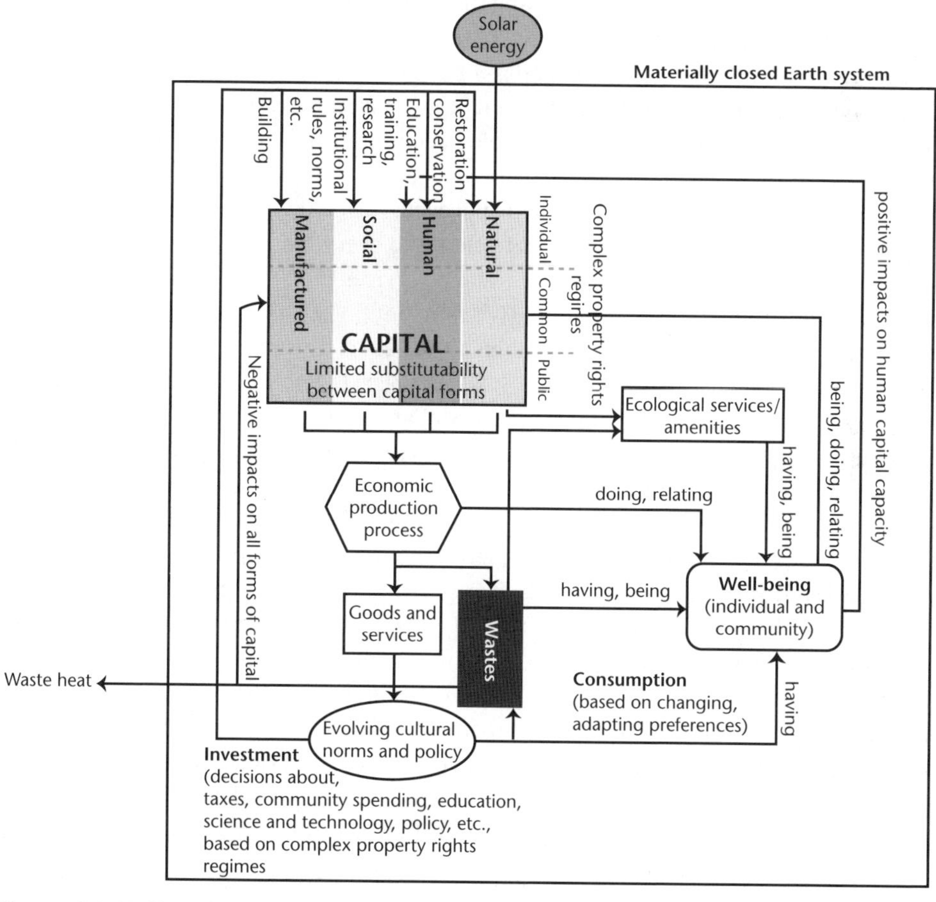

Figure 4.1: 'Full world' model of the ecological economic system (adapted from Costanza *et al.* 2008)

(Source: Costanza, R., Cumberland, J.C., Daly, H.E. Goodland, R. & Norgaard, R. 1997. *An Introduction to Ecological Economics*. Boca Raton: St Lucie Press, p. 275.)

mind, namely a 'materially closed' system, except for solar energy inputs and natural heat losses. It implies a zero-waste economy in which all outputs are inputs, eco-system services have financial values so that they can be factored into the economic equation, and natural capital stocks are maintained over time. This provides the basis for the attempt by Costanza *et al.* to value the world's eco-system services. In the second most cited article in environmental studies over the past decade, they concluded that the world's eco-system services are worth, on average, US$33 trillion per annum, whereas at the time of the publication, annual global gross product was worth US$18 trillion (Costanza *et al.* 1997b).

Although these approaches envisage far-reaching changes to the current economic system (including, for Costanza *et al.*, a break from private ownership as the only type of ownership), they shy away from radical state-led redistributive perspectives. Nor is it ever made clear what this all means for developing countries, where growth is unavoidable when it comes to establishing the basic material infrastructures for poverty eradication (such as urban infrastructures, transport systems and housing) and where an aggressive middle class of up to a billion people aspires to consume like the average Californian. Callously telling them to give up the American dream when a billion others continue to live it is how the notion of 'ecological limits' can be construed as an apology for the status quo.

The alternative may be a rights-based approach, which depicts the natural resources of the world as part of the commons that belongs to everyone, thus limiting our right to exploit these natural resources and eco-system services (what is sometimes referred to as the 'strong sustainability' principle) (Ostrom 1999). Two perspectives are pertinent here—'ecological footprinting' (Chambers *et al.* 2001; Costanza 2000; Gasson 2002; Hammond 2006; Hubacek & Giljurn 2003; Wackernagle & Rees 2004) and 'environmental space' (Buhrs 2007; Carley & Spapens 1998; McLaren 2003).

Both ecological footprinting and environmental space approaches take as their point of departure the notion that there are a finite set of quantifiable natural resources which are globally available to a given population. To this extent they go beyond the *Brundtland Report* which insisted there are no 'absolute limits'.[9] By dividing the resources available by population size, it is possible to derive a per capita 'fair share' that can then be used as the basis for a global social movement—aimed eventually at attaining a new global deal—to equitably redistribute the planet's remaining natural resources. For ecological footprinting, though, all resources are reducible to a productive land value which, in turn, translates into an 'ecological footprint' per capita which now stands at between 1.5 and 1.8 ha/cap, down from over 5 ha/cap a century ago when the population was much smaller (World Wildlife Fund *et al.* 2006; World Wildlife Fund 2008). Ecological footprinting is a surprisingly popular

[9] See World Commission on Environment and Development. 1987. *Our Common Future.* Oxford: Oxford University Press. p. 8—this notion that there are no 'absolute limits' also laid the basis for the notion of 'intersubstitutability', i.e. that one form of capital (natural capital) could be substituted for another (human, financial).

method, even within mainstream business networks, which are fond of connecting it to performance indicators.

For those who use the 'environmental space' argument, resources are not reducible to a productive land factor. Instead, each resource must be treated separately in ways that are appropriate in order to arrive at a per capita allocation (for example, CO_2 t/cap/yr; metals t/cap/yr; or barrels of oil /country/yr). This then makes it possible to determine who is over- and who is under-consuming each resource which, in turn, creates the basis for globally negotiated redistributive interventions (via new combinations of taxation, pricing, licensing, subsidies and commodity trading systems, etc.). The environmental space approach is clearly the most radical and idealistic, but there are already examples of pragmatic proposals which reflect this underlying approach. These include the 'cap-and-share' approach to carbon, which is the alternative to the current 'cap-and-trade' approach (which lies at the centre of the Kyoto Protocol). A 'cap-and-share' approach would value what is left of a particular resource and then divide this by the total population to arrive at an equitable share per person. Each person, therefore, will be allocated some sort of voucher, equal to their inalienable right to a portion of the value of the global commons. Given that people in developed countries will have more than their fair share (that is, their current share of the value of the commons will be greater than the value of their voucher), it follows that further consumption can happen only when people in poorer countries are compensated for the fact that they each have had less than their fair share. So, for example, a global tax could be made applicable which, over a period of time, rectifies the imbalances. A 'cap-and-share' approach to the commons would be the essence of a politically radical agenda that would undermine growth in the developed world and finance accelerated growth in the developing world.

All perspectives within ecological economics share the view that the current mainstream economic growth model, as implemented in most developed and developing economies, is inconsistent with the finite nature of the planet. Ecological economists may use very different conceptual languages and they disagree on whether or not it is necessary to dismantle capitalism to achieve sustainability, but all imply that a time will come when policies will be required that will radically change the economic growth model. As the South African policy community knows, it is one thing to agree that a system must change, but it is quite another to implement changes over time because of the complexities of institutional change, social learning and the capacity for innovation. The MFA approach, discussed in Chapters 2 and 3, will be useful when it comes to implementing change, because it makes it possible to calibrate the modalities and temporalities of relative decoupling and absolute resource reduction with respect to both material inputs ('resource decoupling') and environmental impacts ('impact decoupling'). This needs to be done in ways that take into account institutional capabilities and the learning process.

If the transition to a sustainable socio-ecological regime means anything, it will mean investing in innovations that will demonstrate practical ways of promoting non-material economic growth, which will require minimal quantities of non-renewable resource inputs, the reuse of all waste outputs, and replenishment of renewable resources that have been over-exploited (such as fish, soils, water supplies, air quality, biomass and biodiversity). Although many development economists will find it impossible to reconcile non-material growth with poverty eradication, ecological economists insist that the poor will be the biggest losers in an increasingly unsustainable world.

Rethinking innovation

Much will depend on what we mean by innovation and how serious developmental states are about investing in the human, relational and institutional preconditions for really transformative innovations. The problem with the national innovation systems that have been promoted by many governments around the world over the past two decades is that they have been aimed almost exclusively at promoting economic growth, with very little attention paid to the various dimensions of decoupling (cleaner production being an obvious major exception, plus sector-specific investments in innovation, such as renewable energy in Germany). Running in parallel with this has been the critique of growth by ecological economists, who have argued that growth is responsible for environmental destruction, which is now undermining traditional growth models. In other words, innovation is not in and of itself a good thing from a sustainable resource management perspective. Maybe the word itself is irredeemably tied to creating more rather than less and should be replaced by a new word, such as *exnovation* so that the intention is clear from the outset. What is obvious is that a new conception of innovation is required. Carlos Montalvo, a researcher at TNO in The Netherlands, has proposed the notion of 'sustainable innovations', which deserves further exploration (Montalvo 2008), while others have started to refer to 'sustainability-oriented innovations' (Stamm *et al.* 2009).

Sustainability-oriented innovations (SOIs) are specific interventions (at the technological, institutional and relational levels) that result in the dematerialisation of economies by increasing resource productivity and reversing environmental degradation. Such innovations will take to scale what is already emerging from life-cycle and material flow analysis, ecological design, environmental impact and fair trade thinking: long-life design instead of rapid obsolescence, zero-waste, cleaner production, product stewardship, re-manufacturing, factor 4/5 improvements in productivity, fair trade agreements, green taxation, and so on. As Montalvo puts it: 'The challenge lies in replacing a large proportion of our current technological stock with new technologies underpinned by new science and applied knowledge that do not violate but accommodate the first and second laws of thermodynamics' (Montalvo 2008).

Four key insights can be drawn from the conventional literature on innovation, which are relevant when it comes to understanding SOIs (Lundvall 2007), namely:

1. Innovations are different to inventions — an invention is when a new idea emerges for a new product or process, while an innovation is the synthesis of the idea with a complex set of financial and institutional arrangements to implement the new idea on a broader scale

2. Innovations are not random events, but are the function of specific incentives and investments
3. Innovations do not arise from single individuals or single firms, but rather from well-networked economic agents, working collaboratively with knowledge institutions (such as universities) and in ways that are open, creative, problem-driven and connected to learning from practice
4. Innovations are not about building up stocks of knowledge capital (patented ideas) created for trade in the so-called 'knowledge economy'; innovations are continuous learning processes that are responsive to the fact that in a highly complex, globalised world, fixed bits of knowledge rapidly become obsolete — we live, therefore, in a learning economy, not a knowledge economy.

Innovation, however, is not simply about technological solutions (the so-called 'techno-fix' approach). We need to think of innovation as a process that manifests at three different levels, namely:

1. Technological innovations — specific techniques for managing/processing materials and energy (for example the steam engine, hydrogen fuel cell, micro-chip, or a process that achieves more with less)
2. Institutional innovations — for managing on a society-wide basis — or even globally — incentives, transaction costs, rents, benefit distribution, dispersal, contractual obligations, precautions, and individual obligations
3. Relational innovations — for managing co-operation, social cohesion, solidarity, social learning and benefit sharing.

If growth and development are dependent on the capacity for innovation, what are the implications for developing countries, which lag behind developed economies when it comes to scientific and technological capacity? Many economists, influenced by Homer-Dixon's notion of an 'ingenuity gap' (Homer-Dixon 2000), are pessimistic about the possibility of developing countries 'catching up', precisely because they will never be able to bridge the 'ingenuity gap'. However, this implies that developing countries should catch up to a level of economic development which is now regarded as ecologically unsustainable. If it is not about catching up with what is unsustainable, but about accelerating the spread of a sustainable economic alternative, are developing countries really at a disadvantage? Following Montalvo (Montalvo 2008), developing economies may actually have an advantage over developed economies with respect to SOIs in the following respects:

- Given that multi-stakeholder co-operation is a precondition for innovation, firms that have dominant positions in the market will tend to resist change and protect their technologies (what innovation researchers call 'rigidities'). Emerging firms in developing economies may be less rigid because they do not have dominant positions to defend.
- Key barriers to the diffusion of new technologies are often the dominance of current technologies, which often enjoy protection (via laws, incentives, investments) from regulatory and financial institutions, and consumer cultures are difficult to change.

Regulatory regimes in developing countries are often less restrictive or more permissive.
- From experience it is known that firms and countries can 'leap-frog' stages in the process of developing technological capacity—developing economies have less baggage in terms of non-sustainable production and consumption infrastructures, which means they have an inherent potential for leap-frogging.
- Markets in developed economies are more saturated than in developing economies, thus opening the way for new technologies and products that can meet the same needs in different ways (which presumes a level of protectionism at certain points in the development process).
- Consumers in developed economies are accustomed to mass consumption of products that are designed for rapid obsolescence—a consumer culture which will be hard to change—whereas new consumers could be more adaptable to long-life use if product design moved in this direction.
- Expensive 'high tech' solutions (with costly IP implications for developing countries) will not necessarily result in decoupling—instead, SOIs with the highest impact may involve the reconfiguration of the services delivered by open source or current technologies (whose patents may have expired).

Montalvo concludes: 'Given the above, it is likely that current systems of innovation in developed economies face more rigid structures than those in developing economies, effectively giving a head start to developing economies' (Montalvo 2008). Although much research is needed to substantiate these provocative propositions, this way of thinking does suggest that developing countries may find that sustainability-oriented innovations could be the source of some surprisingly productive growth drivers. If we look at SOIs in places such as Costa Rica, South Korea, Cuba and Sweden, the evidence suggests that it may, in fact, be smaller developing countries that have the most to gain from being first movers. This important conclusion should be considered seriously by developing countries that insist on clinging to the notion that conventional growth must come first, followed by environmental clean-up, despite mounting evidence that resource constraints will undermine this conventional development trajectory.

Towards a synthesis

We can now bring together the key threads of the arguments developed thus far. Firstly, we argued in Chapter 3 that we are witnessing a messy intersection between two transitions—an industrial and epochal transition. This provides a context for the rise to prominence since the onset of the financial crisis in 2007/08 of the notion of a 'low carbon economy' or Green New Deal (Barbier 2009). We are going through a crisis that marks the turning point for our Information and Communication Age, which should culminate in power shifts from finance to productive capital. This, however, is not inevitable—the consolidated power of finance capital may well be an obstacle when it comes to finding long-term investments in the 'real economy'. Substantial state intervention will clearly be required to release investment from the grip of financial capital and to restructure the institutions and norms of society to facilitate the full-scale deployment of ICTs across the entire system of production and consumption. If ways can be found to constrain the scourge of hyper-financialisation, and if the cost of labour in China continues its upward trend (in particular via rising investments in health care, rural consumption and infrastructure), the pre-conditions may fall into place for global (re-)industrialisation as the basis for a new era of redistributive prosperity. But, it has been argued, this will be short-lived if the ecological limits to our current growth models are ignored. The transition to a sustainable socio-ecological regime will be driven by the effects of global warming and the depletion of strategic resources (such as oil, water, atmosphere, soils and certain key strategic metals and minerals). Following the logic of ecological economics and the accepted science of the IPCC and International Resource Panel, radical decoupling will be required to bring down the consumption of materials by some developed economies from over 30 t/cap/yr to 6 t/cap/yr within a matter of decades. This will be painful and entail radical changes to a consumption-driven growth model. Equally, developing economies will be forced to rethink the basis for their growth strategies to eliminate poverty, because there is no way they can do this if it means increasing material consumption per capita to levels similar to those that are assumed to be the norm in developed economies. While developed economies face the painful challenge of radical decoupling, developing countries face the challenge of inclusive equitable growth by investing in consumption and production systems that decouple rates of growth from rates of resource consumption and associated environmental impacts. Put simply, without decoupling, global (re-)industrialisation becomes a scientific impossibility.

Unfortunately, the language of transitions, industrial ages and decoupling is often devoid of a sense of politics and agency. Things don't happen because they should, but because choices are made within specific contexts informed by a particular understanding of what is going on. This is why it is so important to bring the developmental state into the wider discussion of industrial and epochal transitions. Building on the work by Evans, we have argued that knowledge, capabilities and innovation have become the key foci of developmental strategies. To reinforce this, we elaborate a framework for thinking about SOIs as key to developing practical ways of achieving non-material growth.

Transitions to more sustainable ways of using resources are evident across the world at global, national and local levels (as documented by Smith *et al.* 2010; and Von Weizsäcker *et al.* 2009). These transitions have invariably been achieved by reconfiguring institutions (and also often communities and neighbourhoods) around a new set of ideas in response to eco-system breakdowns and socio-economic needs, and the result has often been the release of new investments as new value chains are created out of sources of value that were previously ignored or suppressed. In most cases, the existence of well-networked stakeholders, inspired by visionary/creative leadership, is what made the advances possible. Barriers to these advances which favour the old ways of doing things (for example, subsidies, legislation, marketing power, R&D funding) were ever-present, but not insurmountable. These new value chains often undercut the constricted value chains that are so tightly controlled by the established (often monopolised) corporate sectors and the constraints they impose via their globally protected patent regimes and highly geared financial arrangements.[10] Alternatively, they create new flows of value from which established corporates are absent, because their costly systems prevent adequate returns from these niche markets. And, of course, the impact of new technologies and consumer cultures create new value chains, as the costs of the old technologies rise to critical thresholds (in particular in the energy, built environment, waste, transportation, water and food production fields).

These trends suggest that sustainability challenges could provide a new context for innovations, which could lead to major new valuc chains with significant positive implications for employment creation and market expansion within *developing* economies. Following the logic of institutional economics, this will not happen via private investment alone, because the returns on investments in social learning accumulate largely (and unpredictably) within the public sphere via open systems and knowledge networks, instead of into privately owned, tightly controlled intellectual property regimes which often lack the high-speed innovation capacity that open source systems offer (Weber 2004). If environmental public goods are left to the market, the result will be failure. This was empirically demonstrated by Harriss-White and Harriss who revealed that climate change mitigation in the UK is a government priority, but implementation is left to the private sector. As a result, very little progress has been made (Harriss-White & Hariss 2006). This may suggest that when it comes to measuring up to the task of cataclysmic ecological breakdown that could detrimentally harm billions of people, private capital is rather weak and disorganised. It also suggests the need to move beyond 'ecological modernisation', which is the approach most favoured by global business groups, because they would like to achieve incremental sustainability via market-driven innovations that keep the regulators out (Korhonen 2008). Following the rationale for a developmental state, a weak private sector and the need for public investments in innovation are the classic arguments in favour of state intervention and/or (re-)regulation to prepare the conditions for an adequate response. In particular, it will mean substantial public investments in social learning for sustainable living at all three levels of the innovation process — technological, institutional and relational.

Conclusion

We have attempted to synthesise the new institutional and ecological economics in order to conceptualise what could be called a *sustainability-oriented, innovational developmental state* which has the ideational and institutional capacity to facilitate the ex-/in-novations required for the dual transitions that are necessary at the industrial and epochal levels. We have referred to technological, institutional and relational innovations

[10] For instance, although 'big pharma' are the profit giants of the world, their record of innovation is pitifully poor precisely for this reason.

that will, inevitably, pan out in complex, contradictory and uneven ways. The most ideal scenario is least likely. This would involve moving beyond the October 2008 crash into the deployment phase of the ICT-based industrial transition using technological innovations which ensure substantive decoupling and dematerialisation — or put simply, the 'greening' of global (re-)industrialisation, or what global leaders have started to refer to as the 'Green New Deal'. For this to happen, far-reaching institutional innovations will be required to rebuild developmental states that are, once again, prepared to lead the development process via investments in innovation, which favour well-supported formations of productive capital, bio-economic local diversification, new value chains, capability building and massive labour absorption. This, following Sen, will entail relational innovations which promote open, deliberative, democratic practices for fostering capabilities for development.

The least desirable (and hopefully least likely) outcome would be haphazard global economic recovery, with finance capital still enjoying the upper hand and weak states doing little to support the ascendance of productive capital. Limited (re-)industrialisation will take place. Poverty will deepen and the transition to a sustainable socio-ecological regime will be delayed causing massive suffering for billions of people as eco-systems start collapsing. Bewilderment and narrow nationalism will most likely be strengthened as innovations at all levels get minimised and even suppressed. The number of so-called 'failed states' in the developing world will increase rapidly, and fanatics could well come to power in developed countries as electorates get whipped into a frenzy of fear by reactionary fundamentalists.

A more likely scenario is a series of stop-start contestations as these longer-term transitional trends pan out in both contradictory and complementary ways. Productive capital may well win out across key sectors of the global economy, but only if assisted by developmental states with the necessary capabilities and if the costs of production continue their current upward trend in China (thus making profitable manufacturing more possible elsewhere). This will pave the way for the next long-term development cycle, which will be path-dependent initially with respect to more traditional resource-intensive technologies. Far-sighted developmental states which understand the sustainability challenge may actively invest in sustainability-oriented innovations to reverse this, but this may well end up being too little too late. In short, the result will be a relatively short-lived and incomplete global (re-)industrialisation process that will falter on the rocks of increasingly serious eco-system breakdowns. If China remains a resource predator and artificially keeps its currency devalued, the severity of these breakdowns could be exacerbated. By delaying the much-needed but inadequately acknowledged transition to a more sustainable socio-ecological transition, two things may happen. The first may be the triggering of a sixth industrial transition that will come to be known as the Age of Sustainability, which then simultaneously drives the transition to a more sustainable socio-ecological transition. This could take 15–30 years to reach its turning point and deployment phase. Evidence for this comes from financial analysts who have already started to talk about a 'green-tech bubble' as venture capital pours into the new so-called 'green-tech' start-ups. The problem is that, from an ecological perspective, very few scientists think we have 15–30 years to avoid some of the most severe natural disasters and ecological crises. Conversely, it is precisely these accelerated crises on which the new 'green-tech investors' are banking to render redundant the resource- and energy-intensive socio-technical systems that currently underpin a politically powerful complex of economic interests.

Alternatively, the second option is that the fifth transition may stumble, but get salvaged by just sufficient levels of investment in innovation to stave off the most serious ecological collapse(s) (especially those affecting the rich nations). This will more than likely involve hugely expensive, grand techno-fixes which, in the end, will exacerbate the severity of the crises some decades down the line. An alternative to this, of course, is a messy mix — failures to green the fifth transition will spur the innovations required to trigger and drive up investments in the sixth transition. How this pans out will be dependent on whether a leadership can emerge which is capable of comprehending the dynamics of epochal transitions as they manifest themselves within the seemingly mundane dynamics of everyday crisis management.

It will be impossible to determine at any moment which trajectories are at play, or which forces have the upper hand. Like lightning on a dark night, moments of crisis might reveal key markers here and there. In the meantime, the task will always be to live the future in the present by turning sustainability-oriented innovations into a way of life.

Chapter Five

Rethinking Urbanism

In his painting the city became a demented maze, clogged and vibrant with bright colours, but seen through an effluvial mist. He caught the truckpushers, the carriers of heavy loads, the hawkers, the traffic policemen in their orange uniform tops. He caught the hundreds of feeder roads, the paths, the streets, wild lines that lead only into confusion. He caught the streets of cars jammed and crooked. He even managed to convey the dramatic gestures of Lagosians in their frozen, angled positions. The agonies and the comedies of the city. Framing the haste and frenzy was the lagoon. It was of the same green as his painting of the scumpool. All roads lead into the maze of the city. The chaos and the frustration of the city. But the only ways out lead to the forests of the interior and to the sea.

He looked at his work and, with despair and joy in his soul, he thought: 'Art is a poor approximation, but the best we have.'

(Okri 1996, *Dangerous Love*)

Introduction

Okri's description of Omovo, the Lagosian artist, has a ring of truth about it — art may well be all we have to capture the 'agonies and the comedies of the city'. Nevertheless, admitting to a lesser art for the job, we will attempt to connect the discussions of transition and the developmental state to an equally significant socio-demographic transition, namely the transition to a predominantly urban world — a process driven by the twin forces of 'informational' economic globalisation and the 'second urbanisation wave'. Cities are the geographical spaces in which there are the greatest concentration of overlapping networks of actors engaged in imagining, mediating, contesting and deploying the differentiated modalities and temporalities of these complex industrial and socio-ecological transitions.

Following the work of Erik Swyngedouw, Simon Marvin and others who are using actor-network theory and material flow analysis to reconceptualise cities (Graham & Marvin 2001; Graham 2010; Guy *et al.* 2001; Heynen *et al.* 2006; Hodson & Marvin 2009a; Hodson & Marvin 2010a, 2010b), we extend our understanding of global material flows from Chapters 2 and 3 into an understanding of material flows through the networked infrastructures of the cities, and how urban actors engage to contest the nature, impacts and benefits of these flows. In doing so, we are motivated by the same concerns expressed by Nik Heynen *et al.* who introduce their edited collection entitled *In the Nature of Cities* by arguing that:

> ... *it is surprising, therefore, that in the burgeoning literature on environmental sustainability and environmental politics, the urban environment is often neglected or forgotten as attention is focused on 'global' problems like climate change, deforestation, desertification, and the like. Similarly, much of the urban studies literature is*

symptomatically silent about the physical-environmental foundations on which the urbanisation process rests. Even in the emerging literature on political ecology, little attention has been paid so far to the urban as a process of socio-ecological change, while discussions about global environmental problems and the possibilities for a 'sustainable' future customarily ignore the urban origin of many of these problems. Similarly, the growing literature on the technical aspects of urban environments, geared primarily to planners and environmental policy makers, fail to acknowledge the intimate relationship between the antinomies of capitalist urbanization processes and socio-environmental injustices. (Heynen *et al.* 2006: 2)

Our search to 're-nature the urban' is obviously heavily influenced by our experiences in building the Lynedoch EcoVillage, described in the Introduction and in more detail in Chapter 10. By building in ways that re-established natural systems and resources, we became obsessed with the actual socio-technical connections between urban systems and the eco-systems within which they are embedded.[1] When we decided to use earth worms rather than chemical systems to process our sewage, this connection became a rather intimate affair. But this was not just about an ethical urge to 'live in ecologically sustainable ways': it was also because worms do for free what mechanical or chemical systems do at a financial (and usually ecological) cost.

Our Lynedoch experience has led to the realisation that in searching for connections between 'urban systems' and 'eco-systems' we need to complement the traditional concerns of urbanists with structure, physical spaces, boundary lines, technical constraints, spatial functionality, the architecture of built forms and even the image of interlocking systems, and look in relational ways for *processes* which connect urban living, working and playing with the resources that flow into, through and then out of urban systems—what is referred to as the 'socio-ecological metabolism' of the city (Girardet 2004; Guy *et al.* 2001; Heynen *et al.* 2006). Cities are not fixed, physical artefacts or historical subjects, nor are they simply spaces within which other things happen. Cities are, pre-eminently, emergent outcomes of complex interactions between overlapping socio-political, cultural, institutional and technical networks, which are in a constant state of flux. Vast socio-metabolic flows of material resources, bodies, energy, cultural practices and information work their way through urban systems in ways that are simultaneously routine, crisis-ridden, unpredictable and transformative.[2]

1 It is interesting to note that others with a similar motivation had similar experiences—see Revi *et al.* 2006.

2 By depicting the city this way, we are obviously using the language of complexity theory. However, we want to avoid the increasingly common approach of assuming *in advance* that cities are complex systems, and then proceeding by simply fitting the empirical material into these predetermined categories in the hope that the predetermined relationships between the categories will generate the analysis automatically, so to speak (for examples of this mistaken application of complexity theory, see Allen 1997; Pulzelli & Tiezzi 2009; Rotmans 2006). The basic logic of this type of argument usually runs as follows: 'cities are a complex system', and 'because they are complex systems' (established purely by describing them in the language of complexity), they must now be managed as complex systems (and to this end a set of truisms are drawn down from the mountain of literature on managing complex systems). There is, unfortunately, no substitute for the hard work of doing empirical research on the actual relations that may or may not mean that cities are complex systems—indeed, it is doubtful whether the notion of a 'city' denotes the boundary of a viable system at all. It is the relations and networks that make up a city that are the real (potentially complex) systems.

Seeing cities this way means paying attention to the role that urban infrastructural networks play in the metabolism of the city. Here we mean almost anything that conveys a flow of some sort, such as the roads, pavements, footpaths, electrical cables, canals, sewage pipes and waste sites, dams, aquifers, and water lines, fibre-optic cables and communication lines, servers, vehicles, telephone lines, food supply lines, airports, shipping and railway lines, quarries and soils — all the networks on which the users of the city depend for their well-being and survival. But none of these networks exist outside of everyday life: we take them for granted as we go about our daily business, but every one of them has been imagined, designed, negotiated, funded, constructed and managed by a range of actors along the way, all of whom do what they do with particular outcomes in mind, which cannot be divorced from their socio-cultural backgrounds, value-preferences and aligned interests. We are, as Latour argues, '*many participants* ... gathered in a *thing* to make it exist and to maintain its existence' (Latour 2004: 246). This is the idea at the centre of Swyngedouw's conception of the *cyborg city* — half human, half machine, the city becomes a writhing mix of material, cultural and institutional flows (Swyngedouw 2006).

Contrast Okri's depiction of the flows of an African city with Swyngedouw and Kaika's description of someone at the epicentre of the flows that produce that great icon of urban modernity — Piccadilly Circus:

> *Imagine, for example, standing on the corner of Piccadilly Circus and consider the socioenvironmental metabolic relations that come together and emanate from this global-local place: smells, tastes, and bodies from all nooks and crannies of the world are floating by, consumed, displayed, narrated, visualised, and transformed. The Rainforest shop and restaurant play to the tune of ecosensitive shopping and the multibillion pound eco-industry while competing with McDonald's burgers and Dunkin' Donuts; the sounds of world music vibrate from Tower Records and people, spices, clothes, foodstuffs, and materials from all over the world whirl by. The neon lights are fed by energy coming from nuclear power plants and from coal or gas burning electricity generators. The coffee I sip connects me to the conditions of peasants in Colombia or Tanzania and to the Thames River Basin as much as to climates and plants, pesticides and technologies, traders and merchants, shippers and bankers, bosses and workers. The cars burning fuels from oil-deposits and pumping* CO_2 *into the air, affecting forests and climates around the globe, further complete the global geographic mappings and traces that flow through the urban and 'produce' London's cityscape as a palimpsest of densely layered bodily, local, national, and global — but geographically depressingly uneven — socioecological processes. This intermingling of things material and symbolic combines to produce a particular socioenvironmental milieu that welds nature, society, and the city together in a deeply heterogeneous, conflicting, and often disturbing whole.* (Swyngedouw & Kaika 2000: 568)

Whether we are with Okri in a Lagos traffic jam or with Swyngedouw and Kaika in Piccadilly Circus, what distinguishes each experience is the unique pattern of flows that makes that particular place and moment what it is. What allows us to distinguish one place from another is not reducible to a particular economic structure or spatial configuration or mode of governance, but rather a recognisable *pattern* created by the complex mix of flows that cannot be predicted with any precision prior to the actual experience (and analysis) of the context.

In this chapter we start off by describing the dimensions and dynamics of the second urbanisation wave. We then locate the second urbanisation wave within the wider context of globalisation and the Information Age, which have transformed the connections between cities and parts of cities. With this as background, we proceed to argue, following recent work on the socio-metabolism of cities (Guy *et al.* 2001; Heynen *et al.* 2006; Hodson & Marvin 2010), that urban researchers and the sustainability community have not paid sufficient attention to the significance of the vast networked urban infrastructures that connect everyday living and working to the natural and informational resources on which urban dwellers depend across cities in the developed and developing world.[3] Urban infrastructure is crucial if one wants to come to terms with the challenge of building more 'sustainable cities' by reworking the metabolic flows through the ecological and urban systems which are structured and directed by these infrastructures.

The second urbanisation wave

The first urbanisation wave took 200 years — 1750 to 1950 — and resulted in an increase in the number of urban dwellers in Europe and North America, from 15 million to 423 million people (United Nations 2006). This was also the process that resulted in the iconic images of the 'modern city' as the architects, engineers and planners gave cultural form to this radical transformation of everyday life and work. Marshall Berman's great classic text (which took its prescient title — *All That is Solid Melts into Air* — from a line in the *Communist Manifesto)* revealed the intimate connection between the rise of modernity as the aesthetics of the new middle classes produced by the northern industrial revolutions, and the drive to transform pre-industrial cities into paragons of urban modernity (Berman 1988). This powerful movement — complete with Faust as its mythical hero — has its origins in Haussmann's Paris, Peter the Great's St Petersburg, and the New York that Robert Moses built. All three were archetypal 'developers' and, as extreme reductionists, were inspired by the power of rational planning to impose neatly measured boxes and straight lines onto messy complex spaces and, by the powers of the new technologies of light, water and motion, to build the perfect frictionless modern city using law, cash, art and force. It is a movement that gets mass produced in the incarnations of Le Corbusier's profoundly reductionist, one-dimensional, urban

[3] It is recognised that Stephen Graham and Simon Marvin in their seminal opus *Splintering Urbanism* failed to recognise the ecological significance of natural resource flows. Their focus was on the interfaces between globalisation, urban development, networked infrastructures and urbanism. But this was remedied in the subsequent edited volume *Urban Infrastructure in Transition*, which combined the analytical logic of *Splintering Urbanism* with material flow analysis to generate an extraordinarily fruitful and new perspective on the challenge of thinking about cities, urban infrastructures and green buildings from a sustainability perspective.

modernity which, in turn, inspired the urban design of cities from Atlanta to Brasilia, and from the formalised colonial suburbs of many African cities, to the contrived new capitals like Abuja, Nigeria. In South Africa we had Lord Milner's 'Kindergarten'[4] that planned and built early Johannesburg, and H.F. Verwoerd, the architect of apartheid, who personally planned the township of Mdantsane as a racially exclusive model 'garden city' for black people, outside East London (Swilling 1984).

For nearly 300 years, the image of the modern industrial city became synonymous with what a 'city' is supposed to be. It embodied the meaning of progress, rationality, secularism, universality and all that was associated with the Enlightenment. All other cities were either getting there, or were — as in the case of the great pre-colonial African cities or historic core cities in many Islamic countries — either entirely forgotten or denied the status of being cities in their own right (Malik 2001). The awesome power of this historically constructed lens is what makes it all the more difficult to 'see' the cities being created by the second urbanisation wave.

The second urbanisation wave will take less than 100 years — 1950 to 2030 — and is taking place in developing countries, where the urban population is projected to grow from 309 million to a staggering 3.9 billion people over this period (United Nations 2006). China alone will urbanise more people in 50 years than were urbanised in 200 years in North America and Europe together. As the global population increases from 6 to 9 billion, it is the urban centres of Africa and Asia (some of which don't exist yet) that will be home to the additional 3 billion people expected on the planet by 2050. Consequently, pressures will be greatest where the urban and institutional infrastructure is weakest, and where the cultural memories needed for urban living must still be created.

For some, this confirms the sum of all fears: cities are the irredeemable sites of unsustainable consumption (Low *et al.* 2000) and dystopic social marginalisation (Davis 2005). There is undoubtedly some truth in this requiem for urban modernity, especially given the correlation between levels of urbanisation and GDP per capita growth (Figure 5.1) that may be a proxy for development, but which is also a proxy for a process that has created the billion or so urban-based over-consumers who drive global ecological destruction. However, we share the view that there is the potential for cities to be different because, after all, the concentrations of the intellectual resources for innovation created by an urban-centred science and education system, should provide the ideational, cultural and institutional context to foster imaginaries about more sustainable futures (Dodman 2009; Hodson & Marvin 2009a; Satterthwaite 2009).

Part of the reason for our optimism is that contrary to popular belief, the urbanisation trends across all continents are not creating more mega-cities, but rather a rich, global patchwork of smaller cities of around a million people or less. It is possible to estimate that if current trends continue, by 2015 nearly 60 per cent of the total urban population across high-, low- and middle-income countries will live in cities of less than a million

[4] This is the name of the group of young whizz-kids whom Lord Milner recruited to design and build post-Boer War Johannesburg as an icon of British colonial order.

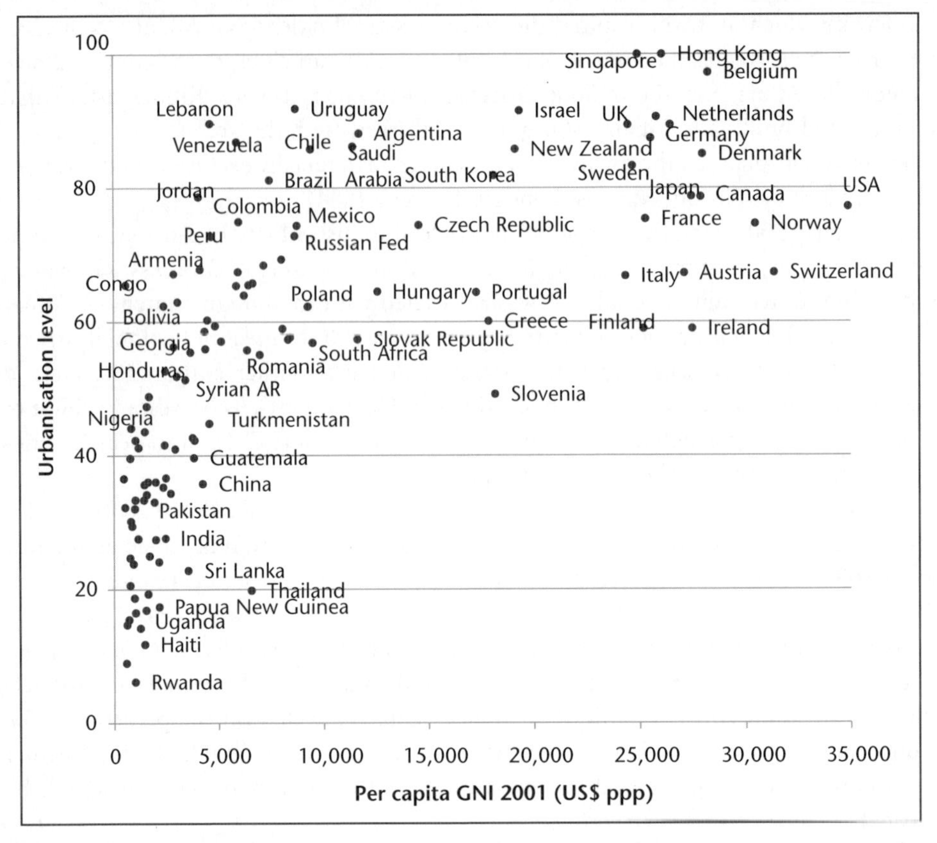

Figure 5.1: The association between nations' level of urbanisation and their average per capita income, 2000/01

(Source: Satterthwaite, D. 2007. *The Transition to a Predominantly Urban World and its Underpinnings.* Human Settlements Working Paper Series Urban Change No. 4. London: IIED.)

people. Approximately 25 per cent will live in cities of 1–5 million people, 5 per cent will live in cities of 6–10 million people, while only 10 per cent will live in cities of 10 million people or more (calculated from National Research Council 2003; United Nations 2006).

Although the bulk of future urban growth will be in the developing world, only 17 per cent of these cities are growing at 4 per cent or more per annum, while the bulk (36 per cent) are growing at a more moderate 2–4 per cent per annum. What is staggering, though, is that 40 per cent of cities in the developed world actually shrunk during the 1990s (UN-HABITAT 2008b: 11), whereas 17 big cities (of 5 million people or more) in the developing world also contracted during the decade leading up to 2000 (UN-HABITAT 2008b: 36). By contrast, between 1990 and 2000, 694 new cities were established in the developing world in locations in which there were fewer than 100,000 people before 1990: 510 of these became small cities (100,000–500,000 people), 132 became intermediate cities (500,000 to 1 million), and 52 grew rapidly into big cities

(1–5 million). These trends, when read together, suggest that it is far too simplistic to depict our urban future as dominated by the (often dystopic) spectre of mega-cities as the logical outcome of a linear, irreversible urbanisation trajectory from 'small to big'.[5] Something else is clearly going on.

Maybe it is out of these smaller, more manageable cities — both old and new, contracting and expanding — that new cultural fusions will emerge which will inspire the innovations needed to realise what Peter Evans has called 'liveability' — more sustainable and more equitable modes of urban living (Evans 2002). After all, if it was the innovations for economic growth that resulted in the extraordinary concentrations of economic, political, communicative and knowledge power in the globally connected cities of the world (Borja & Castells 1997), maybe sustainability innovations will generate a more geographically distributed hierarchy of smaller cities that are embedded within their respective, uniquely configured, bio-economic regional systems of production and reproduction. Undoubtedly, the most significant driver of this outcome will be the declining viability of long-distance transportation of food as transport costs go up and soils collapse. A cursory glance across the world at the cities that are at the cutting-edge of sustainability innovations tends to reinforce this conjecture — names like Curitiba (Brazil), Rizhao (China), Melbourne (Australia), Vancouver (Canada), Freiburg (Germany), San Jose (USA), Bogota (Columbia), and Portland (USA) immediately come to mind. None in Africa has attempted to capture for themselves a similar identity, although Cape Town may well be getting there. In all these cases, very specific local ecological drivers have connected with economic processes and knowledge networks, which have culminated in surprisingly profound bio-economic diversifications, with positive developmental results. In almost every case, state institutions have played a key role, thus reinforcing the argument we developed in Chapter 4.

Although an emerging urban future of smaller rather than larger mega-cities is heartening from a sustainability perspective, the bad news is that the second urbanisation wave has created a global population of slums. According to the path-breaking UN-HABITAT report *The Challenge of Slums*, there were nearly a billion people living in slums by 2001 (United Nations Centre for Human Settlements 2003).[6] In other words, by 2001 one in three people living in the cities of the world lived in a slum.

A slum is defined in the UN-HABITAT report as a settlement made up of households that lack one or more of the following conditions: access to improved water, access to improved sanitation facilities (minimally, a pit latrine with a slab), sufficient living area (not more than three people sharing the same room), structural quality and durability of dwellings, and security of tenure.

[5] The empirical substantiation for this notion of an urban future made up of a hierarchy of smaller (potentially more innovative) cities is derived from the *State of the World's Cities* reports (*State of the World's Cities* reports for 2006/07, and 2008/09), and David Satterthwaite's interpretation of statistics generated by the UN Population Division.

[6] This figure was revised downwards to 810,000 in the *State of the World Cities* report for 2008/09 because of a decision to exclude those households that have access to an improved pit latrine (UN-HABITAT 2008b).

Significantly, although half of all slum dwellers are in Asian cities, it is only in sub-Saharan Africa that one finds cities in which the majority of the population live in slums. No less than 62 per cent of all urban dwellers in sub-Saharan Africa live in slums, compared to Asia where it varies from 43 per cent (southern Asia) to 24 per cent (western Asia), and in Latin America and the Caribbean where slums make up 27 per cent of the urban population (UN-HABITAT 2008b). The large majority of cities in sub-Saharan Africa are, therefore, slum cities. Given the fact that urbanisation rates in Africa are the highest in the world at 3.3 per cent (UN-HABITAT 2008a: 4), the slum cities of sub-Saharan Africa will be with us for decades. Africa is now 40 per cent urbanised and is projected to be 60 per cent urbanised by 2050, which translates into an increase in the *urban* population from the current 373 million to 1.2 billion by 2050 (UN-HABITAT 2008a: 5). If Africa's governments continue to ignore this problem (by stubbornly insisting that slum dwellers are a problem only because they refuse to go back to the rural areas), the additional 800 million urban dwellers will land up in Africa's mushrooming slums.

Africa is becoming a continent of slum cities and, in so doing—as Okri's prose reveals—it is transforming entirely what we mean when we use the word 'city' to describe quite a unique set of urban dynamics and modalities (Pieterse 2008; Simone 2004a; Simone 2001; Swilling *et al.* 2003). Indeed, for many analysts and policy-makers, African cities don't deserve to be called cities at all—a position that is tenable only if you assume that the 'Western city' is the only legitimate template for defining the city. Maybe it is time to realise that the iconic image of the 'Western city', which emerged from the specificities of the first urbanisation wave, has become little more than a mirage. Maybe it's time to find non-Western reference points for rethinking our deepest assumptions about the purpose, meaning and impact of the city (Malik 2001; Swilling *et al.* 2003). And maybe it is also time to realise that industrialisation, modernisation and (from the late 1980s onwards) high-tech informationalism—the traditional economic drivers of urbanisation—have not been the primary driving forces of African urbanisation and the emergent urbanisms we see across the diverse cities of the continent. Again, as we will suggest later, something more complex and difficult to grasp is going on when it comes to the making and shaping of slum urbanism (Pieterse 2008; Simone 2004a; Swilling *et al.* 2003).

Globalisation, restructuring urban space and resource flows

The second urbanisation wave has taken place more or less at the same time as economic globalisation has reconfigured the spatial and temporal relationships between cities and between the globalised networked enclaves within cities in both developed and developing economies, which share a place in the new hyper-mobilities of the Information Age. In this process, new borders have been created between those included and those that were excluded by their disconnectedness. A well-established literature now documents in great detail the combined impact of the Information Revolution, neoliberal economic policies, financialisation (nominally secured by urban assets), privatisation of public services, the relocation of production into low-wage, rapidly industrialising zones (mainly in Asia, but also Latin America and Eastern Europe) and the rise of new, globally networked, urban spaces which are more 'connected' to one another via ICTs than to their respective 'home bases' or national 'hinterlands' (see Amin & Thrift 2002; Borja & Castells 1997; Bridge & Watson 2000; Castells 1999; Graham & Marvin 2001; Sassen 2000; Urry 2000). A key characteristic of this new era of information-based globalisation is what Castells called the 'space of flows', which has, in turn, trumped the 'space of place' in the new networked world that neoliberal globalisation has stimulated and promoted (Castells 1997). Graham and Marvin describe this new space economy as follows:

> *Very broadly, those global and second-tier cities, parts of cities, and the socioeconomic groups involved in producing high value-added goods, services and knowledge outputs, are tending to become intensively interconnected internationally (and sometimes even globally)... Using the capabilities of high quality information, transport, power and water infrastructures, zones of intense international articulation — business spaces, new industrial spaces, corporate zones, airports, new cultural or entrepreneurial zones, logistics areas — are emerging in such cities, albeit to highly varying degrees.* (Graham & Marvin 2001: 305–306)

Those local places that find a role in this globalised 'space of flows' flourish, while those that fail literally fall off the edge and become 'irrelevant and dysfunctional' (Castells 1997: Vol. 1: 147) for a global economy which came to be structured to meet primarily the consumption needs of an expanding urban-based class of about a billion (over-) consumers.

The triple drivers of the second urbanisation wave were the declining capacity of rural areas to support naturally expanding populations, a natural increase in the cities (which is now a bigger driver than rural-urban migration) and the increasing concentration of political, economic and networked informational power in the expanding developing country cities, which globalisation stimulated. For cities (mainly in sub-Saharan Africa) which were incorporated into the global economy as ports for extracting raw materials rather than as sites for industrialisation and urbanisation, the Information Revolution and globalisation have meant little more than expanding slums. While recognising that poor governance contributes to negative outcomes, the reason why the second urbanisation wave has resulted in a billion slum dwellers is because of the uneven way in which the new international division of labour excluded/incorporated all or parts of cities in the developing world. Some sub-Saharan African cities were excluded entirely because they simply did not have the infrastructure to connect up, thus forcing them to depend on a completely different set of flows to those associated with the global 'space of flows' that Castells described.[7] However, many developing country cities were partially incorporated as low-wage factories (often in export-processing zones) and related service industries opened up requiring labour that could be paid very little,

[7] Hence the significance of studies that have refused to assume that just because these cities fall off the edge of the Information Highway, they somehow cease to exist—instead they are also dependent on local, regional and global flows of information that structure very different modes of value to those that flow through the formalised global economy (Simone 2004a; see Swilling *et al.* 2003).

often *because* they lived in slums. As the 'connected' part of developing country cities expanded, the cities needed to be restructured to create spaces for industrial areas and middle-class residential neighbourhoods (in particular in old, densely populated, developing country cities like Manila, Mumbai, São Paulo). As a result, slums were demolished and people forcefully relocated to areas where they were often disconnected from economic opportunities — a process which entailed extensive urban social conflict as communities resisted the demolition of their communities (Cabannes *et al.* 2010; Evans 2002; Peet & Watts 2000). Vast informal economies opened up which gave the unemployed access to the crumbs of value that fell from the circuits of the formal economy and its expanding illegal appendages (loan sharking, child labour, drugs, sex work, organised crime, corruption). In some cases, communities successfully resisted relocation and negotiated their way into at least the outer edges of the space of flows (for instance, as waste pickers), while others accessed localised flows unrelated to the global economy (for example, as urban farmers).

This new era of 'stark techno-apartheid' — to use the words of former EU-Commissioner Ricardo Petrella (cited in Graham & Marvin 2001: 307) — not only created a highly inequitable global geography of accelerated economic accumulation during the quarter decade leading up to the 2008 crash, it was also an era of unprecedented demands on the planet's ecological resources. This included a 70 per cent increase in global greenhouse gas emissions due to human activities from 28.7 Gt/CO_2-eq/yr in 1970 to a staggering 49 Gt/CO_2-eq/yr by 2004 (Intergovernmental Panel on Climate Change 2007: Figure SPM 3); plus a 36 per cent growth in global material extraction from 40 billion t/yr in 1980 to 59 billion t/yr by 2005 (as discussed in Chapter 2). Significantly, extracted materials for construction over this period increased by 40 per cent (compared to fossil fuels at 30 per cent, biomass at 28 per cent and metals at 56 per cent [Behrens *et al.* 2007]), thus reflecting the importance of constructing the built environment (infrastructures and buildings) as a driver of the increase in total used resource extraction and related environmental impacts. This was confirmed in a 2009 report, which concluded that only three activities — all of them primarily urban activities — were responsible for over 60 per cent of all environmental impacts, namely mobility, eating/drinking, and housing/urban infrastructure, with housing and infrastructure making up half of this (31 per cent) (SERI Global & Friends of the Earth Europe 2009).

The current global economic crisis — which marks the mid-point of the Information Age and the end of a long-term development cycle as discussed in Chapter 3 — is also an urban crisis. This is reflected in the role played by debt-financed consumerism, secured against urban property in the West (see Gowan 2009), and massive debt-financed urban infrastructure projects in the cities of the developing world, funded by consortia involving the World Bank and its related financial institutions, the increasingly significant Sovereign Wealth Funds and the large commercial banks. Together, these financial institutions effectively bankrolled globalisation and accelerated resource depletion across the planet, with the worlds' expanding urban assets working as security for the loans.

The global ecological crisis is also, therefore, an urban crisis. The construction industry worldwide is a US$4.2 trillion-plus global industry (Langdon 2004), is responsible for 10 per cent of global GDP, employs over 100 million people globally (International Labour Organisation 2001), uses up nearly 50 per cent of global resources annually and 45 per cent of global energy (5 per cent during construction), 40 per cent of water globally and 70 per cent of all timber products (Edwards 2002).[8] Unless this industry—including the architects and engineers who produce the designs used by construction companies—finds ways of doing a lot more with less, it will exacerbate global warming, resource depletion and eco-system degradation and, in so doing, set these cities up for cataclysmic failure over the medium to long term.[9] The global economic crisis might have been caused by debt that was decoupled from the real underlying value of the assets used as security, but we can only imagine what could happen if the value of the assets themselves are eroded by natural disasters, such as flooding, salinated aquifers, pandemics, toxicity, resource depletions or hurricanes. The first attempt to comprehend the consequences of massive infrastructure failure (Graham 2010) underscores how dependent city dwellers really are on systems that they take for granted as permanent and secure elements of everyday life.

The strategic centrality of cities in the global polycrisis is reflected in the importance given to cities in the 'solutions' embedded in the so-called 'rescue packages'. The evidence suggests that the publicly financed investments to stimulate the global recovery will be targeted primarily at investments to refurbish/extend the ageing urban infrastructures of cities in the developed world, and the under-serviced, over-burdened urban infrastructures of the burgeoning cities in the developing world (many of which have only the original colonial enclaves built for the settler elites during the first urbanisation wave). The first estimates of what this will cost globally are already being published. The US-based, global consulting firm Booz Allen Hamilton, which depends heavily on world-wide contracts to build infrastructures allocated to a handful of the world's largest engineering contractors for its US$4.5 billion turnover, has compiled a detailed estimate of the investment required to meet demand for urban infrastructure over the next 25 years across all the cities of the world. Significantly called *Lights! Water! Motion!* and published in the influential business journal *Strategy and Business,* this report estimates that a total of US$41 trillion is required to refurbish the old (in mainly developed country cities) and build new (mainly in the developing country cities) urban infrastructures over the period 2005–2030—in other words, more than the value of all stocks on the world's stock exchanges in 2007 (Doshi *et al.* 2007). The Boston Consulting Group independently arrived at a similar estimate when it argued that US$35–$40 trillion will need to be invested in infrastructure by 2030 (Airoldi *et al.* 2010). Of the US$41 trillion estimated by Booz Allen Hamilton, over 50 per cent ($22.6 trillion) would be required

[8] We are grateful to Llewellyn van Wyk from the CSIR, Pretoria, for directing us to these references to the global construction industry.

[9] There is evidence that this industry is starting to recognise this challenge—for example the Holcim Foundation, set up by Holcim which is the largest cement producer in the world, runs a global competition for the most sustainable building designs (see www.holcimfoundation.org). Obviously, Holcim assumes that cement production is reconcilable with sustainable construction and urban development.

for water systems,[10] $9 trillion for energy, $7.8 trillion for road and rail infrastructure, and $1.6 trillion for air- and sea-ports (Doshi *et al.* 2007: 4).

In a revealing statement, which demonstrates the sales pitch that these powerful global firms use to land these lucrative contracts (which, in turn, are echoes of the Stern Report's new dictum that it is 'cheaper to fix things earlier rather than later'), the authors of the Booz Allen Hamilton report write:

> *Sooner or later, the money needed to modernise and expand the world's urban infrastructure will have to be spent. The demand and need are too great to ignore. The solutions may be applied in a reactive, ad hoc, and ineffective fashion, as they have been in the past, and in that case the price tag will probably be higher than $40 trillion. After all, infrastructure projects are notorious for cost overruns. But perhaps the money can be spent proactively and innovatively, with a pragmatic hand, a responsive ear, and a visionary eye. The potential payoff is not simply the survival of urban populations,* but the next generation of great cities. (Doshi *et al.* 2007: 4—emphasis added)

To their credit, Booz Allen Hamilton recognise (albeit only in a side box) that the grand retooling of the world's urban infrastructures will mean finding new designs and technologies that will make it possible to use natural resources more sustainably:

> ... *[C]ities that ignore environmental impact will find themselves facing another collapse of infrastructure 30 or 40 years from now, and our children and grandchildren will bear a much higher price tag.* (Doshi *et al.* 2007: 13)

Just as economists look back on the investments in automobiles, roads, petro-chemicals and mass production systems made from the 1930s through to the post-World War II period (including the Marshall Plan in Europe) as the investments that 'resolved' the 1929–1933 economic crisis and paved the way for the deployment period of the fourth industrial transition, we predict that in 10–15 years' time researchers will look back and realise that the investments that helped 'resolve' the crisis of 2007–2010/12 were, in fact, investments in networked urban infrastructures—using, of course, 'Web 2.0'-type ICTs as their operating systems (for instance, smart grids, telecommuting, virtual shopping, remotely controlled intelligence systems, and digitalisation). Retooling the world's cities for (hopefully more equitable) global re-industrialisation is what is at stake here, or, to use the language that the global consulting and development finance community might use, how to create 'the next generation of great cities', which will become the geographical nodes of the next long-term development cycle.

The question is, of course, what kind of networked urban infrastructures will be built? Will they set cities up for sustainable socio-ecological metabolisms, or will they set cities up against nature's resource-bases and eco-systems? Will they reinforce the 'stark

[10] Although not specifically defined, we presume this means 'water and sanitation' infrastructure because the one without the other does not make much technical sense.

techno-apartheid' that is splintering cities around the world or will they create the basis for greater equity, reduced levels of poverty and greater opportunities to build a sense of community? Will more sustainable modes of resource-use reinforce or undermine the search for greater equity and a sense of place?

Networks, flows, urbanisms

Which city will be the first to configure their metabolic flows so that citizens can enjoy a decent quality of life and sense of community while emitting no more than 2.2 tonnes of carbon and consuming a maximum of 6 tonnes of extracted materials per annum? This convergence point — where the poor get more to have enough and the rich make do with less — is what is increasingly referred to as 'sufficiency' (first clearly defined by Revi *et al.* 2006; see also a discussion of this term in Von Weizsacker *et al.* 2009). For about a billion of the wealthier urban dwellers this will entail drastic consumption reduction (from between 15 and 30 t/cap/yr to 6 t/cap/yr) and for the billion who live in slums it will mean significant increases in resource consumption. But for about a billion or so urban dwellers who do not live in slums and who are connected into the mainstream socio-metabolic flows via a set of networked infrastructures that deliver adequate basic services (such as water, energy, sanitation, and waste services, and access of some sort to public and/or private mobility), this is more or less how they live now.

In very practical terms, our focus will be the set of socio-technical *systems* and associated socio-metabolic *flows* listed in Table 5.1.

Building on the small, but growing, literature on the political ecology of socio-metabolic urban flows (Coutard *et al.* 2005; Girardet 1996; Girardet 2004; Graham & Marvin 2001; Guy *et al.* 2001; Heynen *et al.* 2006; Hodson & Marvin 2009a;

Table 5.1: The set of socio-technical systems and associated socio-metabolic flows

System	***Flows***
Water supply (including dams, pipes, pump stations) and sanitation (in particular the sewage treatment works).	Water from catchment areas and sewage usually into large treatment works (noting that sewage includes useful ingredients such as nutrients, methane and water).
Energy generators and grids, and biomass supply lines	Electricity generated usually from fossil fuels, but also hydro, nuclear and biomass, plus other forms of energy including the burning of biomass for cooking purposes, and renewable energy
Mobility, such railways, air- and sea-ports, roads	Bodies and goods in vehicles, trains, airplanes, ships
Solid waste, including landfills, transfer stations, incinerators, etc.	All kinds of solid waste, including nutrients, recyclables, biogas
Communications, from traditional land-lines, to fibre-optic cables and satellite systems	Data, voices, images, etc.

Food as a socio-metabolic flow?

The food supply system should be regarded as the sixth networked infrastructure, with food and drink as the socio-metabolic flow that it delivers into the city. This is also a vast networked infrastructure in which the state has traditionally had a much more limited role (for instance, the provision and operation of a centralised fresh-food market at which farmers sell to wholesalers, who then on-sell to retailers). In many cases in the developing world the state plays no role in the food supply system, so food markets sprawl informally along the main mobility routes. Nevertheless, in both cases the food supply system comprises highly complex, entrepreneurially organised, logistical paraphernalia for managing the massive socio-metabolic flows—and biochemical processes—involved in connecting foodstuffs to urban dwellers and, where sanitation systems exist, managing the outflowing sewage (for how this perspective can help to understand urban hunger, see Heynen 2006; for urban obesity, see Marvin & Medd 2006; for the best empirical review to substantiate this argument, see Steel 2008). Although policy and research interest in urban agriculture has accelerated since the early 1990s (see Mougeot 2006), it has yet to be conceptualised as part of the wider system of socio-metabolic flows and biochemical processes through a specific set of networked infrastructures, despite the obvious (well-documented) linkages to related socio-metabolic flows (for instance, capturing nutrients from sewage for reuse as fertiliser, reusing grey water for toilet flushing and irrigation, and composting solid organic wastes for reuse in agriculture and biodiversity restoration).

Hodson & Marvin 2010a; 2010b), we can now describe in more conceptual terms the key elements of the urban infrastructure system. These urban infrastructures are often vast technical networks which can be configured in many different ways, depending on levels of financial investment, politically determined roles of the state relative to markets, institutional capacities, geographical boundaries, technical know-how, capacities of civil society to engage the managers, and user demand (which, of course, is fractured by class, race, space and gender). This also implies that different networked infrastructures channel and process the range of socio-metabolic flows through the urban system in different ways. It is obvious, for example, that in developing country cities, where a formal sanitation system may be non-existent, human excrement will not circulate through the urban system in the same way as in cities which do have a formal sanitation system. Anyone who enters central Mumbai for the first time can literally smell what it means when large numbers lack access to water-borne sanitation. Similarly, cities which are not hardwired with fibre-optic cables will not be populated by businesses that depend on 24/7, high-speed, low-cost connections to global information flows. It is, therefore, not hard to imagine that different configurations of networked infrastructures and their associated socio-metabolic flows will, in a non-deterministic way, influence and shape the nature, mobilities and, indeed, cultures and subjectivities of everyday urban living — an ensemble of practices and perceptions that is referred to here as *urbanism.*[11]

In summary, we are talking about three key concepts here: the *networked infrastructures* which are the physical and technical systems that are fixed in space (such as roads, cables, satellites, pipes and rail) and managed by specific sets of actors embedded within public and/or private institutions which, in turn, must operate within specific regulatory environments. These networked infrastructures, in turn, provide conduits for the *socio-metabolic flows* through ecological and urban space in time (such as, vehicles, water, data, energy and food). Excluding, for the moment, information flows that have unique properties (such as data-systems), these flows can be linear metabolic flows of virgin materials with minimal reuse, or circular metabolisms where outputs are seen as potential inputs (Girardet 1996). But not all inputs translate into outputs in the short term because a large proportion of the input resources get fixed in space over very long time frames as buildings and infrastructures.[12] Without being reductionist, *urbanism* refers to the 'ways of life' associated with these built forms, infrastructures and flows — they shape through design how these infrastructures and flows evolve, and they are conditioned by the infrastructures and flows on which everyday life depends.[13]

Despite the dangers of setting up typologies in advance and then applying them to a wide range of different contexts, we think it is useful to identify four generic *urbanisms,* each of which displays what Guy and Marvin call a different 'logic of network management' (Guy & Martin 2001: 32) and associated flows and practices. Each of the four is characterised by specific interventions to structure and extend urban infrastructures in order to access and direct the socio-metabolic flows. We hasten to add, however, that none exists in a pure form, especially in today's sprawling, developing country cities in which all four can co-exist in complex and contradictory ways. The logics of the four approaches are defined as *inclusive urbanism, splintered urbanism, slum urbanism* and *green urbanism.*[14] Green urbanism is rapidly becoming the spatial discourse of ecological modernisation. If the next long-term development cycle turns out to be an *unjust transition* inspired by ecological modernisation, then green urbanism will, in all likelihood, be the spatial expression of this logic. In line with our normative commitment to a *just transition,* we propose a fifth perspective which we have called *liveable urbanism,* explored via a case study of the Lynedoch EcoVillage in Chapter 10.

[11] This is our reading of the argument developed by Pile and Thrift (Pile & Thrift 1996), but also an interpretation of our earlier work on African cities with Maliq Simone (Swilling *et al.* 2003) and Simone's other writings (see Simone 2004a; 2004b).

[12] Birkeland, for example, has suggested that 90 per cent of extracted materials remained fixed in use as buildings and infrastructures (2008: 26).

[13] This is our interpretation, using slightly different language, of the argument that Guy and Marvin have developed (2001).

[14] The first two, inclusive and splintered urbanism, have been thoroughly described and analysed in the path-breaking book by Graham and Marvin (Graham & Marvin 2001). The notion of 'green urbanism' was coined by Beatley in his classic review of Western European cities (Beatley 2000). And slum urbanism is substantially drawn from Davis (Davis 2005), but qualified by the work of Pieterse (Pieterse 2008), Bayat (Bayat 2000) and Swilling (Swilling *et al.* 2003) who do not share the foreboding absence of agency that pervades Davis' work (for a critical review, see Satterthwaite 2006).

Inclusive urbanism

The booming Victorian era in Britain (the deployment period of the Age of Steam and Railways) that followed the economic crisis of 1848–1850 was not simply about knitting the British Empire together with railway networks to speed up the flow of raw materials to feed the fast-evolving manufacturing sector of post-slavery Britain. It also kick-started a *century-long* effort by a small Faustian coterie of 'city builders' who managed to capture the imaginations of generations of intellectuals, politicians and financiers to build support for their efforts to systematically centralise and standardise the design and delivery of road, water, waste, energy and communications grids in Western cities (as well as the colonial/settler enclaves in the New World and colonial outposts). In so doing, the engineers put in place the infrastructures and institutions that created the urban preconditions for the third (Age of Steel, Electricity and Heavy Engineering starting in 1875) and fourth (Age of Oil, Automobiles, and Mass Production starting in 1908) industrial transitions — what Graham and Marvin call the 'modern networked city' (Graham & Marvin 2001: 40–89). This was an awesome vision that fixed in steel and concrete (the highest of all aspirations of modernity) the desire to make progress, and to include everyone equally.

The crisis of 1929–1933 not only gave birth to Keynesian social democracy as the all-encompassing framework for fostering the Age of Oil and the post-World War II consumer society, it was also inspired by the democratic vision of an *inclusive urbanism* in which (almost) every urban dweller had 'rights to cheap, good-quality and accessible

Baron Haussmann — pioneer of modern urbanism

Baron Haussmann (1809–1891) must surely be one of the great pioneers of modern urbanism, and the first grand master of the art of debt-financed urban infrastructures. One can only imagine the urban future that he must have contemplated in 1852, just before implementing Napoleon III's mandate to transform Paris into a 'modern city'. It was a job that was his obsession for 20 years and which entailed forcefully ramming his 'boulevards, gardens, railways, gas pipes and aqueducts' through the teeming slums of Paris. The massive debts he ran up to finance it all most certainly catapulted Paris into the Victorian Boom, but the risks he took and the debts he incurred lost him his job in 1870. The destruction of Old Paris was so brutally disruptive that, a year after he lost his job, Paris was consumed by the revolutionary movement that resulted in the setting up of the Paris Commune (which, in turn, inspired Karl Marx's concept of 'communism'). In reality, the revolutionaries tried to recreate the sense of community that Haussmann had decimated and they reoccupied the spaces that had so forcefully been cleared to build the grand boulevards for the preening bourgeoisie to show off their new-found wealth. It is surely most remarkable how similar these dynamics are to the contemporary post-2007 crisis brought on by debt-financed urban assets within increasingly unequal cities, but now writ large as a global phenomenon.

infrastructure services and its associated obligations of prompt payment and respect of technical boundaries' (Guy & Martin 2001: 29). Presided over by increasingly large, vertically-integrated, public monopolies delivering uniform cross-subsidised services during the prolonged post-World War II growth period, massive resource-intensive socio-metabolic flows were created that coursed through the 'consumption cities' that emerged during the first urbanisation wave. The end result was remarkably equitable, highly unsustainable, and only (almost) fully realised in the developed industrial economies. By the end of the 1960s, these massive, state-run, ecologically destructive, cash-guzzling, networked infrastructures had reached their apogee. As major contributors to the stagflation crisis that set in after the first Oil Crisis in 1973, they became the prime targets for the neoliberal reformers, who despised cross-subsidised, inclusive urbanism and the public monopolies that made this possible.

Splintered urbanism

The assault on Keynesian economics and the dismantling of social democratic governance that followed the conservative electoral victories in the late 1970s/early 1980s in Western Europe/North America unleashed the forces of economic globalisation and hyper-financialisation and fostered the launching of what we, following Perez, have called the fifth industrial transition or, following Castells, the Information Age (Castells 1997). The three key investment strategies to revive the global economy involved incentives to move industry into developing countries in search of cheap labour and new markets, financial deregulation to enable the higher-risk debt-financing of expanded consumption of globally traded goods (manufactured mainly in the new, fast-industrialising economies), and the drastic regulatory measures to privatise and liberalise the delivery of urban services to break the power of the large public-sector monopolies in nearly every developed, developing and post-communist society. This 'commodification' of urban services (McDonald & Ruiters 2005) using new debt and intellectual property regimes—that effectively transferred 100 years of publicly funded physical and intellectual capital into private hands (usually) at a massive discount—opened up the global market for extraordinary, highly profitable, private sector investments, which reconfigured globally networked cities in ways that were discussed earlier in this chapter.[15] More significantly from a sustainability perspective, this globally implemented privatisation of networked urban infrastructures gave profit-seeking corporations access to resource flows (especially water, energy and mobility) during a period that was also characterised by accelerated GHG emissions and resource extraction. Given that cities are such huge resource consumers and energy users (hence GHG emitters), and given that corporations were licensed to exploit these business opportunities as the second urbanisation wave gathered momentum, with no regard for ecological limits, it is unsurprising that the period of neoliberal globalisation

[15] For a personal insider account of how this was managed in practice, see Perkins 2004.

has ended not just in a financial crash rooted in the limits of debt-funded splintered urbanism, but in an unprecedented ecological crisis.

Urban infrastructures were rapidly 'unbundled' and the new 'informational cities' that emerged as enclaves within — but disconnected from — the 'old cities' were now connected globally through the new, privately delivered ICTs. An entirely new global industry emerged to deliver high-tech 'premium infrastructures' (Graham & Marvin 2001), which ensured that these enclaves (whether they were large cities such as Manhattan, or islands of modernity in the slum cities of the world, or entirely new cities with 'connected' cores) were kitted out with privately supplied water, sewerage, waste, energy, mobility and communication services paid for via user-charges collected to a large extent automatically via new electronic systems. Gone was the democratic vision of an *inclusive urbanism* held together by an integrated, publicly owned and standardised networked infrastructure. Gone was the possibility of the southern cities, created during the second urbanisation wave, benefiting from the vision of a more inclusive urbanism. In the 'public choice' language of individualised needs, market segments, competition and the user-pays principle, what you got was what you could afford.

The result is what Graham and Marvin called *splintered urbanism* (Graham & Marvin 2001). While it failed in developmental terms because it fostered the unprecedented social fragmentation of cities across the developed and developing world, it also created the space for innovations, choice, access and decentralised solutions more appropriate for the emergence of large, complex urban systems (such as mega-cities), with which centralised public systems may never have been able to cope. This is why it was so central to the second growth period of the long-term development cycle. But like *inclusive urbanism, splintered urbanism* depended on highly unsustainable, linear, socio-metabolic flows — the only difference was that access to these flows became far more inequitable.

Slum urbanism

One in three urban dwellers today live in slums, and yet only 10 per cent of these are seen as potential 'beneficiaries' of the programmes that aim to achieve the UN Millennium Development Goal of improving the living conditions of a 100 million slum dwellers by 2015. Assuming for a moment that these plans work, it follows that little will be done about the 900 million who are already living in slums! Add to this a good proportion of the additional 3 billion expected to be living in cities in Africa and Asia by 2050, and a shocking spectre emerges. And so it is not an over-statement to say that slum urbanism is probably here to stay for foreseeable generations as the urban poor struggle to resolve their own problems, often resisting forced relocations to make way for the 'premium networked infrastructures' of the globally connected enclaves (Cabannes *et al.* 2010). This is why splintered urbanism and slum urbanism are two sides of the same coin.

Slum urbanism is not a 'passing phase' as cities move along the linear trajectory from a primitive 'pre-modern' urban form to the 'modern networked city', as imagined by so much of the urban development literature and associated aid programmes (see the World Bank's *Cities Alliance* for a good example of this). Slum urbanism is an urbanism in its own right, with its own comprehensible networked infrastructures, flows and 'ways of living', which can be understood as readily as any other form of urbanism. However, to do so means jettisoning the rational planning logics at the centre of *inclusive* and *splintered urbanism* and accepting that what we are trying to analyse is a 'mess', which cannot be cleaned up by explaining the mess as an absence of a pre-ordained order (Law 2004) — in this case, the absence of the rational order of the urban planner. It entails accepting what Pieterse calls a 'conceptual inversion', which makes it possible to 'explore the city from the bottom up, or rather through the eyes of the majority of poor denizens who appropriate the city for their own ends' (Pieterse 2008: 109). If 'universal access' was the centre of *inclusive urbanism* and 'commodification' the centre of *splintered urbanism*, then Asef Bayat's notion of 'quiet encroachment' is probably the best candidate for capturing the essence of *slum urbanism:*

> *The notion of 'quiet encroachment' describes the silent, protracted and pervasive advancement of ordinary people on those who are propertied and powerful in a quest for survival and improvement of their lives. It is characterised by quiet, largely atomised and prolonged mobilisation with episodic collective action — open and fleeting struggles without clear leadership, ideology or structured organization.* (Bayat 2000: 545–546)

But what is this 'quiet encroachment' about? What is being achieved (and quite often lost again) over time? In almost every case it is either to somehow get connected to wider (sometimes only partially) networked infrastructures (in particular water, materials for building shelter and food), or to establish autonomous, localised, networked infrastructures for self-managing the flows needed for survival. But it is worth sticking with Bayat's answer to this question — for him the urban poor are inexorably achieving:

> *'[T]he redistribution of social goods and opportunities in the form of the (unlawful and direct) acquisition of collective consumption (land, shelter,* piped water, electricity, roads*), public space (*street sidewalks, intersections, street parking places*), opportunities (favourable business conditions, locations, and labels), and other life changes essential for survival and a minimal standard of living.* (Bayat 2000: 548 — emphasis added)

Such statements are typical of the entire (and now massive) literature on urban renewal, urban social movements and slum upgrading (Cabannes *et al.* 2010; Cities Alliance 2008; Community Organisation Development Institute 2008; De Cruz & Satterthwaite 2005; Menegat 2002; Mitlin 2008; Mitlin & Satterthwaite 2004; Pervaiz *et al.* 2008; Samuels 2005; UN-HABITAT 2008a). The focus is always on the (often highly innovative) community and/or state initiatives to connect the urban poor to what we have called networked urban infrastructures, in particular water and sanitation services. Development policy focuses on this more technical (invariably top-down) task of getting the poor connected to services in order to 'meet basic needs'. The alternative is to emphasise 'quiet encroachment' where non-formalised, hodge-podged, hybridised and contested social orders and territories ambiguate any clear reading of what is really going on. No matter what lens is used, the urban poor are, in one way or another, effectively building and extending a wide range of (connected and autonomous) networked infrastructures, most often piece-by-scrappy-piece as families translate toe-holds into footholds, into full-scale (albeit often fragile) inclusions (even if it takes a generation or more to get there). If one could imagine adding together the millions of everyday actions of slum dwellers to build, connect, repair, clean and protect their tiny spaces, it must surely add up to an effort that is commensurate in scale to the formal investments in networked infrastructures taking place to connect those who can afford to be part of the formal systems. If this were not true, how does one explain the sprawling slums that continue to expand across the developing world as the second urbanisation wave gathers momentum within an increasingly unfair world? The reinterpretation offered here—of slums being an integral part of building the networked infrastructures of the cities of the future, and of slum urbanism as the urban culture of 'quiet encroachment' that animates these activities—opens the way to see slum dwellers as active manipulators of socio-metabolic flows. They are not 'cut-off' or 'excluded' from socio-metabolic flows simply because they lack access to formally constructed and managed networked infrastructures. In one way or another they find ways to access water, energy, food, mobility, building materials and even locations for their sewage and solid waste. In this way they are doing their fair share to exploit the city's ecological resources (rivers for conveying water and waste, soils for planting food and building materials, biomass for fuels and building materials, forests for charcoal, and fossil fuels to access mobility and services such as lighting, the Internet or recharging mobile telephones). As they quietly encroach or actively protest in 'defence of their gains' in ways that are 'collective and audible' (Bayat 2000: 547; see also Cabannes *et al.* 2010; Swilling 2005), they slowly build connections to networked infrastructures, which incrementally secures their access to these flows over the long term (even if they do this illegally or fail to pay taxes or service charges)—or as Bayat puts it, for the urban poor 'modernity is a costly affair' (Bayat 2000: 549).

The global social movement Shack/Slum Dwellers International (SDI) has turned 'quiet encroachment' into a purposive and active strategy of engagement and reform (for the best account by a sympathetic analyst see Mitlin 2008). SDI is a confederation of movements in nearly 30 developing countries, which has demonstrated in practice over the past two decades that 'quiet encroachment' can be accelerated and even formalised by authorities as a legitimate delivery system which is either alternative or complementary to the traditional state- or market-based delivery systems (see www.sdinet.org as well as various editions of *Environment and Urbanization* which tends to carry the best reviews of this work). Accessing resource flows is obviously key to the future survival and prosperity of the urban poor, but how sustainable these resource flows are over time depends on conditions that go way beyond the physical and perceptual purview of the localised spaces that get transformed by the quiet encroachment of slum urbanism.

Green urbanism

Since about 2005/06, *green urbanism* has rapidly evolved as the legitimating ideology for escalating public sector investments in the kinds of networked urban infrastructures that will restructure socio-metabolic flows, although the extent and significance from a sustainability perspective will differ drastically from place to place (for example, the greening of suburban golf estates in the USA is very different from Swedish towns wanting to terminate the use of fossil fuels). Where *inclusive urbanism* was about universal access, *splintered urbanism* about commodification, and *slum urbanism* about quiet encroachment, *green urbanism* is about 'minimising damage' to the environment.[16]

The aim of green urbanism is usually to reduce the environmental impacts of cities (GHG emissions, water, wastes, pollution) and reduce dependence on increasingly costly and insecure long-distance inputs (mainly fossil fuels, but also food supplies, building materials) (Hodson & Marvin 2009b; Hodson & Marvin 2010). The Clinton Climate Initiative, for example, is funding the C40 Cities Climate Leadership Group, which lists 45 major cities as affiliates, including 22 from developing countries (see http://www.c40cities.org). Although the primary focus of this particular alliance is on carbon mitigation, it addresses a broader green urbanism agenda,[17] which is an uneasy mix of urban ecological modernisation (efficiency, 'natural capitalism', greening), 'resource security' (with respect, in particular, to oil dependence, water supplies and food), aspects of inclusive urbanism (renewed appreciation for state involvement in ensuring universal access, especially in developing countries), and the retention of key elements of splintered urbanism (for example, private delivery of services such as rail and telecommunications, in particular the new generation of Web 2.0 infrastructures and 'smart grids').[18]

The top priorities, for example, of President Obama's Green New Deal are all about massive public sector investments in networked urban infrastructures to retool US cities — retrofitting buildings to make them energy efficient, expanding mass transit and freight rail, constructing 'smart grids' to manage electrical grids, and huge investments in renewable energy (wind, solar, second-generation biofuels and bio-based energy) (Barbier 2009; Rotman 2009). Green urbanism has also become the hallmark of China's strategy to make its fast-tracked, state-driven, inclusive urbanism more environmentally sustainable, with a strong focus on reducing the extremely negative environmental impacts of solid, airborne and liquid wastes (China Council for International Cooperation on Environment and Development 2007; Yong 2007). Green urbanism is not just a developed world phenomenon — besides China, it is also taking root in Brazil (Schwartz 2004), India (Revi *et al.* 2006), Costa Rica (Wilde-Ramsing & Potter 2008) and South Korea (Ansems 2009).

---

[16] We have borrowed the term 'green urbanism' from the title of a seminal text by Beatley (see Beatley 2000).

[17] There is no real difference between what we call green urbanism and what Hodson and Marvin call 'urban ecological security' (Hodson & Marvin 2009b).

[18] A similar eclectic mix of discourses is what Kiel and Boudreau have identified in the make-up of what they call 'urban ecological modernization' (2006) with reference to their reading of the environmental politics of Toronto.

Although the recent success of green urbanism has resulted in the taming of its vision so that it can be turned into grand-scale 'techno-fixes' divorced from the realities of social process, culture and power (Guy *et al.* 2001; Hodson & Marvin 2009b; Hodson & Marvin 2010), the ideational roots of this perspective are a strange and diverse mix of movements and aspirations which extend back at least four decades. These include the ecovillage movements of the 1960s, which were anti-urban and autarkic (for example Auroville in India, Crystal Waters in Australia, Gaviotas in Columbia and Findhorn in the UK, but which now extend globally with many affiliated to the Global EcoVillage Network) (Van Der Ryn & Cowan 1996); the German and Dutch cities that, from the 1980s onwards, invested in the greening of the remarkably equitable, inclusive urbanism that evolved during the post-war period (such as Freiburg, Germany) (Beatley 2000; Guy *et al.* 2001); the extraordinary stories of how greening was used in cities such as Curitiba and Bogota to avoid subsidising superfluous middle-class consumption (in particular, the private car) in order to fund a more inclusive Third-World alternative to slum urbanism (Campbell n.d.; Schwartz 2004; Worldwatch Institute 2007: 80–81); the environmental movements that emerged across the world to protest against the negative social and environmental consequences of splintered urbanism (Evans 2002, see also various editions *Environment and Urbanization*; Hawken 2007); the programmatic — and somewhat technocratic — prescriptions for greater social and environmental sustainability that emanated from formal, multi-lateral initiatives such as the Sustainable Cities Programme established in 1995 by UN-HABITAT/UNEP, and the global network of local governments known as ICLEI — trading as Local Governments for Sustainability (see http://www.iclei.org/); extraordinary city-wide partnerships to completely reinvent the city from a sustainability perspective — the best examples being Seoul (South Korea) where the highway through the city was replaced with the river that used to be there, Rizhao (China), Melbourne (Australia) and San Jose (USA) (Ansems 2009; Reed 2007; Wassung 2009; Worldwatch Institute 2007: 88–90, 108–110); and the increasing number of 'habitat awards' for sustainable design that usually feature a vast array of community-based and privately funded projects.

More recently, splintered urbanism has been given a new green sheen by proposals for a new generation of 'green mega-projects' by the world's design glitterati who want to design autonomous 'sustainable cities' for the globally connected elites who want to secede from unsustainable cities and live in safe, carbon-free cocoons (some of them with a 'One Planet Living' stamp of approval from WWF). These projects include Norman Foster's new city called Masdar planned for the Abu Dhabi desert (financed by petro-dollars); Dongtan, the 'first sustainable city', which Arup (the largest consulting engineering firm) has designed in partnership with the Chinese state for Chongmin Island just off the coast of Shanghai; and Lennar Corporation's plan for transforming Treasure Island in San Francisco Bay into another iconic 'sustainable city'. Green urbanism has become big business for the global property development community, with signature architectural egos blazing the trail.

Despite their diversity, all these currents and initiatives have contributed in various ways to the body of practice and knowledge that we have called *green urbanism*. But not all the elements of these antecedents are embodied in contemporary mainstream green urbanism. In particular, the commitments to social justice and the restoration of nature have been conveniently extirpated because they do not fit comfortably with the kind of ecological modernisation that lies at the centre of green urbanism, which is about reconciling sustainability and over-consumption across a range of green 'enclaves' rather than within the context of retrofitted cities (Hodson & Marvin 2010).

Nevertheless, we must not underestimate how fundamentally these investments in more ecologically sustainable socio-metabolic flows (now popularly known as 'green-tech') could change what it means to live a middle-class urban life, and in so doing redefine the kind of urbanism to which others may aspire. These changes could include any combination of the following: ending travel by privately-owned cars; compulsory high-density living in buildings which generate more energy than they consume; enforced waste separation at source; creating potable water from sewage plants; penalties for long-distance travel (by car and air); bans on toxic pollution which could result in the disappearance of a number of consumables (certain kinds of plastics and compounds); decentralised and home-based telecommuting, which will reduce demand for separate commercial and industrial districts; re-engagement with local food production in private allotments or peri-urban farms, either directly or via farmer's markets; rapid price increases for energy-intensive food items such as meat and for consumer goods such as mobile phones which rely on rare non-renewable metals; regulatory and market interventions to restrict carbon-intensive imports and enforce zero-waste; massive new skills training programmes for 'green collar jobs' in the new 'clean tech' industries; and heavy investments in the restoration of urban eco-systems services (such as forests, aquifers, wild areas, rivers and soils). All these changes are already underway in one city or another, and if added up and extrapolated into the future, they represent an image of urban futures that are as different today as the vision of a 'modern city' for the new bourgeoisie that Baron Haussmann must have envisaged in 1852.

Towards a synthesis

Cities will shape the way the dual industrial and socio-ecological transitions unfold over the next 20 to 30 years as they shape the dynamic of the next long-term development cycle. We have discussed four urbanisms ('inclusive', 'splintered', 'slum' and 'green') which can now be mapped against the industrial and socio-ecological transitions that were discussed in detail in Chapter 3 (see Table 5.2). We want to demonstrate that spatial reorganisation has always been an intrinsic manifestation and mediator of the wider socio-ecological and economic transitions that pan out at the landscape level. Although this is not sufficiently acknowledged by those who study transitions (as argued by Hodson & Marvin 2010), it is impossible to understand the contemporary dynamics of transition without grasping the spatial recomposition of urbanism.

Green urbanism has managed to focus attention on what is so ecologically unsustainable about cities, namely the ever-expanding, resource-intensive, environmentally destructive socio-metabolic flows that support the prevailing urban production and consumption

Table 5.2. Four urbanisms mapped against industrial and socio-ecological transitions

	Industrial socio-ecological regime, 1770–present				**Sustainable socio-ecological regime?**
	Long-term development cycle, 1950s–2009/10				***Next long-term development cycle***
	3rd industrial transition: steel/heavy engineering–1875 onwards	*4th industrial transition: cars, oil, chemicals, mass production–1908 onwards*	*5th industrial transition: information & communication technologies, globalised value chains–1971 onwards*		*'greened' deployment phase of 5th transition or 6th industrial transition: sustainable resource use; cleaner production; biotechnology; low carbon; hyper-efficiencies*
Inclusive urbanism, 1850s–1960s	Emerging modern cities in developing countries become nodes for new industries	Spreads out across the developed world, colonial enclaves, and socialist societies	Partially survives in Western Europe and China, and in certain enclaves	**CRISIS**	Returns in countries with a strong commitment to social justice and equity, or—as in China—a political requirement to secure stability through inclusion
Splintered urbanism, 1970s–present/ future			From early 1980s, implementation starts in developed countries, spread through developing countries, and post-communist countries, excluding China—muted in parts of Western Europe (esp. Germany and Holland)	**CRISIS**	Has a role to play, especially with respect to innovation, flexibility, adaptability and choice—but only if properly regulated to prevent rent seeking behaviour; big danger of elite secession in autonomous 'sustainable cities'
Slum urbanism: coterminous with the 2nd wave, 1950–2030		Accelerated expansion in developing countries as the second urbanisation wave gathers momentum from the 1950s onwards	Explosive growth as second wave takes hold, coupled to NIDL, with only pockets remaining in developed countries	**CRISIS**	Unlikely to disappear quickly, positive elements such as self-help and autonomy have a role to play, thousands of projects underway to find low-tech affordable solutions
Green urbanism, 2008–present/ future			Origins in the 1970s/80s, pioneer projects in the 1990s, acceleration from the start of the mid-point crisis (2007) onwards		Origins in 1990s, expanding into consumption and production, key role for cities, mega-projects/techno-fixes, but also new configurations that can bridge green and slum urbanism

systems. Networked urban infrastructures have gradually evolved into highly complex systems which source and deploy these flows from across increasingly vast natural catchments and value chains. Up until recently, these infrastructures have been designed and operated as if there are no limits and those who manage them have not developed the skills — or been incentivised — to think any differently. Green urbanism, therefore, goes up against the entire history of urbanism by suggesting that cities can grow and urban livelihoods improve without increasing — indeed, even reversing — demand for resources. In the words of Hodson and Marvin:

> *Cities have usually sought to guarantee their reproduction by seeking out resources and sinks from locations usually ever more distant and connected through huge socio-technical assemblages. Yet this traditional approach is now being challenged as cities seek to 're-internalise' and 're-localise' resource endowments by creating 'closed loops' and 'circular metabolisms' as they seek to withdraw from reliance on international, national and regional infrastructures.* (Hodson & Marvin 2009b)

The C40 Cities Climate Leadership Group is the most explicit contemporary expression of this approach, but ICLEI shares this vision and has been at it for longer (see http://www.iclei.org/). Related to this is a search for a new generation of green mega-projects which will mobilise giant dollops of funds to stimulate economic growth and prepare for a sustainable future. The most popular grand 'techno-fixes' are mass public transit systems (especially urban rail, but also BRT); closed-loop water and sanitation systems; large-scale renewable energy systems such as Desertec — the biggest solar power plant planned by a consortium of German companies for the Sahara Desert to supply Europe via new DC ('direct current') cables; smart grids; a new generation of buildings that generate more energy than they use; and as mentioned earlier, the new autonomous sustainable cities for the elites such as Masdar, Treasure Island and Dongtan.[19]

There are two problems with this bold green urbanism vision. The first is that it assumes that the next long-term development cycle really is coterminous with the transition to a sustainable socio-ecological regime. Former UK Prime Minister Gordon Brown was correct: if a real deal had been struck in Copenhagen in December 2009, a key condition for the dual transition would have been put in place (*Newsweek*, 28 September 2009). India and China require only one condition to be met to make the deal, namely a guarantee that massive funds will be available from developed countries to finance the kinds of infrastructures that will prepare them for low-carbon futures. This, above all else, is why networked urban infrastructures are such a key factor in determining whether we are going to make the transition to a more sustainable socio-ecological regime now, or wait for the sixth industrial transition to take its course as the global ecological crisis gets much worse.

---

[19] All these projects are referred to in one way or another in a special eight-page supplement in *Newsweek* sponsored by Siemens, which also happens to be investing in most of these mega-projects — see *Newsweek*, 28 September 2009.

The second problem, of course, is that it is difficult to reconcile the capital-intensive techno-fixes envisaged by green urbanism with the messy quiet encroachments of slum urbanism. Although they hardly ever make the headlines, hundreds of thousands of initiatives are underway within slum communities around the world, which are slowly but surely stitching together ingenious ways of connecting slum dwellers to flows of water, energy, food, mobility and building materials. These initiatives run contrary to the opposite logic: daily (usually violent) attacks on slums to relocate people who are 'in the way of progress' (Cabannes *et al.* 2010). In their remarkable review of these initiatives, Cabannes *et al.* document the strategies of engagement that these communities use to contest evictions, which include negotiation (SDI is the champion of this strategy), 'occupy-resist-live', legal challenges via the courts, open struggle and resistance, building rights and pragmatic policies, and campaigning to influence public opinion (Cabannes *et al.* 2010). Where relocation is resisted and connections made to flows that make survival in the city possible, slum dwellers join the ranks of the billion or so people who already live modestly connected to the basic flows that networked urban infrastructures can provide, and who have sufficient to live decent, hard-working lives. Herein lies the historic significance of the movements that SDI gives voice to. This quiet encroachment through countless organisational initiatives is also an unacknowledged driver of the creation of massive domestic markets that are, in turn, catalysing the sonic booms in 'emerging markets' in virtually every developing country outside of China.[20] Maybe this means that slum urbanism is making possible a new kind of *bottom-up inclusive urbanism,* in which state interventions to provide services are largely reactive responses to grassroots power rather than a proactive project of inclusive urbanism enhanced by the kinds of Keynesian economic policies that emerged after the 1929 Depression.

There are, however, strategic perspectives that bridge the challenges of slum urbanism and the perspectives that inspire green urbanism. A rather dramatic example was offered by Michael Rouse, President of the World Water Association and experienced manager and policy-maker. Talking in March 2003 about how to achieve the target set by the 2002 World Summit on Sustainable Development (WSSD) to provide water and sanitation for 1.2 billion people worldwide, he said this would be impossible if this were to be done using the traditional resource-intensive technologies. He said that sewage pipes are too expensive and traditional sanitation systems drain nutrients required for food production out of the urban system. And then, remarkably:

> *If we started sanitation again from scratch in Britain, we would not do it the way we do now. Instead of flushing and piping all the waste away, we would collect the solids once a week like household rubbish, take it to a central depot and compost it. Eventually it would be used as fertiliser, itself a bonus in the developing world, which would be able to cut down on expensive chemical fertilisers.*[21]

[20] Although even China now has its 'quiet encroachments'—(Liu & Vlaskamp 2010).

[21] Cited in 'Keeping sewage on home soil', *Mail & Guardian*, 20–27 March 2003.

This is a graphic image of how a very different kind of networked infrastructure can be designed and operated to direct and channel the same socio-metabolic flows with radically different outcomes for both rich and poor alike. A century ago it would have been unthinkable. But threats to the resource flows that were just assumed when the old system was designed are what force the consideration of alternatives today. More of these kinds of ideas and fewer Masdars is what will connect green urbanism to the quiet encroachments of the slum.

Towards liveable urbanism

During the first growth period of the post-World War II, long-term development cycle, city-wide coalitions for economic growth and extension of networked infrastructures was what urban politics (in the West) was all about—hence the sub-title that Logan and Molotch gave to their classic text on this period: the 'political economy of *place*' (Logan & Molotch 1987). For social movements, this entailed engaging local political and economic elites to secure a greater share of the 'means of collective consumption'. An entire generation of researchers and activists was brought up within this paradigm, complete with its own lexicon and library of training manuals. Then came neoliberalism and globalisation and that classic statement from Borja and Castells that defined the research agenda of the next generation of urbanists: 'The global city and the informational city are also the dual city' (Borja & Castells 1997: 44). In short, the city bifurcated: globalisation and the Information Revolution created one set of spaces (the 'space of flows') while the old space of places was emptied of all power and significance. Hence splintered urbanism and the pessimistic conclusion that Castells promoted—that place had become merely the space of identity, and was no longer the space of economic and informational power. In one conceptual blow, the space of place as the *raison d'être* for decades of urban struggles by the poor—and economic strategies by local businesses—was obliterated. Somehow, 'identity politics' has not really filled this vacuum. In brave attempts to rescue tiny platforms for grassroots struggles from this conceptual wreckage, some writers optimistically argued that although this may all be true, urban communities that suffered the harsh consequences of splintered urbanism still managed to wage campaigns to make their cities a little more liveable (Douglass & Friedman 1998; Evans 2002). But these struggles were, on the whole, depicted as marginal ameliorative engagements to contest social and environmental *impacts*, not struggles about the core logics of resource flows and informational power. Even the new work on the socio-metabolism of the city cited extensively here (Guy *et al.* 2001; Heynen *et al.* 2006) is in general (with some exceptions) more interested in a critique of technicism than the implications of flow analysis for a radically new conception of urban politics.

Based on our reading of the global context and the history of urbanism analysed in this chapter, our conclusion is that successive generations of urbanists have failed to recognise the profound strategic implications of the ecological limits to both inclusive and splintered urbanism. Castells, in particular, mesmerised by the blinking razzmatazz of the Information Age led an entire generation of researchers astray by focusing exclusively on only one set of globalised 'flows', namely informational flows. Once one includes material flows into the 'space of flows' (by using, for example, material flow analysis), the 'space of place' returns not just as a disempowered space of identities, but as the locale of contested sources and sinks for the socio-metabolic flows on which *everything*—the entire web of life—depends. The remarkable report by Cabannes *et al.* (Cabannes *et al.* 2010) provides a glimpse of this new *politics of spatial resources* because their focus is access to land—the source of nearly all the key resources on which contemporary economies and livelihoods depend.

If our analysis in Chapter 3 is correct, many of the key resources will become increasingly costly to procure, and environmental impacts within particular places will intensify. What happens then? The biggest casualty will be the logic of locality specialisation, which has been a key element of splintered urbanism. As Korhonen argued, globalisation of production has entailed the geographical separation of all the key nodes in the product life cycle: raw material extraction, mass production of components at the lowest cost, component assembly in the most efficient ways (as automated as possible), storage and end consumption where the highest prices can be secured—each of these nodes could be in a different country. Each region—and, indeed, even each district within each region—is required to find its 'niche' within this geographically dispersed production and consumption life cycle, and invest accordingly to compete in the global economy (Korhonen 2008). This is what has triggered massive investments in infrastructures, human skills development and cultural capital—all in the name of securing a particular city's 'niche position' within the global 'space of flows'. This, in essence, is what the global obsession within the development policy community with 'local economic development' was all about. However, to make this work, huge quantities of energy are required to transport massive amounts of raw materials and manufactured goods around the globe, often with ludicrous results such as South Africa's agreement to allow Indian multinational ArcelorMittal to export local iron ore to China for processing and then to re-import steel products that could have been manufactured locally. All this works as long as oil and the costs of CO_2 mitigation/adaptation remain relatively cheap. As these factors change (which they will), niche specialisation could change from being a key strength to being a key weakness.

The future may well lie in *regionalised bio-economic diversification,* which is when regions[22] build on all their *ecological, economic* and *social* strengths by maximising the returns from regionally integrated, locally driven value chains, rather than depending *entirely* on globalised supply and distribution chains that are becoming increasingly unstable and unreliable.

South Africa's Western Cape region (with Cape Town as its primary node), for example, has limited energy supplies and sees tourism and agricultural exports as key economic drivers, but both stand to suffer from rising transportation costs. The

22 By region here we mean the functionally integrated locales that make up many urban formations, whether they are smallish cities or large metropolitan areas or even clusters of smaller locales within much wider 'megapolitan' areas (for a useful characterisation, see Clark, Dexter & Parnell 2007).

alternative is catering for the leisure needs of local communities, selling more food into local markets and taking advantage of the best solar radiation levels in the world by investing in renewable energy. Exports would, then, become knowledge-intensive systems and technologies that other regions would require, plus imports that add value (for example, machinery and high-end technology) rather than merely consumables (such as peas from Kenya).

This is what bio-economic diversification — or what we would call *liveable urbanism* — could be all about. Interestingly, it is strongly driven by both learning and identity formation as a 'sustainable city', which in turn translates into the localisation of a new set of informational and resource flows as companies realise that they have more to gain from being embedded in culturally functional communities which can offer significant intellectual capital resources, trust-based transactions across networked supply chains that cut costs, and places in which scarce, high-cost, managerial and technical labour (dissatisfied with corporate towers, malls, traffic congestion and security villages) would prefer to live. In short, liveable urbanism reconnects the 'space of place' and 'space of flows' and establishes, in turn, the basis for a new era of city politics about how to ensure the sustainability of the socio-metabolic flows on which the web of all life depends.

Liveable urbanism is related to, but also substantively different from, green urbanism. It shares with green urbanism the assumption that the cities of the future will need to be low-carbon, more resource efficient and less damaging of the environment. However, the danger with green urbanism is that it is fast becoming a techno-fix for greening the elite residential enclaves and commercial parks without facing the inescapable need to reverse over-consumption and end urban poverty by bringing back the 'universal access' ethos of inclusive urbanism.

Following the remarkably fresh, critical work by Janice Birkeland, who argues that we need to go 'from design for sterility to design for fertility', we concur that it is time to go beyond the 'minimising damage' ethos of green urbanism (expressed most clearly in this movement's most vocal manifestation — the Green Building movement). As Birkeland quips: 'Indeed, if we labelled cigarettes the way we label buildings, people might start smoking more 'light' cigarettes to get healthier.' More seriously, she argues that:

> *[F]ew appreciate that the 'built environment' (cities, buildings, landscapes, products) could generate healthy ecological conditions, increase the life-support services, reverse the impacts of current systems of development and improve life quality for everyone ... For development to become the solution instead of the problem, it must provide the infrastructure for nature to regenerate, flourish and deliver ecosystem goods and services in perpetuity ... This is not only possible, but arguably easier than what we are doing now. The only impediment is fear of change itself.* (Birkeland 2008: 3–4)

Liveable urbanism is inspired by the many examples from around the world where development restored nature (Auroville in India and Gaviotas in Colombia are probably the best pioneer examples), and by the billion or so urban dwellers who have sufficient and therefore live largely within the carrying capacity of the planet. Liveable urbanism can find common ground with those quiet encroachments of slum urbanism that empower the urban poor to build, from below, local economies of inclusive self-sufficiency, which contest the relocations that splintered urbanism tends to foster as developers build more shopping malls, resorts, export processing zones, highways and security estates. To this extent, liveable urbanism is closer to the ethos of inclusive urbanism than the callous selfishness of splintered urbanism. However, liveable urbanism has much to learn from the latter's faith in entrepreneurship as the driver of a diversity of delivery systems instead of a return to the rigidities of public sector monopolies, which gave inclusive urbanism a bad name. The flowering of autonomous, sustainable communities (EcoVillages) as manifestations of ecological entrepreneurship is perfectly compatible with a liveable urbanism, which depends on emerging modes of bio-economic diversification that actively includes the urban poor into new networks of production and consumption.

Conclusion

Sufficiency is what everyone (including slum dwellers and over-consumers) should aspire to if they are concerned about the consequences of an increasingly unfair world. This is not measurable via material metrics such as GDP per capita, but by the proposed indices for measuring 'genuine progress', a sense of community and 'happiness' (Stiglitz *et al.* 2009; Talberth *et al.* 2007; Talberth 2008).[23] GDP measures everything except what is most important. This is why it is not just about (material) sufficiency. It is also about the restoration of life, not only because our survival depends on abundant and thriving eco-systems, such as good soils, a stable climate and clean water, but also because we cannot be expected to be able to live creative, meaningful, fulfilled lives if we participate in the mass genocide of other species, living systems and natural resources. When we realise that investments aimed at restoring our eco-systems — along with investments in social cohesion and innovation — is a key element of the future economy of networked, ecologically thriving, socially integrated, bio-economic regions, we will have embarked on a new kind of *liveable urbanism*. As Okri says: 'All roads lead into the maze of the city... But the only ways out lead to the forests of the interior and to the sea.'

[23] See also the report commissioned by President Sarkozy that was compiled by a commission of eminent academics led by two Nobel Prize winners, economists Joseph Stiglitz and Amartya Sen (Stiglitz *et al.* 2009).

Chapter Six

Soils, Land and Food Security

The Law of the Land[1]

Now this is the law of the land, son
as old and as new as the hills
And the farmer who keeps it may prosper
but the farmer who breaks it, it kills.

Unlike the law of the man, son
this law it never runs slack,
What you take from the land for your own, son,
you've damn well got to put back.

Now we of the old generation took land on the cheap and made good
We ploughed, we stocked, and we burnt, son, we took whatever we could,
But erosion came creeping slowly, then hastened on with a rush,
Our rooigras went to glory, and we don't relish steekgras much.

The good old days are gone, son
when those slopes were white with lambs,
The lands lie thin and straight, son,
and the silt has choked our dams.

Did I say those days were gone son? For me
they are almost gone
But for you they will come again, son,
when the task I set you is done.
I've paid for this farm and fenced it,
I've robbed it, and now I unmask —
You've got to put it back, son,
And yours is the harder task.

Stock all your paddocks wisely, rotate them as you can,
Block all the loose storm-waters, and spread them out like a fan.
Tramp all your straw to compost and feed it to the soil,
Contour your lands where they need it, there is virtue in sweat and toil.
We don't really own the land, son, we hold it and pass away.
The land belongs to the nation, to the dawn of judgement day.

---

[1] Poem by an anonymous poet found hanging on the wall in a roadside cafe near the small town of Heidelberg, Southern Cape, South Africa.

And the nation holds you worthy, and if you are straight and just
You'll see that to rob the land is betraying a nation's trust.

Don't ask of your farm a fortune,
True worth ranks higher than gold;
To farm is a way of living, learn it
before you grow old.

So this is the law of the land, son,
to take, you've got to put back.
And you'll find that your days were full, son,
when it's time to shoulder your pack.

Introduction

In one of his many contributions to the influential website *India Together* (www.indiatogether.org) on the crisis of farmer suicides, award-winning Indian journalist Jaideep Hardikar told the story of Vijay Jawandhia, a local leader of the farming community in Wardha, central India. Lamenting the fact that a European cow gets an average *daily* subsidy of US$2 from government while World Trade Organization (WTO) regulations prevent Indian farmers getting the same benefit, Jawandhia is reported to have said: 'If I were given a choice, I would like to be born a European cow, but certainly not as an Indian farmer, in my next birth… [I]n India, a farmer is a debtor all his life. Post his death, his son inherits his debts and has to borrow money for his funeral' (Hardikar 2006). There have been 166,000 farmer suicides in India since 1997, a death every half hour *(Mail & Guardian* 2008).

Whereas the shock of farmer suicides in India has focused attention on the challenges of agricultural production, the 2008 International Assessment of Agricultural Knowledge, Science and Technology for Development (IAASTD) opened its lengthy report by focusing on the inequalities in consumption: '[g]lobally, over 800 million people are underweight and malnourished, while changes in diet, the environment and lifestyle worldwide have resulted in 1.6 billion overweight adults; this trend is associated with increasing rates of diet-related diseases such as diabetes and heart disease' (Watson *et al.* 2008: Ch. 3: p. 3).

During the first three months of 2008, the real prices for all major food commodities reached their highest levels in 30 years (FAO 2008). These price spikes prefigured the October 2008 financial crash and, according to the United Nations Environment Programme (UNEP), they also forced an additional 110 million people into poverty, adding another 44 million people to the ranks of the malnourished (Nellemann *et al.* 2009: 6). By 2010 the official figure for the number of people who were starving edged over the one billion mark (OECD & FAO 2010). Unsurprisingly, this triggered protest in many parts of the world—by mid-2008 protests had broken out in over 61 countries including Morocco, Guinea, Mexico, Egypt, Burkina Faso, Indonesia, Mauritania, Senegal, Uzbekistan, Haiti, Argentina and Yemen. By the end of 2008 food prices and the related issue of food security had become a permanent feature of the mainstream global policy repertoire (Bailey 2011:7; World Bank 2008).

Although the price spike ended because the recession resulted in shrinkage in demand and a partial (temporary?) withdrawal of financial speculators from the food commodity markets, the 2010 OECD–FAO Report projected that the average price of all food commodities over the coming decade to 2019 will be higher than the preceding decade (OECD & FAO 2010: 18). Confirming this prediction, on 5 January 2011 the FAO announced that its food price index for December 2010 hit an all-time high. So it is not surprising that food protests have continued. Government troops in Mozambique, for example, resorted to force in September 2010 to disperse groups of food protesters, many of whom were children. By 2011 food protests spread across Algeria and broke out in India. Rising food prices fuelled the Egyptian revolution as declining oil exports reduced the funds available to subsidise wheat imports in a country that is the largest importer of wheat in Africa. This was also the year in which Russia imported grain to feed its cattle in the wake of a heat wave that drastically reduced food reserves, and China became an importer of massive quantities of wheat and corn (Brown 2011: 1).

At the same time, there is mounting evidence that the eco-systems that make agriculture possible are steadily deteriorating as the levels of extraction and exploitation intensify. Although agriculture depends on a complex matrix of interlinked eco-systems (such as water, soils, stable climates, nutrient cycles and pollination), it is the degradation of the soils themselves that is the most worrying. Unlike in the water, biodiversity, energy and climate sectors, there are no recent global assessments of soils — the last one, which was done in 1990, reported that 23 per cent of all soils were degraded (Oldeman 1994). Nevertheless, 20 years later no other major global soils assessment has been planned, and the IAASTD and researchers are still referring to the 1990 assessment (Scherr 1999; Watson *et al.* 2008; World Resources Institute 2002).

What do farmer suicides, obesity, starvation, food prices and soils have to do with linkages between the polycrisis, transitions and development? Interestingly, it is possible to correlate the passage of different 'food regimes' with the transitions discussed in Chapter 3. As already argued in that chapter, the post-World War II, long-term development cycle that ended with the 2007–2010 global crisis started off with a growth period (1950s/60s) made possible by the deployment phase of the Age of Oil. Following McMichael's work on 'food regimes' (McMichael 2009a; 2009b), we will show that this 'golden age' was underpinned by the 'second food regime', which effectively ended for economic reasons with the stagflation crisis of the 1970s. The 'third food regime' was made possible by the neoliberal revolution, globalisation and the space this all created for the start of the Information Age, which fuelled the second growth period (1980s/90s) of the post-World War II, long-term development cycle. The food riots of 2008 effectively mark the end of the 'third food regime' and open the way for contestations about what will follow. But our argument will be that the end of the 'third food regime' has as much to do with deeper underlying ecological factors such as soil degradation as it does the usual set of economic factors. Whereas new technologies and the GM revolution are seen by many mainstream policy and business networks as the operating system for the 'fourth food regime',[2] we propose that the 'agro-ecological' alternative is more appropriate if the 'fourth food regime' is to be compatible with the more *sustainable long-term development cycle* envisaged in Chapters 3 and 4.

We argue that there is a link between soil degradation and rising food prices, which is not given sufficient attention in the global 'food security', 'land-use change' and the more radical 'food regime' discussions. If we do not find ways to reverse soil degradation, global food security will be unattainable. We believe, however, that 'agro-ecological innovations' are able to restore soils and thus potentially provide the basis for long-term food security as the core foundation for a more sustainable long-term development cycle.[3] Or, to put it more colloquially, the 'ever-green revolution' may well be a real alternative to the 'green' and the 'gene' revolutions. The logic of the argument is as follows:

- Consistent with our argument about resource prices in Chapter 2, the long-term historical trend in declining food prices has finally bottomed out and there are indications that we are now facing a long-term trend of rising food prices.
- Food prices are rising because yield growth is declining. This, together with rising demand to satisfy more resource-intensive diets (meat, milk products) and biofuel requirements, is the primary reason for food price increases.
- Declining yield growth, however, is directly related to soil degradation. Amazingly, soil degradation is hardly ever identified as a cause of the problem, despite long-standing, reliable, scientific evidence that soil degradation is a major problem and that it does undermine agricultural output.
- The call to mainstream agro-ecological approaches made by, among many others, the IAASTD and UNEP marks a turning point, because it suggests a new kind of restorative agriculture that has the potential to ensure food security by rebuilding the global eco-systems on which agriculture depends. Could this become the basis for a 'fourth food regime'?

Food prices

The massive and rapid rise in food prices between 2004 and 2008 forced the global challenge of food supplies and food security onto the global agenda. Figure 6.1 reveals the close link between the prices of key food commodities and oil prices. Besides contributing to the financial crash by adding yet another financial burden to over-indebted households in developed and developing economies, millions of the poorest people in the world were forced deeper into poverty. For every 1 per cent increase in

2 This is expressed most clearly in Africa by Juma (2011).

3 We would like to gratefully acknowledge the work of our colleagues Prof. Tarak Kate (who is based in India, but who joins us every year for a month or two to teach an agro-ecology module on the masters programme), Gareth Haysom and Candice Kelly, whose work over the years has shaped our thinking and given us courage to pursue the arguments developed in this chapter. We have borrowed from their ideas and teaching, but we take full responsibility for what we have written here (see Haysom 2010; Kate 2010; Kelly 2009).

the price of food, food consumption expenditure in developing countries decreases by 0.75 per cent (Von Braun 2007: 5). Social movements were galvanised into action, reinforcing long-standing critiques of the global food system. Various governments responded with protectionist measures, including bans on food exports and lowering tariffs on imports, and land grabs accelerated, especially in sub-Saharan Africa.

As is apparent from Figure 6.1, the decade that started with the turn of the millennium is characterised by slow and steady price increases, ending off with a spike that quite quickly corrected to a point that is consistent with the overall trend of steadily rising prices since 2000, that is, the correction did not take prices down below where they were in 2000 or, for that matter, before 2005. However, when this 10-year period of rising prices is placed within a wider context of agricultural prices since 1900, it is clear that, excluding the last decade, real prices of agricultural products (like resource prices) have steadily declined over the previous century (see Figure 6.2).

This long-term decline in real prices is largely due to massive increases in agricultural productivity and output that has, with key exceptions that mark moments of crisis (post-World War I, 1929 crash, post-World War II, 1973/74 oil crisis), kept ahead of rapidly rising demand due to population growth, and diet changes caused by modernisation and urbanisation. The obvious question is whether the current spike will end at a point

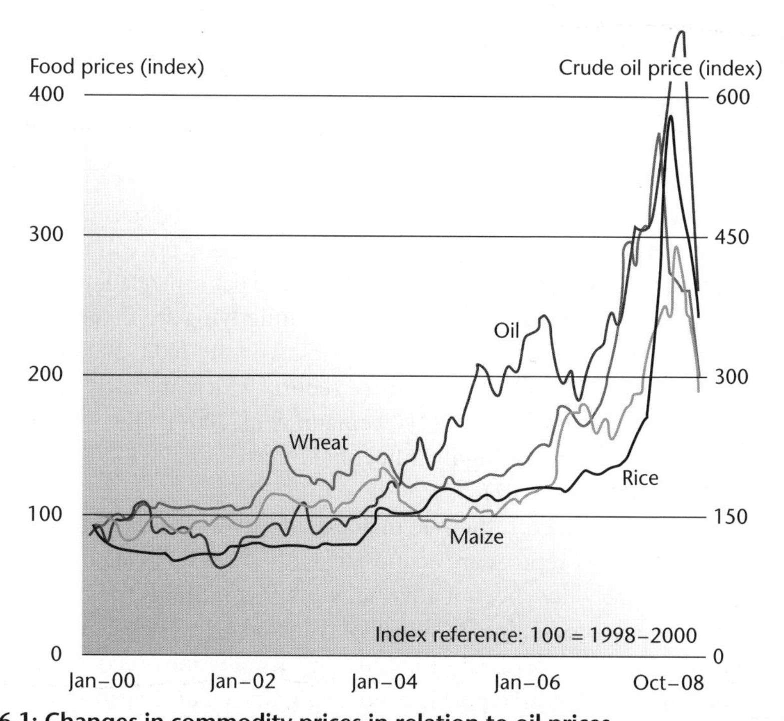

Figure 6.1: Changes in commodity prices in relation to oil prices

(Source: Nellemann, C., MacDevette, M., Manders, T., Eickhout, B., Svihus, B., Prins, A.G., Kaltenbom, B.P. (eds), 2009. *The Environmental Food Crisis: The Environment's Role in Averting Future Food Crises.* United Nations Environment Programme, Brikeland Trykkeri AS, Norway.)

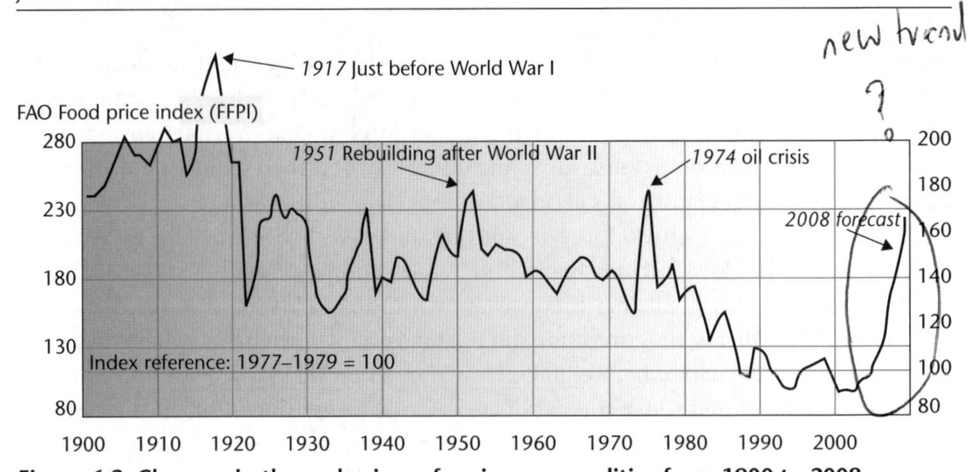

Figure 6.2: Changes in the real prices of major commodities from 1900 to 2008

(Source: Nellemann, C., MacDevette, M., Manders, T., Eickhout, B., Svihus, B., Prins, A.G., Kaltenbom, B.P. (eds), 2009. *The Environmental Food Crisis: The Environment's Role in Averting Future Food Crises.* United Nations Environment Programme, Brikeland Trykkeri AS, Norway.)

which replicates the long-term downward pattern, or whether we are at the start of a long-term increase in food prices driven by a matrix of factors that have not been present in this form before. No other decade, except possibly just after World War II, exhibits a pattern of such steady and steep price increases. So what is going on?

The authoritative 2010 *OECD–FAO Agriculture Outlook Report* predicted that the nominal and real prices of crop and livestock products for the period 2010–2019 will be lower than 2007–2008, but higher than the average price for the decade 1998–2008.

If this prediction turns out to be true, it means there will have been two decades of steadily rising prices — something that has not happened since the start of the twentieth century. Using this and other data, Oxfam has predicted that food prices during the period 2010–2030 will increase by between 120 per cent and 180 per cent (Bailey 2011: 8).

This raises important questions about some of the underlying drivers of steadily rising food prices. Unfortunately, most accounts tend to focus on the 2007–2008 spike and not on the long-term trend. We will focus on three reports which are fairly representative of mainstream thinking, namely UNEP (Nellemann *et al.* 2009), OECD–FAO (OECD & FAO 2010) and a report by the highly influential International Food Policy Research Institute (IFPRI) (Von Braun 2007).

As can be seen from Table 6.1, the vast majority of drivers that have been identified by these three mainstream reports are economic. Only three are related to underlying natural resources, namely the effect of adverse weather conditions (possibly related to climate change) on crop production, the effect of rising oil prices (which for us is related to oil peak, but none of the reports make this link) and degradation of eco-systems (which should include soils). The UNEP report identifies a significant decline in available arable land for food production caused by degraded soils, polluted and limited water supplies, pest infestations, climate change and land-use change (as arable land is reallocated for urban development or non-food agricultural production). The report predicts that if

Table 6.1: Causes of food price increases in 2007–2008 according to major reports

Causes of rising food prices	*OECD-FAO*	*IFPRI*	*UNEP*
Drought/extreme weather causing crop failure, with a climate change link	X	X	X
Declining stocks as demand outstrips supply	X	X	X
Demand for non-food crops for biofuels	X	X	X
High energy prices causing increased fuel and input (fertilisers, pesticides, etc.) costs	X	X	X
Entry of financial speculators into agricultural markets	X		X
Long-term decline in investment in agriculture due to long-term decline in real prices which caused a decline in stocks	X	X	
Depreciation of the dollar thus improving the purchasing power of importing countries buying commodities denominated in dollars	X		
Government interventions in response to price increases to secure food supplies which made matters worse	X		
Economic development in developing countries creating new demand for higher value crops		X	
Vertical and horizontal integration in the corporate value chain (i.e. increased concentration and monopolisation) which reduces competitiveness and pushes up prices		X	
Failure of the Doha Round of trade negotiations resulting in the persistence of US and EU subsidies which reduces investment in and output from developing countries		X	
Degraded eco-system services			X

current trends continue, available cropland could reduce by between 8–20 per cent by 2050 which, in turn, could reduce current yields by 5–25 per cent by 2050 (Nellemann *et al.* 2009: 33). This, in turn, could reduce yield growth to 0.87 per cent per annum by 2030 dropping to 0.5 per cent by 2030–2050. According to Von Braun, global warming by 2 °C or more could reduce global agricultural GDP by 16 per cent by 2020 which, in turn, would result in price increases of up to 40 per cent (Von Braun 2007: 3). Given that agricultural production needs to increase by a minimum of 1.2 per cent per annum to keep up with demand (Nellemann *et al.* 2009:77), a long-term decline in yield growth must be a key explanation for why we are looking at an unprecedented two-decade period of steadily rising real prices for food.

Agricultural yields and projected demand

By the end of the twentieth century, there were approximately 437 million farms in developing countries which sustained the livelihoods of 1.5 billion people and provided food for two-thirds of the human population (Madeley 2002). Up until the 1950s, by far the vast majority farmed using natural methods on lands that were increasingly marginal as a result of centuries of violent land dispossession in favour of imperial powers and their settler offshoots. Until World War II, UK-based capital dominated investments in the 'first food regime' (1870s–1930s) which relied upon imports of grains and livestock from settler family farms in settler colonies and tropical fruits, vegetables, spices, etc. from colonies in general to feed the mushrooming European and North American industrial workforces (McMichael 2009b) created by the third and fourth industrial transitions (discussed in Chapter 3).

The 'second food regime' (1950s–1970s), co-ordinated by the US government and dominated by US-based capital, transformed the US into a global agricultural power, which it achieved by subsidising the deployment of the new Green Revolution technologies (chemical inputs, hybrid seeds and mechanised irrigation systems) across its vast tracts of high-value soils to massively boost agricultural productivity. The US then rerouted huge amounts of surplus food through its network of informal colonies (which emerged after decolonisation) as 'aid' in return for payments into 'counterpart funds' held at local banks, which were then used to finance (mainly American) agri-business expansions into the developing world, using the Green Revolution technologies (McMichael 2009b). As Figure 6.3 demonstrates, the result was massive increases in the use of chemical inputs to drive equally significant increases in agricultural output. Global value chains of cheap subsidised food were created which helped expand the politically useful, bloated, agricultural labour-forces in the US and fed the expanding, urbanised workforces that catalysed the second urbanisation wave across the developing world (discussed in Chapter 5).

The 'second food regime' was also driven by 'developmental states' to generate substantial profits within developing countries to finance modernisation through industrialisation (as discussed in Chapter 4). By the start of the twenty-first century, 40 per cent of the 437 million farms in the developing world were dependent on Green Revolution technologies (Madeley 2002: 21), and many of these generated the surpluses required to finance urban-based modernisation through industrialisation (especially in Asia and Latin America; less so in Africa, but certainly in South Africa). The rest were small farmers on marginal land, often the victims of land dispossessions to make way for cities and massive agri-business operations on the best land. Those who were no longer on the land had migrated to the burgeoning cities as part of the second urbanisation wave.

In essence, the Green Revolution 'package' entailed hybrid seeds (seeds developed by specialist seed companies — usually large multi-national corporations — and sold to farmers, thus replacing the age-old traditions of seed banking and exchange); chemical inputs derived mainly from oil, but also from rock phosphate (fertilisers and pesticides produced and distributed globally by multi-national corporations); large-scale mechanised irrigation systems (often funded and installed by the state using loans provided by the World Bank); and micro-credit facilities. This is why this package is often referred to as 'high external input' (HEI) agriculture. It was a package which

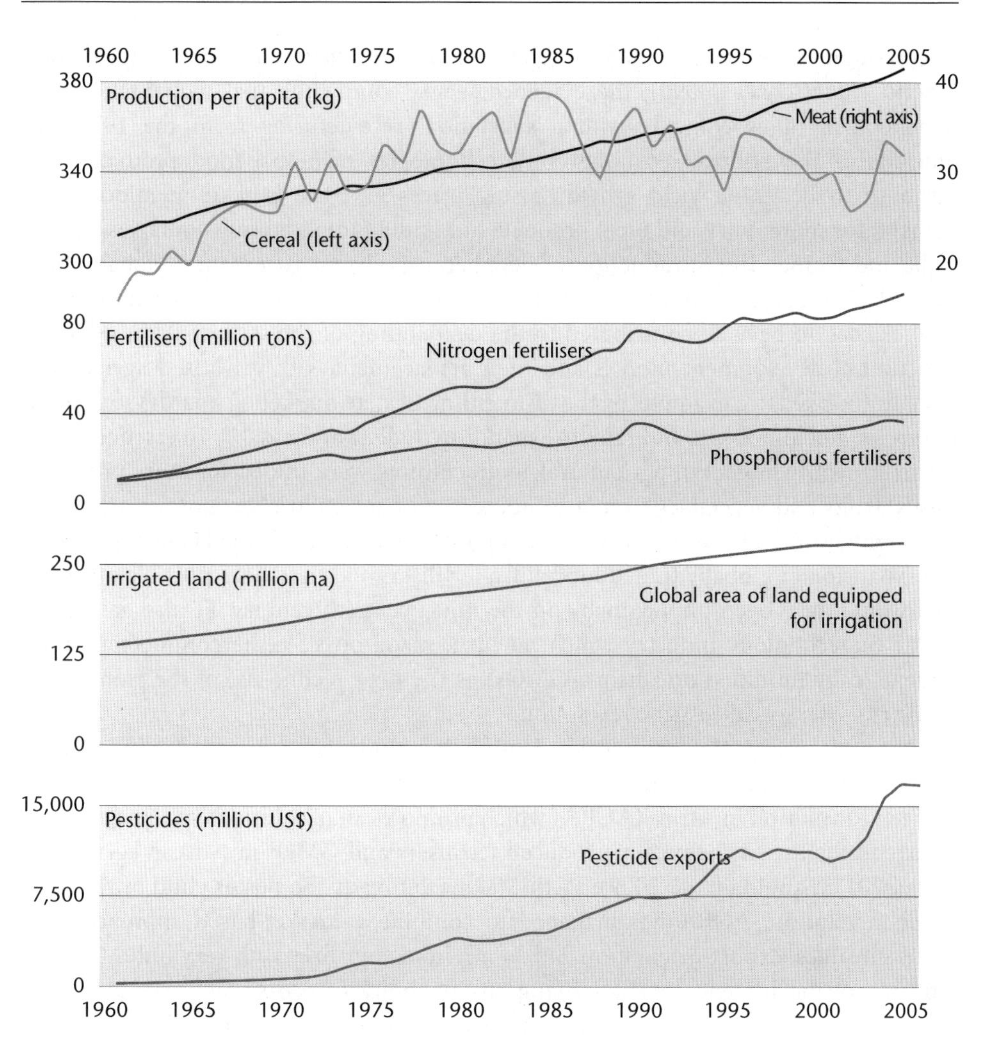

Figure 6.3: Global trends (1960–2005) in cereal and meat production, use of fertiliser, irrigation and pesticides

(Source: Nellemann, C., MacDevette, M., Manders, T., Eickhout, B., Svihus, B., Prins, A.G., Kaltenbom, B.P. (eds), 2009. *The Environmental Food Crisis: The Environment's Role in Averting Future Food Crises.* United Nations Environment Programme, Brikeland Trykkeri AS, Norway.)

became central to what 'development' was all about after decolonisation in Asia and Africa, and was key to the Latin American 'modernisation' drive. Outputs doubled and even quadrupled wherever the package was applied, in particular in strategic pockets of India during the 1960s (encouraged by lavish grants from US foundations to establish the scientific knowledge base, practices and supply lines). Without the Green Revolution global food production would have been half of what it was by 2000, and the environmental impacts of moving food production into environmentally unsuitable areas would have been far more serious than they are today (Uphoff 2002a).

Although there is some debate about whether we have entered the 'third food regime' (McMichael 2009b), the evidence seems compelling that globalisation and neoliberalism does coincide with a substantial restructuring from the 1970s/80s onwards of the political economy and technologies of global food production in response to declining yield growth, rising prices and the expanding middle-class demand for more dairy and meat products in rapidly industrialising countries such as India and China. The introduction of neoliberal modes of governance, globalisation, deregulation, privatisation, the establishment of the WTO rules for agriculture, and financialisation have all contributed to the dismantling of the state-centred, national, agricultural-development models and their replacement with privatised agricultural systems (marked, for example, by the dismantling of state marketing boards) structured to service global markets and rapidly expanding trade (Barker 2007). In essence, stable grains grown in northern agricultural monocultures were traded for mass-produced meats, fruits and vegetables from a complex mix of agricultural economies across the developing world. The information technology revolution transformed logistics, making the expansion of globally traded foodstuffs, fertilisers and pesticides possible on scales that would have been unimaginable in the mid-twentieth century. IT also gave birth to the biotechnology industry which, in turn, made possible the commercialisation of genetically modified organisms (GMOs) as the new 'techno-fix' of the global food industry — the so-called 'gene revolution'.

In recent years, the Rockefeller Foundation and the Bill and Melinda Gates Foundation have provided the core funding for the increasingly influential Alliance for a Green Revolution in Africa (AGRA) which promotes an innovations system approach to extend the use of chemically produced fertilisers and GMOs in African agriculture (Juma 2011; Sangina *et al.* 2009). With Malawi acting as the poster child of this new 'green revolution', AGRA has managed to combine a market-based approach with input subsidies, export-orientation and a faith in techno-fixes with very little attention paid to soil health and the underlying sustainability of eco-system services (Kelly 2009). This is not surprising, because instead of embracing the whole systems science that inspires the agro-ecological approach, the underlying science in AGRA remains technicist and input-oriented, with its innovation systems approach seen primarily as a communications tool.

By 2005 the largest 10 seed corporations controlled 50 per cent of all commercial seed sales; the top five grain trading companies controlled 75 per cent of the market; the largest 10 pesticide manufacturers supplied 84 per cent of all pesticides; and when it comes to vegetable seeds there is only one company that completely dominates the market and this is Monsanto, which controls roughly 30 per cent of the seed market for beans, cucumbers, hot peppers, sweet peppers, tomatoes and onions — the remainder is divided between a large number of small operations (Barker 2007:7).

It was at the level of everyday consumer culture that the dynamics of this 'third food regime' became most apparent. Supermarket chains rapidly increased their grip on retail food sales between 1992 and 2002, with South Africa leading the world in 2002 when nearly 60 per cent of all food was sold through supermarket chains, followed by

East Asia (excluding China) and South America (at just over 50 per cent) and China (at just below 50 per cent). Only 10 years earlier South America had been the highest at 15 per cent, and South Africa had been below 10 per cent (Reardon *et al.* 2003). This remarkable rise to dominance of the supermarket chains would not have been possible without the 'just-in-time' logistical systems that ICT made possible.

Some analysts believe that price increases are related to this new global role of supermarkets (Von Braun 2007: 1). As these powerful supermarket chains have striven to secure year-round supplies of 'food from nowhere' to supply mass consumer and specialist niche markets, they have integrated their value chains horizontally with some now directly controlling food production in remote locations around the world. This includes directly or indirectly participating in land grabs to secure access to food production (Cotula *et al.* 2009: 81; Nellemann *et al.* 2009), supporting a concerted global bid funded by US foundations (Gates, Rockefeller, Ford) to accelerate fertiliser use in under-exploited regions such as sub-Saharan Africa, and promoting GMOs as the new panacea to 'feed the world'.

By the end of the 1990s, virtually without exception every major report and researcher expressed concerns about declining rates of growth of agricultural yields. As reflected in the calculations using FAO data, yield growth rates for cereals have declined since the 1970s, with sub-Saharan Africa experiencing the most severe declines (see Figure 6.4). Declining yield growth rates are obviously a serious challenge for governments, but rising prices clearly suit the global food corporations. On the positive side, rising prices have helped to reverse the longer-term decline in levels of investment in agriculture.

When measured against what agricultural yield growth levels should be to meet rising demand, these declining rates of growth are even more alarming. In the most optimistic scenario, which ignores ecological constraints, the World Bank has estimated that yield growth of at least 1.5 per cent per annum to 2030 would be required to meet demand, dropping to 0.9–1 per cent per annum between 2030 and 2050 (World Bank 2009). The UNEP report, which factors in ecological constraints, concludes that yield growth could drop to 0.87 per cent per annum up to 2030 and drop to 0.5 per cent per annum between 2030 and 2050 (Nellemann *et al.* 2009: 91). In a detailed modelling exercise published in *Crop Science*, Hubert *et al.* concluded that yield growth 'will continue to slow': yield growth for cereals is expected to drop from an average of 1.96 per cent per annum for the period 1980–2000 to 1.01 per cent in 2000–2050, with even slower growth rates for developed countries (0.9 per cent) and slightly faster growth rates for the Middle East and North Africa (1.16 per cent), Latin America and the Caribbean (1.24 per cent) and sub-Saharan Africa (1.59 per cent) (Hubert *et al.* 2010: 41). As a result, all these reports conclude that food prices are set to rise steadily in response to declining yield growth in the context of rising demand through to 2050, thus confirming the argument that we have reached the end of an era of long-term decline in food prices. However, even if higher yield growth was attained in certain exporting regions, this would not necessarily translate into lower prices for local consumers, because prices will inevitably be determined by international markets affected by rising demand.

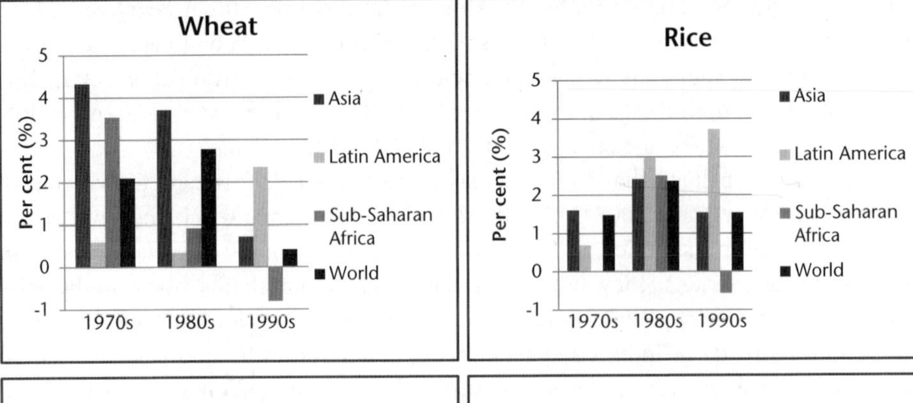

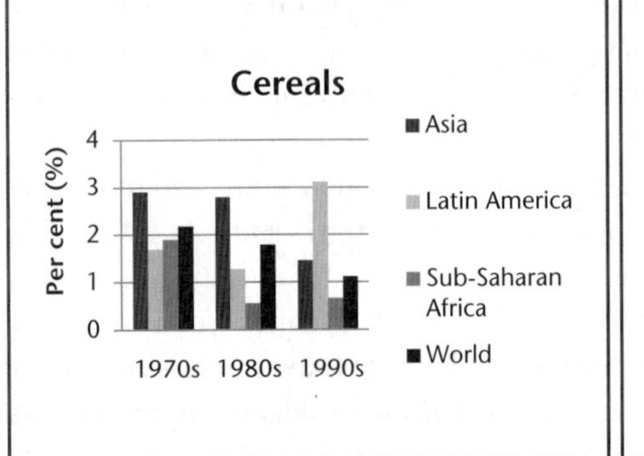

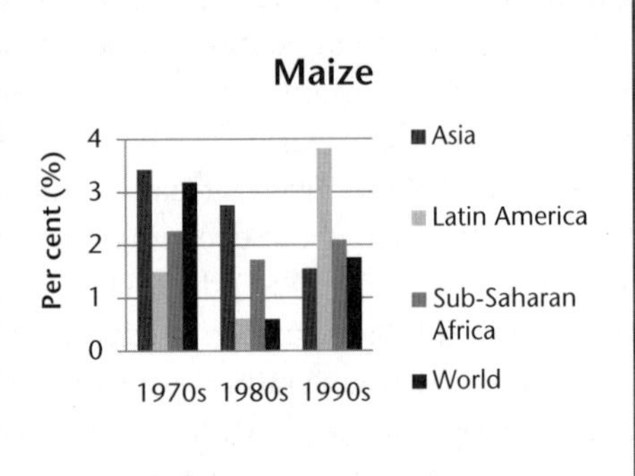

Figure 6.4: Annual cereal crop yield growth rates
(Source data from Gruhn *et al.* 2000: 7)

The problem, of course, with the discussion thus far is that it is cast entirely in terms of quantities and ignores unequal distribution and the question of quality. It assumes yield growth generates more food for everyone, including the poor. Some have questioned the assumptions underlying the numbers themselves — for example, the projections cited above assumed that everyone will be adopting Western diets (along with all the health problems) (Soil Association 2010). Raj Patel and others have argued convincingly that the real problem is that the global food industry produces energy- and resource-intensive food (which contains too much fat and sugar) to supply the world's billion over-consumers via the supermarket chains; that the large majority of poor, urban households can afford only nutritionally poor, mass-produced, cheap foods (hence the new correlations between obesity/diabetes and malnutrition); and that rural households suffer the effects of increasingly degraded soils and land dispossession caused by HEI farming and agribusiness practices (Patel 2008). This line of argument leads to the conclusion that the solution does not lie in higher yield growth, but rather a complete restructuring of the way food is produced, distributed and consumed: if food production restored rather than depleted the soils and produced healthy, low-carbon foods in sufficient quantities for all, maybe the discussion should not be about quantities and prices, but about qualities and sources (Friedman 2003). There is sufficient evidence that there is enough food produced in the world today to supply every person on the planet with 2,720 kilocalories (Kcal) per day which is what the average person needs to live well (Erb *et al.* 2009). The question, in short, is not whether there is enough food for everyone, but who gets it and who produces it for whom? But we are getting ahead of ourselves. We need to first find out what is happening to the world's soils which are, ultimately, what the 'first', 'second' and 'third' food regimes have taken for granted. Maybe the 'fourth food regime' will be the first that results in farming practices which restore rather than degrade the soils.

Land and soils

The total ice-free land surface of the Earth is 13 billion hectares (Bha) of which 1.5 Bha is unused 'wasteland' and an additional 2.8 Bha is unused and inaccessible. This leaves 8.7 Bha which humans in the *anthropocene* can choose to 'use' for a wide variety purposes, including pasture, forests and cropland. Of this, 3.2 Bha are potentially arable, the rest being marginal land from a cultivation perspective and covered by forest, grassland and permanent vegetation.[4] Of the potentially arable land, only 1.3 Bha is deemed to be moderate to highly productive. Just under half of the 3.2 Bha of potentially arable land (1.47 Bha) is cultivated as cropland (that is, just over 10 per cent of the ice-free land surface of the Earth is the resource on which humans depend for the bulk of their food). This 1.47 Bha of cropland, plus approximately 3.2 Bha of permanent pasture and 4 Bha of permanent forest and woodland are what makes up the 8.7 Bha of 'usable' land (Scherr 1999). It is noticeable that the only African countries with very extensive or moderately extensive arable land resources are Nigeria, Ethiopia, South Africa and Sudan. The majority of African countries have limited arable land resources with high population pressures. Yet African countries are earmarked by all the models of the future for substantial yield increases — they are also where most of the land grabs are taking place (Cotula *et al.* 2009).

As the population has grown, the amount of agricultural land per capita has shrunk while the increase in the amount of land brought into agricultural production has levelled off (see Figure 6.5).

As already indicated, many assume that cropland needs to expand by at least 120 million hectares (Mha) by 2030 to produce enough food. But they are also concerned by the fact that cropland area seems to be declining in absolute terms for a combination of reasons, but mainly environmental and reduced investment in agriculture over the long term (Nellemann *et al.* 2009: 35). An eminent group of scientists and ecological economists have estimated that cropland could be safely expanded by 400 Mha (Rockstrom *et al.* 2009) — an area roughly the size of India! Most of this land is Latin America's cerados and grasslands (Brazil, Argentina) and in the African savannas (Sudan, Democratic Republic of

[4] Lambin and Meyfroidt estimate that there is approximately 4 Bha available for 'rain-fed agriculture' (Lambin & Meyfroidt 2011: 3466).

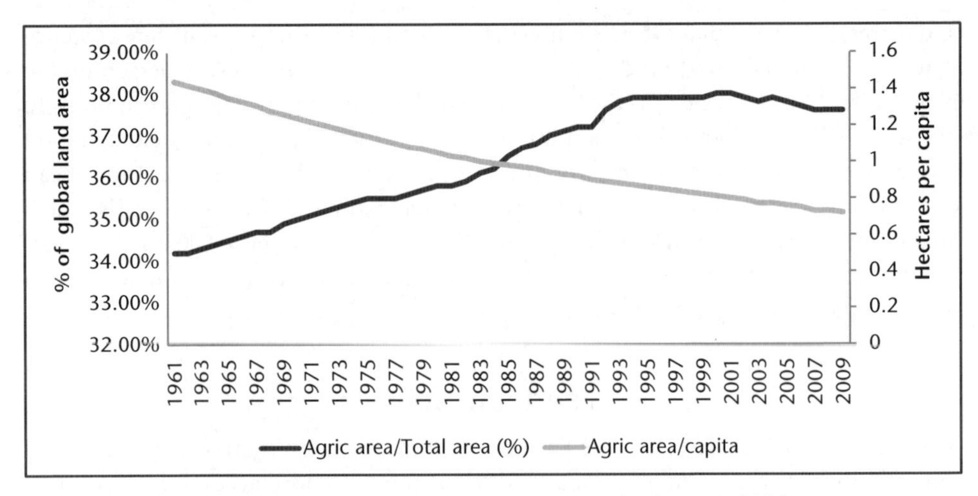

Figure 6.5: Share of land devoted to agriculture has peaked, 1961–2008
(Source: Calculated from FAO, http://faostat.fao.org/site/377/default.aspx)

Congo, Mozambique, Tanzania, Madagascar) (Lambin & Meyfroidt 2011: 3466). Lambin and Meyfroidt estimate additional land requirements by 2030 for all agricultural outputs plus urban expansion[5] to be between 285 Mha and 792 Mha (Lambin & Meyfroidt 2011: 3466). The problem is that expanding cropland for food production can be achieved only by converting potentially arable land currently supporting permanent pasture, forest and woodlands, or moving into marginal land (the least productive land that can, nevertheless, be farmed). Either way, the consequences from a carbon storage, water use or biodiversity perspective would be drastic, to say the least.

To reconcile the need to increase food production by 68 per cent without massively increasing the demand for cropland, the FAO has constructed a scenario to 2050, which envisages an 'intensification' of agricultural production on existing cropland without expanding global cropland area by more than 9 per cent (which is approximately 120 Mha, see Figure 6.6). Significantly, this FAO scenario and all the other scenarios that have been compiled envisage the most substantial increases in cropland area taking place in sub-Saharan Africa (Erb *et al.* 2009; FAO 2006; OECD & FAO 2010; Watson *et al.* 2008). As will be argued later, the evidence suggests that the arable soils in sub-Saharan Africa are too degraded to justify such optimism.

Although most international reports envisage increased production and limited expansion of cropland area (for example, Erb *et al.* 2009; Hubert *et al.* 2010; Lambin & Meyfroidt 2011; Watson *et al.* 2008), exactly how to achieve this level of 'intensification' and how to bring an additional 120 Mha into production remains unclear. There is no global agreement on whether significant cropland expansion to extend food production using conventional farming methods within the context of the 'third food regime' is practically

[5] Lambin and Meyfroidt estimate that urban expansion could require between 48 Mha and 100 Mha by 2030 (Lambin & Meyfroidt 2011: 3466). This is highly problematic because normally the most fertile soils are located in regions surrounding cities.

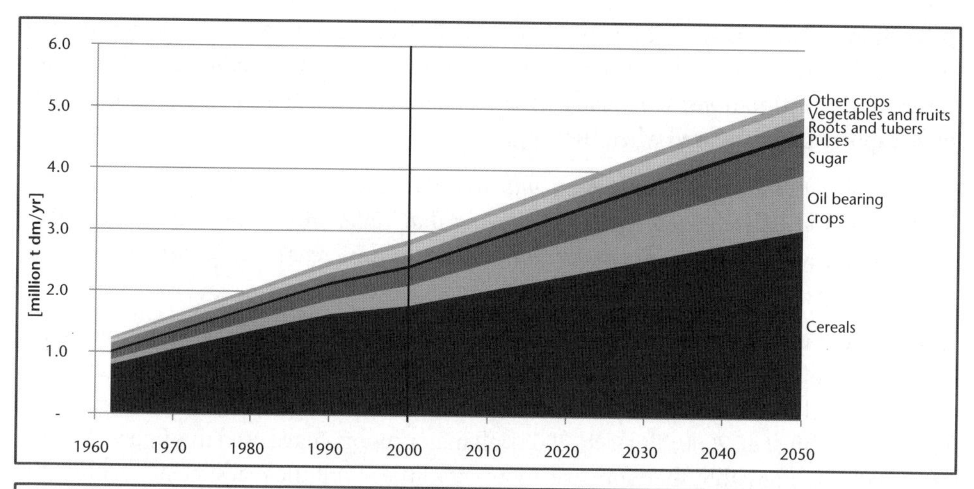

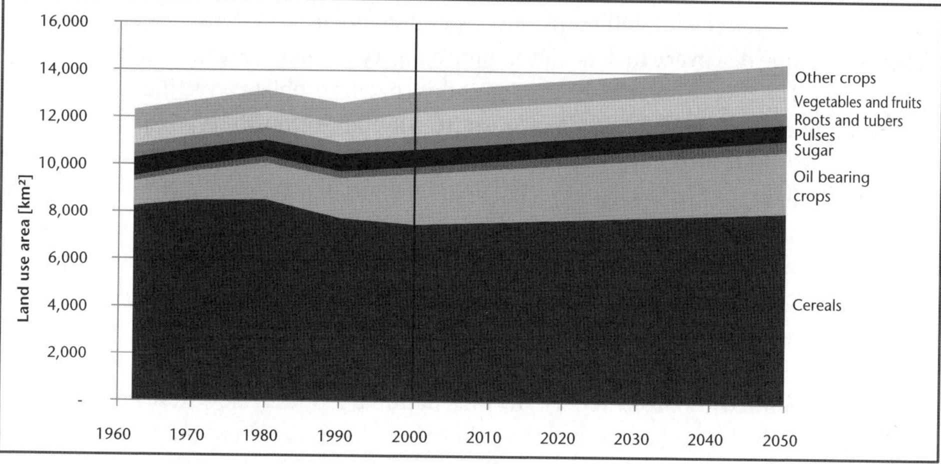

Figure 6.6: Cropland production 1961–2050 in the 'FAO intensive' scenario with respect to (a) production of food (Btonnes of 'dry matter'[6]/yr) and (b) amount of arable land required for food crops (km^2)

(Source: Erb *et al.* 2009: 15.)

Note: 1 km^2 =100 ha

possible, socially just or ecologically sustainable. There is, however, an alternative to land expansion which is not often considered,[7] namely the restoration of previously arable land which now suffers from degraded soils. But before this is discussed, we need a better understanding of soil degradation.

The global discussion summarised above is limited to land use (geographical space in hectares) and largely ignores soils and nutrients. The problem with this is obvious: if future

6 'Dry matter' is a standard term used when quantifying the weight of agricultural produce—it refers to weight excluding the estimated or actual water content.

7 Inexplicably, not even the authoritative work by Lambin and Meyfroidt (2011) consider this option. Lambin himself was unaware of the work cited here by Scherr (personal communication with Lambin, Paris, May 2011).

projections include land that has been degraded and is, therefore, no longer productive or as productive as it should be, conclusions about what can be produced will obviously be gross over-estimations. We must, therefore, factor soil health into the equation. Gruhn *et al.* capture this issue well when they argue:

> *Because agriculture is a soil-based industry that extracts nutrients from the soil, effective and efficient approaches to slowing that removal and returning nutrients to the soil will be required in order to maintain and increase crop productivity and sustain agriculture for the long term.* (Gruhn *et al.* 2000: 1)

The problem is that we do not have an up-to-date understanding of the conditions of the Earth's agricultural soils because the last global assessment was done in 1990 (Oldeman 1994). Since then, despite plenty of evidence of accelerated degradation (Den Biggelaar *et al.* 2004; Gruhn *et al.* 2000; Tan *et al.* 2005), all major reports have used this figure (Watson *et al.* 2008). Remarkably, therefore, we know very little about the real state of health of the fragile resource on which 6 billion people depend for their food. Most worrying of all is Scherr's stunning discovery that 'no developing country has in place a national monitoring system for soil quality' (Scherr 1999: 7). Unless this specific problem is rectified, no effective management of the most significant soil resources in the world will be possible.

Since 1990 there have been calls—although not many—for a new assessment or for analysis that takes soils seriously (Bai *et al.* 2008; Ballayan 2000; Gruhn *et al.* 2000; Henao & Baanante 1999; Scherr 1999; Tan *et al.* 2005). Unfortunately, while assessments of other resources abound (especially water, but also energy, fisheries, atmospheric carbon, biodiversity, eco-system services, even metals), a new comprehensive global soils assessment had not been initiated by the time of writing.[8] Instead, in order to find out the state of health of our soils we must rely primarily on the remarkable (although admittedly dated) report by International Food Policy Research Institute (IFPRI) researcher, Sara Scherr, read together with the most recent published research by renowned soil scientists R. Lal and colleagues (Scherr 1999; Tan *et al.* 2005).[9]

Sherr reviewed 26 global and regional studies and 54 national or local studies in 26 developing countries (Scherr 1999: 3) in order to arrive at conclusions that make it possible to link soil degradation and declining yield growth. Her point of departure, drawing on Lal's work, was to describe the soil characteristics that affect yield, namely nutrient content, water-holding capacity, organic matter content, soil reaction (acidity), topsoil depth, salinity and soil biomass. 'Changes over time in these characteristics', she wrote, 'constitutes degradation' or 'improvement' (Scherr 1999: 5). Degradation is not, however, a static state; it is ongoing with a specific combination of causes. The

8 The most significant set of activities that address soil degradation is UNEP's Land Degradation Assessment in Drylands (LADA) initiative funded by the Global Environmental Facility. The focus of this initiative is to develop 'tools and methods' for assessing land degradation on dryland eco-systems. It includes local case studies in Argentina, China, Cuba, Senegal, South Africa and Tunisia. One of the key outputs of the LADA project is the Global Land Degradation Land Information System (GLADIS) which maps the six main aspects of land degradation: biomass, soil, water, biodiversity, economy and social/cultural aspects (see http://www.unep.org/dgef/LandDegradation/LandDegradationAssessmentinDrylandsLADA/tabid/5613/Default.aspx).

9 All future references in the text to 'Lal' will, in fact, be referenced as Tan *et al.* 2005.

processes include erosion, compaction and hard setting, acidification, declining soil organic matter, soil fertility depletion, biological degradation and soil pollution. The *causes* of soil degradation include some combination of water erosion, wind erosion, soil fertility decline due to nutrient mining, waterlogging, salinisation (often caused by irrigation systems), lowering of the water table and over-use of chemical inputs causing soil pollution (Scherr 1999: 5–6).

The 1990 assessment concluded that 23 per cent of global soils were degraded. Although Asia has the largest amount of degraded land, this is a relatively smaller proportion of its total arable land area than is the case for Africa which has the second highest level of degraded soils in the world (30 per cent) after Central America (see Table 6.2).

What matters for our argument is the fact that 38 per cent of all agricultural land was degraded by 1990, and that we can safely assume that since then it has got worse. If the focus is exclusively on yield growth by finding an additional 120 Mha to 400 Mha for cropland, it means using up some of the 3.2 Bha of permanent pasture and 4 Bha of forests/woodland for cropland (both of which are far less degraded than the agricultural lands). Given the negative ecological consequences of doing this, the obvious alternatives would be the restoration of the 750 Mha of 'lightly degraded soils' and/or, even better, the more costly restoration of at least some of the billion or so hectares of 'seriously degraded soils' that are not yet wastelands. As far as Africa is concerned, what really matters is the fact that 65 per cent of its agricultural soils are degraded and that 321 Mha are 'seriously degraded'. What this means is that *Africa has nearly 321 Mha of previously arable land, possibly as much as half of which is today irrecoverable wasteland,* which will be (possibly too) costly to be rehabilitated. But this means it has 170 Mha of 'lightly degraded soils' which will be relatively cost effective to rehabilitate. Unless ways are found to increase investments in soil restoration, projections for increased yields in Africa will remain pipe dreams.

Given that these are 1990 estimates, it becomes important to know whether these trends have continued. Using the data from the 1990 GLASOD assessment, some researchers estimated that the annual rate of loss of *cultivated agricultural land* due to soil degradation was — and therefore continued to be — 12 Mha, while others estimated the losses at closer to 6 Mha per annum (Scherr 1999:21). In other words, if the rate is 12 Mha per annum, in the short space of a decade we will have lost an amount equal to the 120 Mha UNEP and FAO reckon is needed to meet future demand. However, loss of area is probably much less significant than the loss of soil nutrients over time.

Conventional wisdom since the Green Revolution in the 1960s has been rather straightforward: to increase agricultural output you have to apply NPK (nitrogen, phosphorus and potassium) fertilisers. Figures 6.7 and 6.8 seem to confirm this assumption: as the application of fertilisers per hectare has increased, so has output per hectare. But on closer inspection, something else is happening.

Figures 6.7 and 6.8, when read together, confirms the popular belief that in order to increase yields it is necessary to increase the application of fertiliser. This appealing

Table 6.2 Global estimates of soil degradation by 1990, by region and land use[10]

	Agricultural land			***Permanent pasture***			***Forest and woodland***			***All used land***				
Region	*Total*	*Degraded*		*Total*	*Degraded*		*Total*	*Degraded*		*Total*	*Degraded*		*Seriously degraded*	
	(million hectares)		%	(million hectares)		%	(million hectares)		%	(million hectares)		%	(million hectares)	%
Africa	187	121	65	793	243	31	683	130	19	1,663	494	30	321	19
Asia	536	206	38	978	197	20	1,273	344	27	2,787	747	27	453	16
South America	142	64	45	478	68	14	896	112	13	1,516	244	16	139	9
Central America	38	28	74	94	10	11	66	25	38	198	63	32	61	31
North America	236	63	26	274	29	11	621	4	1	1,131	96	9	79	7
Europe	287	72	25	156	54	35	353	92	26	796	218	27	158	20
Oceania	49	8	16	439	84	19	156	12	8	644	104	17	6	1
World	1,475	562	38	3,212	685	21	4,048	719	18	8,735	1,966	23	1,216	14

(Source: Calculated from data from FAO and 1990 assessment, as cited in Scherr 1999: 18.)

10 Note: the last two columns refer to 'seriously degraded' soils with one column which states the total quantity involved and the other which reflects this as a percentage of the total amount of land suffering from degraded soils. Scherr includes in her definition of 'seriously degraded' soils the following three categories of degraded soils: 'moderately degraded soils', which refers to soils that are still suitable for farming but with 'greatly reduced productivity' because 'original biotic functions are partially destroyed' and for which restoration is 'beyond the means of local farmers'; 'strongly degraded soils', which refers to soils that are no longer productive and are no longer suitable for local farming systems because the 'original biotic functions are largely destroyed'—restoration would entail major investments and engineering works; 'extremely degraded soils' which are a 'human-induced wasteland', that is 'beyond restoration' because the biotic functions have been 'fully destroyed'. The only soils not included in the definition of 'seriously degraded soils' are those defined as 'lightly degraded' which are soils that have a slightly reduced productivity, but are suitable for local farming because their original biotic functions are still intact and restoration to full productivity is affordable for local farmers and practical if they modify their farming practices. The difference between the amount of land that is 'seriously degraded' and the 'total degraded' amount is equal to the amount of land that is 'lightly degraded'.

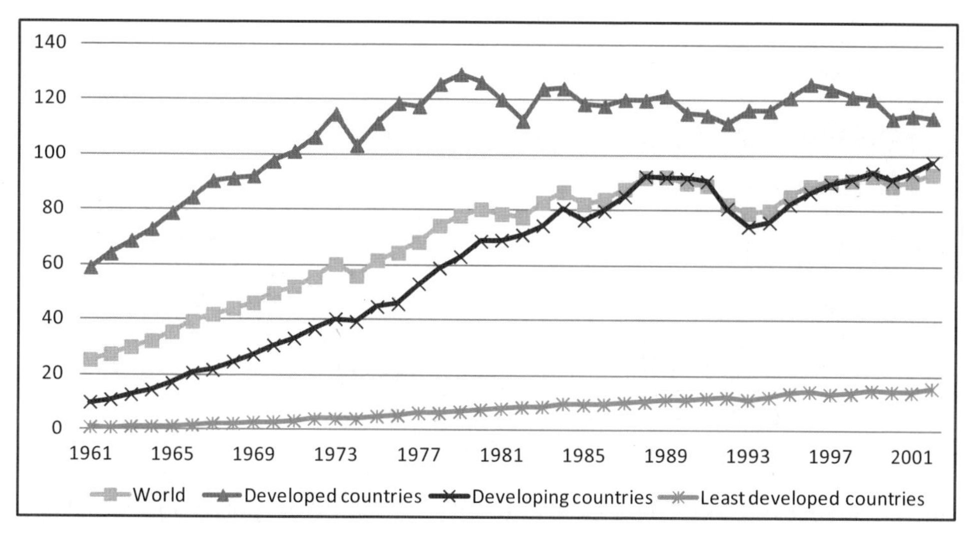

Figure 6.7: Average rates of NPK fertilisers applied on arable land and permanent crop areas since the 1960s (kg/ha)

(Source: FAO)

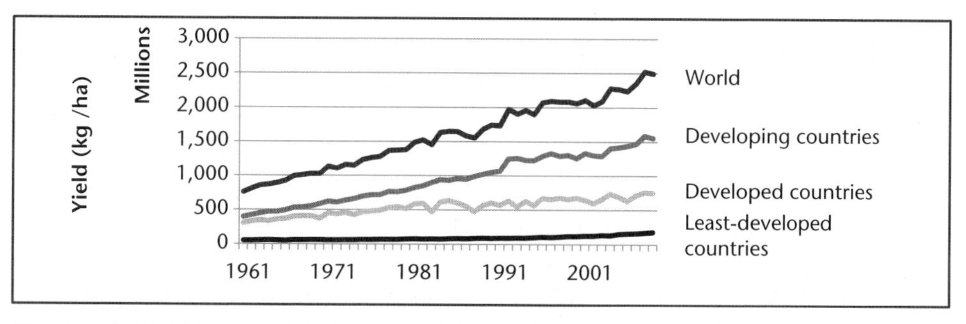

Figure 6.8: Global mean yield change of all cereal crops since the 1960s

(Source: FAOSTATS)

Note: developing countries excluded from count for least-developed countries; Yield (Kg/ha) means kilograms of cereal produced per hectare per year; cereal includes wheat, rice, maize, barley, oats, rye, millet, sorghum, buckwheat and mixed grains and excludes crops harvested for animal feed or grazing.

equation, however, ignores another simple and obvious calculation: How many kilograms of crop can you grow with a kilogram of fertiliser? How has this changed over time? Figure 6.9 shows that on average it was possible to produce 161 kg of cereals from 1 kg of NPK in 1961, but by 2000 this had dropped to 93 kg. The decline was most severe in the poorest countries (from 3,150 kg to 272 kg), but with serious declines in developing countries (from 494 kg to 71 kg).

What is the explanation for the fact that in *developed* countries more could be produced from 1 kg of NPK in 2000 than in 1961? The clue to the answer lies in Figure 6.8, which shows that since 1980, the application of fertilisers per hectare has *levelled*

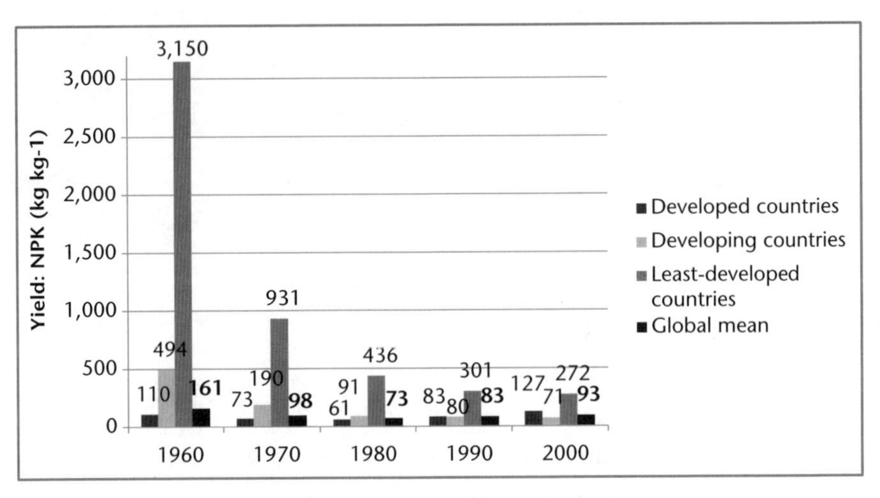

Figure 6.9: Ratio of cereal crop yield (kg) to NPK fertilisers (kg) applied since the 1960s
(Created from data in Tan *et al.* 2005: 140)

off without adversely affecting yields per hectare. This can be explained with reference to two factors: firstly, unlike in developing countries, farms in developed countries have had access to energy infrastructures which have meant crop residues and manures could be returned to the soil rather than used for fuel (residues and manure), fodder (residues) and building materials (manure and certain residues); and secondly, the massive scientific research infrastructures in developed countries gave farmers ready access to real-time research which allowed them to understand what was happening to their soils and, in turn, empowered them to carefully calibrate and harmonise the application of fertilisers and irrigation (eventually using sophisticated computerised techniques such as soil moisture meters for triggering intermittent 'just-enough' irrigation volumes), crop residues, and manuring to retain a healthy nutrient balance. Real-time information available to the farmer meant less reliance on the advice of fertiliser agents on how much and when to apply NPK. Other factors which could have played a role include the rise of the sustainable farming movement (and its various interpretations from integrated pest management through to organic farming, to biodynamic farming) in developed economies (in particular Europe) which questioned the chemical input-output concept of soils; and the fact that in developed countries fertilisers may have been subsidised in various ways, but they were not delivered to farmers as aid, which meant that farmers had to make market-based calculations which correlated expenditures on fertilisers with returns on their investment. Nevertheless, we must not forget that by 1990, 20 per cent of Europe's soils were 'seriously degraded'[11], a level second only to Central America.

So if the assumption is correct that the level of output from soils is determined by the quantity of NPK inputs, how is it possible to explain the diminishing returns on fertiliser

[11] Degraded soils in Europe are found mainly in Southern European states such as Spain, Greece and southern Italy.

use? Crops derive their nutrients from three sources from a conventional farming perspective: the soil, manure (which is added), and fertilisers (which are also applied). Tan *et al.* quote estimates that by 1970 (on a global scale) crops derived 48 per cent of their nutrients from the soil, 13 per cent from manures, and 39 per cent from fertilisers. By 1990 soils provided only 30 per cent of the nutrients required by the crops, followed by 10 per cent from manures and 60 per cent from fertilisers. If agricultural practices remain unchanged, by 2020 the contribution of nutrients by the soils will have more than halved from their 1970 level to 21 per cent with only 9 per cent provided by manures and a whopping 70 per cent coming from fertilisers (supplied by a few multinational companies). As Tan *et al.* then conclude, 'the increase in crop production has, to a large extent, been at the expense of the decrease in soil fertility...[C]ontinuing soil nutrient exhaustion is leading to an increasing dependence of crop yields on fertilisers' (Tan *et al.* 2005: 138).

This decline of nutrient supply from the soils is what is generally referred to as 'nutrient mining' and is usually reflected in quantifications of NPK deficits. Analysing only land cultivated for wheat, rice, maize and barley, Tan *et al.* calculated that by 2000, 56 per cent of global cropland used for these four crops was affected by N deficits averaging 17.4 kg/ha/yr; 80 per cent of cropland used for these crops was affected by P deficits averaging 5 kg/ha/yr; and 56 per cent of cropland used for these crops was affected by K deficits averaging 38.7 kg/ha/yr. This made it possible for Tan *et al.* to calculate that by 2000, the total global NPK deficit was 20 Tg[12] across the 562 Mha of cropland used to cultivate wheat, rice, maize and barley.[13] This NPK deficit, however, was unevenly distributed, with 75 per cent of the deficit occurring in developing countries, 14 per cent in developed countries, and 11 per cent in the least-developed countries (Tan *et al.* 2005: 133). The low NPK deficit in developed countries correlates with, as already indicated, declining use of NPK per hectare in developed economies, which is consistent with the overall argument that nutrient provision by soils goes down as NPK applications go up. From a resource management perspective, it is this missing 20 Tg of NPK which Tan *et al.* argue is a key explanation for yield decline.[14] Furthermore, there is no evidence that expansion of cropland will resolve this problem, especially if this means clearing land which currently holds biomass that generates nutrients for soils (for instance, nitrogen-rich woodlands or wild biomass that provides fodder for manure-producing cattle) or if it means moving onto marginal soils with limited natural nutrient content.

Nor is there much evidence that it can be resolved by applying more agrochemicals because this is clearly making matters worse. As of 2003 the total amount of synthetic N fertiliser in circulation had doubled compared to pre-industrial times, to 82 million tonnes (FAO data cited in Badgley *et al.* 2007: 91), while the N deficit had risen to

---

[12] Tg = Teragram = 10^{12}g

[13] If we use the average price in 2010 for NPK, which was $400/tonne, the 20 Tg deficit can be valued at $80 billion.

[14] Bob Howarth from Cornell University—an expert in global nutrient flows—points out that 20 Tg is actually such a tiny amount compared to total global nutrient flows that it is difficult to prove that this is statistically relevant to any discussion of yield decline—personal communication, Paris, May 2011.

5.4 million tonnes by 2000 (Tan *et al.* 2005: 133). As far as P is concerned, it has been estimated that 22 Tg lands up in the oceans each year after travelling through soils, rivers, food, bodies and sewage-treatment plants — roughly triple the rate from pre-industrial times. Although massive increases in nutrient flows have many different negative consequences for eco-systems, eutrophication of rivers and dams which results in 'dead zones' is the most problematic outcome (Howarth *et al.* 2005). This has led to calls for a very different approach to the sustainable management of the nutrient cycle and agricultural eco-system resources. This would need to include the harnessing of the largely under-utilised naturally produced 140 million tonnes of N that could be made available by nitrogen-rich leguminous cover crops which can be planted without displacing production (Badgley *et al.* 2007: 91).

In short, nutrient mining and the consequent oversupply of fertilisers to compensate for NPK deficits had the (unintended) consequence of inducing nutrient imbalances (and the related problem of soil pollution) which, in turn, explains declining yields per kilogram of NPK and, ultimately, declining yields per hectare. The empirical fact that nutrient mining can occur as the external input of NPK per hectare increases calls into question many of the scientific assumptions about soils that inspired the Green Revolution, in particular assumptions that privileged soil chemistry (that is, a narrow focus only on NPK) over soil biology — or, to anticipate the next section by quoting Tan *et al.*'s conclusions: 'Soil fertility decline includes the deterioration in the chemical, physical, and *biological properties* of the soil that affect plant nutrition, and is a result of specific processes such as reduction of SOM [soil organic matter] and *soil biological activity*, adverse changes in soil nutrient resources and development of nutrient imbalances, and build-up of toxicities and acidification through incorrect fertiliser use, etc.' (Tan *et al.* 2005: 137). This is a remarkable indictment of a chemistry-centred soil science paradigm that has been dominant since at least the 1960s.

We need now to return to Africa's degraded soils to find out how this happened. Gruhn *et al.* cite various studies that confirm large-scale nutrient mining has been taking place in Africa for decades. They cite a World Bank study which estimated that sub-Saharan Africa has suffered a net loss of 700 kg of N, 100 kg of P and 450 kg of K per hectare of *cultivated land* since the 1970s. Another study of sub-Saharan Africa found that 22 kg of N, 2.5 kg of P and 15 kg of K were lost per hectare per annum during the period 1982–1984 (Gruhn *et al.* 2000: 11). Henao and Baanante estimated that 86 per cent of African countries lose more than 30 kg of NPK per hectare per annum (Henao & Baanante 1999).

One explanation for the rate of nutrient mining in Africa may lie in the fact that more than half of all NPK applied to African soils is imported into Africa in the form of aid and handed out for free. By 1990, 22 of 40 sub-Saharan African countries received *all* their fertiliser imports as aid (Gruhn *et al.* 2000). Given the argument that increased fertiliser application can lead to soil degradation if incorrectly or wastefully applied, it is logical to assume that if fertiliser is provided to farmers for free with little scientific support regarding the appropriate balance between irrigation, nutrient mix, plant uptake of nutrients and seed selection, there is little incentive for the farmer to invest in the soils while the going is good, which is usually for the first 10 to 15 years.[15] After that soil degradation is so severe, that even if farmers do realise that soil nutrients need to be rebuilt, they will face considerable costs to make this happen. As Table 6.3 suggests, although the costs of restoring the nutrient balance might be relatively low (presumably because it is assumed that more NPK can be added at relatively low cost), it gets costly to remedy more structural problems such as polluted soils, reduced biological activity, salinisation or compaction. Scherr concludes:

> *Many water, nutrient, and biological problems in soils can be reversed over 5–10 years through soil-building processes and field- or farm-scale investments and management changes. Some types of physical and chemical degradation, such as terrain deformation and salinization, are extremely difficult or costly to reverse.* (Scherr 1999: 8)

And yet surprisingly, although overall yields across Africa declined over the last three decades of the twentieth century, yield growth *per hectare* in sub-Saharan Africa *increased* between 1960 and the mid-1990s. The only explanation for this is that African farmers were able to abandon land that had become unproductive due to nutrient mining and move on to available virgin land, where they probably had a smaller piece of land with reasonably good soils to start off with which, in turn, explains the rising yields per hectare. The data supports this line of argument: between 1973 and 1988, arable and cropped land increased in sub-Saharan Africa by 14 Mha, forest and woodland areas shrank by 40 Mha, and pasture land remained stable. This means that during this period 26 Mha was abandoned, most likely left behind as wasteland during this period (Gruhn *et al.* 2000: 11). This provides an insight into a much wider process which explains how at least some of Africa's 321 Mha of once-productive arable land was degraded. However, it is obvious that there is a limit to the amount of available, viable, virgin land. Once this limit is reached, three things inevitably follow: firstly, people can leave the land and migrate to the cities (which partly explains the urbanisation patterns described in Chapter 5); secondly, resource conflicts increase, such as the one that triggered the Rwanda genocide and the ongoing Sudanese conflicts (as described in Chapter 7); and thirdly, people slowly starve to death. In short, soil degradation may well be a root cause of many of Africa's biggest problems and yet it is hardly ever mentioned in any of the accounts about Africa's socio-economic and political challenges.

To take the next step in the logic of the argument, it is necessary to relate nutrient mining to yield decline. Scherr refers to various studies that used the old data from the 1990 assessment, one of which estimated in 1998 that global cropland production was 12.7 per cent lower and pasture production 3.8 per cent lower than they would have

[15] This estimate is based on a series of interviews with Indian farmers conducted by Mark Swilling in 2005. All these farmers had converted to Green Revolution technologies in the 1960s and experienced a doubling of yields during the first 10 years, then a decade during which yields plateaued out, followed by steadily declining yields that persisted no matter how much NPK was added.

Table 6.3: Relative reversibility of soil-degradation processes

Type of degradation	Degradation process	Largely reversible, low cost	Reversible, significant cost	Largely irreversible/ very high cost
Physical	Clay pans, compaction zones		x	
	Surface sealing and crusting	x		
	Subsidence			x
	Topsoil loss through wind or water erosion		x (if active deposition)	
	Terrain deformation (gully erosion, mass movement)			x
Waterholding	Reduced infiltration/ impeded drainage		x	
	Reduced waterholding capacity		x	
	Waterlogging		x (farm scale)	x (landscape scale)
	Aridification		x	
Chemical	Organic matter loss		x	
	Nutrient depletion/leaching	x		
	Nutrient imbalance	x		
	Nutrient binding		x	
	Acidification		x (if liming feasible)	x
	Alkalinisation/salinisation			x
	Dystrification			x
	Eutrophication		x	
Biological	Reduced biological activity due to soil disturbance		x	
	Reduced biological activity due to agrochemical use		x	
Pollution	Contamination		x	
	Pollution (accumulation of toxic substances)			x

(Source: Scherr 1999: 10.)

been without soil degradation. This corroborated a 1995 study that found that global production would have been 12–13 per cent higher if only 15 per cent of the 1.2 Bha of 'strongly degraded soils' (see Table 6.2) could be restored to full productivity (Scherr 1999: 19). This would bring back into production 180 Mha of land, which is more than the 120 Mha that FAO and UNEP estimated would be needed to meet future demand.

Tan *et al.* have argued that soil-nutrient deficiencies have directly undermined agricultural yields of wheat, rice, maize and barley. By working out what would have been produced if nutrient deficits were not present, they calculated that 'average yield decline (equivalent rice yield) due to the deficit in K, P, and N was 1,372, 1,093, and 670 kg/ha/yr, respectively. These reductions were equivalent to 27 per cent of the average crop yield in the year 2000' (Tan *et al.* 2005: 134). This is more than double the reductions estimated by the earlier studies reviewed by Scherr. It was much worse for the least-developed countries where nutrient deficiencies reduced the annual crop for 2000 by 35 per cent with, of course, devastating socio-economic consequences. Yield reduction for 2000 due to nutrient deficiencies in developing and developed countries was estimated to be 27 per cent and 11 per cent respectively (Tan *et al.* 2005: 134).

Half of the developing world's arable and perennial cropland is in just five countries — Brazil, China, India (with 22 per cent), Indonesia and Nigeria. The fact that China and India, home to the largest populations, have similar problems to sub-Saharan Africa is of major concern (Scherr 1999: 24). China has 96 Mha of arable and permanent cropland while India has 168 Mha. It has been estimated that up to 30 per cent of China's land was degraded by 1990 and that agricultural yield growth in the 1980s and 1990s would have been 12 per cent higher without environmental degradation. A major part of the problem was overuse of N, reflected in the fact that grain production increased by 71 per cent between 1997 and 2005, whereas the application of N fertiliser increased by 271 per cent (Bindraban 2009: 14–15). Similarly for India, up to 43 per cent of arable soils were regarded as suffering from 'high degradation', while 5 per cent was so degraded it was unusable (Scherr 1999:24).

In short, there is a clear causal link between soil degradation and yield reduction per hectare. If yield reduction is a significant cause of rising prices, it follows that long-term soil degradation needs to be seen as an important cause, not of the price spike of 2007/08 (which had more to do with financial markets), but rather of the long-term rise in prices since 2000. As already demonstrated, the OECD–FAO anticipates prices rising well into the second decade of the twenty-first century and the IAASTD has argued that 'real world prices … are projected to increase in the coming *decades*' (Watson *et al.* 2008, Ch. 5: 3 — emphasis added). The economic and ecological evidence, therefore, seems to confirm once again that the era of declining food prices is over because, in part, yield decline caused by soil degradation will continue if science paradigms and policies remain unchanged.

It follows that declining yield growth and rising prices can only be resolved by recognising the significance of soil degradation. Surely this resolves the inelasticity puzzle, that is, why yield growth did not improve as prices increased over the past decade. It follows from all this that yield growth will depend on investments to reverse soil degradation. If we accept the World Resources Institute estimate that the annual value of global agricultural production is US$1.3 trillion (World Resources Institute 2002), and if we accept the more conservative estimate that soil degradation reduces annual yields by 10 per cent, then we are talking about an annual loss of US$130 billion. Maybe what the world needs is a fund that will invest just 10 per cent of this opportunity cost per annum (that is, US$13 billion) to restore the degraded soils of the planet as an alternative to the potentially riskier and more costly venture of finding an additional 120 Mha of arable land in an increasingly densely populated and ecologically endangered world. The question this raises, of course, is how does one restore soils? We answer this question by supporting the argument put forward by the IAASTD that this will require a new body of knowledge — an agro-ecological approach — which has hitherto had no place in the agricultural science that has driven mainstream HEI agriculture for many decades.

Agro-ecology: Towards a 'fourth food regime'?

We have made clear that by 1990 there was 750 Mha of 'lightly degraded land' that could be restored to full productivity over a period of 5–10 years at relatively low cost. We were also able to show that a significant portion of the 1.2 Bha of 'seriously degraded land', which was not yet unrecoverable wasteland, could also be restored, but at a much higher cost. We can only assume that over the two decades since the 1990 assessment some of what used to be 'lightly degraded land' has degraded further, thus increasing the amount of 'seriously degraded land', and the amount of wasteland has more than likely increased. It has also been shown that soil degradation is a key cause of yield decline, especially in the five countries in which 50 per cent of the arable soils of the developing world are located (Brazil, India, China, Indonesia and Nigeria). It therefore follows that instead of looking for virgin soils, we should be focusing on how we restore degraded soils to increase yields. But we also argued that farming systems, which focus on applying increasing amounts of NPK to restore nutrient imbalances, may well be having the opposite effect. It follows, therefore, that the primary question we now face is the following: is there a food production system that makes it possible to increase total yields and feed the poorest people *by restoring the soils*?[16]

Significantly, there is a high degree of agreement (which excludes most of the biotechnology industry and some who retain a faith in conventional HEI farming) that agricultural knowledge does need to incorporate what Bindraban usefully calls 'agro-ecological intelligence' (for an authoritative argument by the Director of World Soil Information at Wageningen, The Netherlands, see Bindraban 2009). The IAASTD was the first significant global assessment of agricultural practices and science since the global application of the Green Revolution technologies in the 1960s.[17] Although even

reframing

[16] Note that this is a different question to the usual one, which is 'how do we increase agricultural yields without damaging the environment?' A minimising damage approach in agriculture will not reverse soil degradation.

[17] The major sponsoring partners were the FAO, Global Environment Facility (GEF), United Nations Development Programme (UNDP), United Nations Environment Programme (UNEP), United Nations Educational, Scientific and Cultural Organization (UNESCO), World Bank and World Health Organization (WHO). What was called the 'Bureau' of the initiative included representatives from 28 governments from around the world, 15 major civil society groups, 5 from private sector companies and 6 from major research institutions. Major GMO companies, Monsanto and Syngenta, were originally part of the core group but formally withdrew because they concluded that the IAASTD report was not going to endorse unqualified support for GMOs. Fifty-eight countries approved the Executive Summary that was released for policy-makers at the launch conference in Johannesburg, excluding Australia, Canada and USA who stated they could not approve the final text.

this report devoted surprisingly little space to the question of soils, the publication of the IAASTD report marked a key turning point because, despite the enormous complexity of the terrain that it covered (and its various internal inconsistencies), it reached two fundamental conclusions of major significance: that HEI agriculture may have been responsible for the doubling of yields since the 1960s, but the ecological damage for which it continues to be responsible is undermining the eco-systems (including soils) on which future production depends; and that increased yields in future will depend on the inclusion of an agro-ecological understanding into mainstream agricultural science, knowledge and technology development. In one of the very few references to soils in the IAASTD report, it was argued that:

> *The consequences of population growth and economic expansion have been a reduced resource base for future agriculture: now there are pressing needs for new agricultural land and water resources. In recent decades the development of integrated pest/water/nutrient management practices, crop/livestock systems, and crop/legume mixtures has contributed greatly to increased agricultural sustainability, but further progress is needed, especially to combat declining* soil fertility. (Watson *et al.* 2008: Ch. 3: 4 — emphasis added)

Since at least 1995, with the publication of Altieri's seminal work entitled *Agroecology: The Science of Sustainable Agriculture* (Altieri 1995), an alternative body of agricultural science, knowledge and practice has emerged to that which underpinned the HEI approach that emerged from the Green Revolution. Although agro-ecological methods can get incorporated into HEI farming, proponents of the agro-ecological approach break from the HEI paradigm in two important respects: to farm sustainably means working *with* rather than *against* nature and this can only happen if agro-ecological systems are understood as complex systems (as discussed in Chapter 1); and that small farms are generally more productive per hectare than large farms and, therefore, must become the focus of policy (also partly confirmed by the IAASTD report). These are, actually, inter-dependent and need to be understood in relation to one another, rather than as separate stand-alone factors.

Agro-ecology is a useful term because it does not refer to any specific farming system as such, but rather to a 'body of principles' which are applied along a wide continuum of farming traditions, including systems that use agrochemical inputs where soils are particularly deficient (for example, the West African Sahel region that has limited phosphorus content). The farming traditions that reflect the application of these principles in one form or another include *permaculture* associated with the ecologist Bill Mollison; *biodynamic farming* associated with the anthroposophist Rudolf Steiner; the *one-straw revolution* founded by the Japanese farmer Masanobu Fukuoka, which so inspired the Indian organic farming movement; the *bio-intensive* farming system popularised in the USA by John Jeavons; the *No Tillage* movement in Brazil; and a wide range of 'sustainable', 'biological', 'organic' and 'natural' farming systems that in one way or another apply the basic principles of agro-ecology. The same principles can be found in traditional African agricultural systems and the so-called Globally Important Heritage

Systems that have survived into the present day (Altieri & Koohafkan n.d.; Dlamini 2007).[18] Although most of these specific systems would eschew any use of agrochemicals, the agro-ecological approach does not necessarily imply that agrochemicals should be excluded at all times in every context (especially in areas with phosphorus-deficient soils). As Uphoff suggested, it is 'more useful to consider practices and technologies along a *continuum* between likely-to-be-sustainable and unlikely-to-be-sustainable, rather than to categorise practices and technologies—and their proponents—into separate and opposing camps.' (Uphoff 2002a: 8)

Breaking with the reductionist science that defined soils purely as a medium for carrying chemically defined nutrients needed to grow crops (usually on large commercial farms with access to irrigation, roads and extension support), agro-ecology adopted a complex systems approach to understand agricultural systems as indivisible wholes 'supported by interactions and synergies between and among biological components that enable these systems to sponsor their own soil fertility, productivity enhancement and crop protection' (Altieri 2002: 41). Instead of focusing on isolated factors to increase productivity through targeted technical interventions (for instance, application of agrochemicals, irrigation or biotechnology), agro-ecology advocates a knowledge-intensive focus on the health and co-evolution of the entire indivisible social and ecological system as it pertains within specific unique contexts. This context-specific know-how can either be generated in rational scientific ways through research and learning using advanced transdisciplinary, quantitative and qualitative techniques, and/or it can be embedded in indigenous knowledge systems, which have learnt about what works through trial and error over long periods of time. In most cases, it is a combination of the two. From this perspective, Altieri continues:

> *[a]n area used for agricultural production, such as a field, is regarded as a system in which ecological processes that are found also under natural conditions are occurring: e.g., nutrient cycling, predator/prey interactions, competition among species, symbiosis and successful changes. Implicit in agroecological research is the idea that, by understanding these ecological relationships and processes, agroecosystems can be enhanced to improve production and to produce food, fibre, etc. more sustainably, with fewer negative environmental and social impacts, and using fewer external inputs.'* (Altieri 2002: 41–42)

Halweil captured the significance of this merging of indigenous and scientific knowledge when he wrote that 'organic farming is a sophisticated combination of old wisdom and modern ecological innovations that help harness the yield-boosting effects of nutrient cycles, beneficial insects, and crop synergies. It's heavily dependent on technology—just not the technology that comes out of a chemical plant' (Halweil 2006: 19).

[18] See also the extraordinary case studies captured on the Globally Important Agricultural Heritage Systems website at http://www.fao.org/nr/giahs/en/

Following Altieri (Altieri 2002: 42), the five key principles of agro-ecology that get applied in different ways appropriate to each context are as follows:

1. Recycle and reuse all available biomass (for instance, crop residues, cuttings from surrounding trees/shrubs, manures) in order to replenish and constantly restore soil nutrients
2. Grow the plants by building the soils, focusing in particular on soil organic matter and soil biotic activity by, for example, adding manures and promoting the growth of earthworm populations
3. Minimise losses of growth factors above and below ground by protecting the soils from direct solar radiation, strong winds and erosive water flows — this by ensuring constant soil cover by way of companion planting in densely packed rows, contouring to control water flows, and wind protection measures
4. Maximise species and plant diversity in order to build up the resilience of the system, which in practice means above all avoiding monocultures
5. Enhance beneficial biological interactions and synergies so that natural ecological processes can work to enhance, rather than undermine, agricultural production, for example, boundary planting that attracts beneficial insects, birds and small animals which feed off potential threats to crops.

For the proponents of agro-ecology, however, these principles are particularly suitable for small farms for two reasons: firstly, small farmers tend to develop a deep knowledge of the intricacies of their micro-ecologies, which is essential for the implementation of agro-ecological approaches; and secondly, there is growing evidence that small farms are more productive *per hectare* than large commercial farms (Altieri 2008; Cousins & Scoones 2010; Cousins n.d.; Lahiff & Cousins 2005; Wiggins 1995). For these reasons, the agro-ecological approach may be particularly suited for resolving problems where the challenge is greatest, namely, the approximately 250 million farms in the developing world that have not adopted Green Revolution technologies (mainly because they could not afford agrochemical inputs and irrigation systems), plus a significant proportion of the 200 million or so who did adopt these technologies but — like the Indian farmers who have been committing suicide — are experiencing rising input costs (due to peak oil and rising phosphorus prices) and declining yields because of soil degradation. Given that small farms dominate the five countries in which 50 per cent of the developing world's arable soils are located, it certainly does make strategic sense to focus on ways that can increase yields on these farms without tying them to external inputs, which are steadily going up in price, or assuming they need to be aggregated into large farms to be productive.

A wide range of practical applications flow from these principles that have become part of the everyday practices and technologies of millions of farms around the world, including mulching, green manuring, worm farming, contouring, tree planting, zero tillage, species diversification and the application of cattle manure and urine as natural fertilisers (for the best recent overview of the different types of agro-ecological farming, see Buck & Scherr 2011). The application of agrochemicals to enhance NPK content in certain contexts is not inconsistent with this approach, but as Altieri puts it 'there is a *burden of proof* that these will actually add to economic and environmental net benefits over multiple years, and that such benefits cannot be attained by other, less costly means' (Altieri 2002: 45 — emphasis added). But, he warns, agro-ecology is not a menu of technologies, but rather an entire paradigmatic approach that redefines the relationship between humans and nature:

> *Just adding or subtracting certain practices or elements within present production practices will not produce a more self-sufficient and self-sustaining agriculture. This transformation requires deeper understanding of the nature of agro-ecosystems and of the principles by which they function.* (Altieri 2002: 41)

Indeed, where this paradigmatic difference is most apparent is when it comes to soil management and related research. Although they cannot refer to a systematic research survey, Fernandes *et al.* estimated that 60–70 per cent of soil research since the 1950s in the USA and elsewhere has focused on soil chemistry (mainly NPK modalities), 20–30 per cent on soil physics (structure and composition), and only 10 per cent on soil biology (microbes, worms, soil organic matter, water and entrapped air content, etc.). Furthermore, they estimate that, at most, 10 per cent of soil research has addressed 'sub-surface processes and dynamics' (Fernandes *et al.* 2002: 31–32). This focus on plant health and NPK is not only much easier to do (a soil test for chemical content is quick, relatively cheap and generates a fixed quantitative result), it is consistent with a much wider reductionist conception of soils that devalued the importance of biological processes and, of course, it is a knowledge set that was consistent with — and reinforced by — the agrochemical companies, whose business was to sell NPK,[19] even though the economic and environmental case against these practices was becoming increasingly clearer (Wilson & Tisdell 2001). This is a classic case of technological 'lock-in' that goes a long way towards explaining one of the central concerns of this chapter, which is the extraordinary absence of soils from the global discussion of food prices, yield growth and food security. By elevating soil biology to a central place in its overall transdisciplinary approach to soils as a complex system, agro-ecology has clearly gone up against this path dependency, which reductionist soil science is responsible for perpetuating.

In short, our conclusion is that agro-ecological science provides a knowledge set which defines a set of practices that can, in turn, tackle the question posed at the outset of this section, namely how to increase yields and food security for the poor by restoring the soils. The survival of future generations may well depend on how the highest quality soils are protected by deploying these practices across key regions, including the temperate zones of South America, the fertile deltas of South and South-East Asia, and the deep volcanic soils scattered throughout the tropics. There is no space here to review the burgeoning research on agro-ecological systems since the 1990s, all of which confirm in one way or another that yield growth, higher incomes from farming and improved food

[19] Interestingly, by contrast, companies which sell pesticides and not fertilisers have in recent years supported research into agro-ecological zero/low tillage systems because these systems reduce external NPK requirements, but remain fairly high users of pesticides.

security are being achieved by restoring soils and protecting water resources (Badgley *et al.* 2007; Halberg *et al.* 2005; Lampkin & Padel 1994; Pimentel *et al.* 2005; Pretty *et al.* 2003; Stanhill 1990; The Worldwatch Institute 2011). Journals such as *Journal of Sustainable Agriculture* and *Renewable Agriculture and Food Systems* repeatedly publish useful case studies which confirm how soil restoration and related ecocentric measures result in quantitative increases and qualitative improvements in yields.

However, it needs to be noted that the bulk of the evidence cited by the above references is also highly contested. Bindraban brings together all the most authoritative studies that question many of the assumptions made by the proponents of 'organic farming', which is nevertheless only one strand of the agro-ecological approach. This body of empirical research questions claims that 'organic agriculture' is less exploitative of natural resources than conventional agriculture, that yields equal to 80 per cent of yields on conventional farms can, in fact, be attained without N supplementation from additional land, that environmental pollution is less in organic systems, that N losses are minimised on organic farms, that losses per unit of product in organic systems are less than in conventional systems, that the quality of organic food is better, that organic products are less toxic given that manures result in toxic releases, that there is greater biodiversity on organic farms, and that nutrient use is more efficient in soils on organic farms than on conventional farms (Bindraban 2009: 17). No doubt each of these empirical studies can be countered with evidence produced by those who support organic farming. But this is primarily a debate that relates to certified organic farming systems in developed world contexts. It is important that this is not conflated with the wider meaning of agro-ecology, as defined by Altieri, Uphoff and others, which does acknowledge that in certain contexts external chemical inputs will be necessary.

Case studies cited in Uphoff's remarkable edited collection all refer to production increases of 50–100 per cent, and in some cases even 200–300 per cent (Uphoff 2002b). These cases are drawn from the following areas: smallholders in Kenya and Nigeria; agroforestry projects in Kenya and Zambia; integrated aquaculture in Malawi; Senegal's Peanut Basin where yield increases were achieved by merging organic and inorganic inputs; Mali's Sahel region where yield increases and food security were achieved using soil conservation, seed banking and market gardening; low-external input methods that were introduced to small farmers in Honduras and Guatemala; combined crop and livestock farming in the Andean mountains; no-till agriculture in Parana State, Brazil; rice farming using the Farmer Field-school approach in Bangladesh; integrated pest and crop management in Sri Lanka; and contour farming in the Philippines.

In a background report prepared for the compilation of the World Bank's *World Development Report*, Pretty provides a detailed account of the largest study of sustainable agro-ecosystems involving analysis of 286 projects in 57 countries covering 12.6 million small farmers, farming 37 million hectares of land. For the 360 reliable yield comparisons from 198 projects, the mean relative yield increase was 79 per cent across a wide variety of systems and crop types. A study of 144 projects has shown that water efficiency was enhanced. Under rain-fed conditions, the water use efficiency was improved by 70.2 per cent, 102.3 per cent and 107.5 per cent for cereals, legumes, roots and tubers respectively, and by 256.6 per cent for vegetables and fruits when compared to the pre-intervention stage. Further, Pretty reports on 'positive side-effects' which include improved 'natural capital' such as increased water retention, reduced soil erosion and more agro-biodiversity; improved 'social capital', including better internal social organisation and connectedness to external institutions; and improved 'human capital', including better health, reversed urban migration, improved status of women and advances in decision-making and problem-solving capabilities (Pretty 2006).

The above cited studies refer to developing world contexts in which yield growth has, in general, not been pumped up to the maximum using tightly managed NPK and irrigation programmes. It is therefore not surprising that agro-ecological approaches will generate better results than the conventional approach in these regions. However, in developed countries where yields have been pumped up to the maximum using conventional methods, one would expect yield growth from agro-ecological methods to be lower in comparison. This is, in fact, not the case. Halweil cites several scientific studies from Europe and North America by major recognised scientists that show that organic yields are only slightly lower (about 5–20 per cent) or equal to conventional yields (Halweil 2006: 19).

In recent decades the agro-ecological approach has influenced a number of policy-driven conversions of agricultural production (Buck & Scherr 2011). One of the more better known is the conversion of Cuban agriculture after the collapse of the Soviet Union that resulted in a massive drop in oil supplies, which forced Cuba to invest in organic farming to secure food supplies (Funes *et al.* 2002; Wright 2008). A more recent example is the passing of the Law of Productive, Communal and Agricultural Revolution in Bolivia in 2011, which aims to reverse the neoliberal model (exports of primary products, imports of processed food) by re-orienting agricultural production to meet the needs of Bolivians using agro-ecological methods. This programme will be supported by an annual government investment of US$500 million for 10 years (Cabitza 2011). The *China Daily* reported that Chinese government statistics show that organic food production quadrupled between 2005 and 2010 (Woke 2011). A United Nations Environment Programme report found that an agro-ecological approach in sub-Saharan Africa would be the most effective way to ensure food security (United Nations Environment Programme 2008).

One of the pet discussion topics that always comes up when agro-ecological agriculture is compared to conventional HEI agriculture is whether farming methods organised along agro-ecological lines 'can feed the world'. This is a rather strange discussion because it is often conducted on the *assumption* that conventional HEI agriculture *can* feed the world. Somehow it is only agro-ecological agriculture that needs to prove that it can achieve the *same results*, with many—such as Nobel Prize-winner, Norman Borlaug—predicting global disaster if everyone converted to organic farming (Borlaug 2000; see also Smith 2000; Trewavas 2002). As suggested in this chapter, setting the benchmark at what HEI agriculture can achieve is totally inadequate, especially if soil degradation is factored into the equation. Nor—as the IAASTD concluded—is there sufficient evidence that the GMO solution can help resolve the problem of soil degradation, although there are those who see no reason why agro-ecological methods cannot be used in conjunction with GMOs — a position which ignores the fact that GMOs are owned by a few, powerful, global corporates who enforce their intellectual property rights in ways that contradict the self-empowerment values that underpin the agro-ecological approach. Much more has to be achieved than conventional industrial farming can in terms of output, eco-system restoration and human capability development. With regard to output, it is worth reviewing two major reports that have tried to establish whether, indeed, enough food would be produced if it was produced using agro-ecological methods (Badgley *et al.* 2007; Erb *et al.* 2009).

The University of Michigan study (Badgley *et al.* 2007) developed a global dataset based on 293 (largely) peer-reviewed studies of comparative yield ratios between what they called 'organic' farming (with a similar definition to Altieri's 'agro-ecological principles') and conventional farming. Significantly, 160 were from developed countries and 133 from developing countries. Table 6.4 presents the results, with the *N* column referring to the number of studies, and the *Average* column referring to the ratio between organic and non-organic production for particular products that were described in the case studies. For example, a ratio of 0.928 for 'grain products' in developed countries means that under organic production the yield was 92.8 per cent of the yield from conventional production.

This remarkable table reflects well-known trends: yields from organic production in developed countries are slightly below those from conventional production (because in these countries yields are high due to well-supported and resourced HEI farming practices), and in developing countries they are substantially higher than conventional production. In order to then calculate whether organic production could feed the world, the researchers made a few necessary conservative assumptions: that the amount of land available was equal to the amount of land devoted to crops and pasture as of 2001 (in other words, the FAO assumption that more land is needed was not adopted); that diets would not change; that waste levels would remain the same (10 per cent); that the same amount of foodstuffs would be produced for animal feed (36 per cent of global grain production); and they excluded the problem of unequal food distribution (that is, they follow the prevailing — albeit factually misleading — practice in food supply studies of dividing global supply by the population to derive 'average caloric intake'). They used standard FAO procedures for calculating yields and constructed two models: one using the yield ratios for developed countries ('Developed, countries' column in Table 6.4), which were then applied globally, and the other using the yield ratios for developing countries, which were then applied globally ('Developing countries' column in Table 6.4).

According to standard FAO figures for 2003, the world food production system can supply 2,786 kcal/pp/day[20] which compares well with the average caloric requirement

[20] kcal/pp/day = kilocalories per person per day.

Table 6.4: Average yield ratio (organic:non-organic) and standard error (S.E.)

	World			**Developed countries**			**Developing countries**		
Food category	*N*	*Average*	*S.E.*	*N*	*Average*	*S.E.*	*N*	*Average*	*S.E.*
Grain products	171	1.321	0.06	69	0.928	0.02	102	1.573	0.09
Starchy roots	25	1.686	0.27	14	0.891	0.04	11	2.697	0.46
Sugars and sweeteners	2	1.005	0.02	2	1.005	0.02			
Legumes (pulses)	9	1.522	0.55	7	0.816	0.07	2	3.995	1.68
Oil crops and vegetable oils	15	1.078	0.07	13	0.991	0.05	2	1.645	0.00
Vegetables	37	1.064	0.10	31	0.876	0.03	6	2.038	0.44
Fruits, excl. wine	7	2.080	0.43	2	0.955	0.04	5	2.530	0.46
All plant foods	266	1.325	0.05	138	0.914	0.02	128	1.736	0.09
Meat and offal	8	0.988	0.03	8	0.988	0.03			
Milk, excl. butter	18	1.434	0.24	13	0.949	0.04	5	2.694	0.57
Eggs	1	1.060		1	1.060				
All animal foods	27	1.288	0.16	22	0.968	0.02	5	2.694	0.57
All plant and animal foods	293	1.321	0.05	160	0.922	0.01	133	1.802	0.09

(Source: Badgley *et al.* 2007: 88)

of a healthy adult, which is between 2,200 and 2,500 kcal/pp/day (if, of course, food was available equally to all). If global food production is converted to organic production, and if the ratios for developed countries are applied globally, the system would be able to supply 2,641 kcal/pp/day which is below the current level of 2,786 kcal/pp/day, but above the accepted standard of 2,200–2,500 kcal/pp/day. (Put another way, if the average caloric consumption rate was, in fact, 2,641 kcal/pp/day, global food poverty would have ended.) If, however, the ratios for developing countries are applied globally, the supply would shoot up to 4,381 kcal/pp/day, which is 57 per cent greater than current global availability (or, alternatively, 157 per cent of current global availability).

The Michigan University researchers conclude: 'This estimate suggests that organic production has the potential to support a substantially larger human population than currently exists' (Badgley *et al.* 2007: 92). But what is important for the purposes of this chapter, is that this conclusion is based on the assumption that no additional land is required to achieve this objective — indeed, the researchers even go so far as to suggest 'the possibility that the agricultural land base could eventually be reduced if organic production methods were employed ...' (Badgley *et al.* 2007: 94). This is possible purely because soil degradation is reversed without high external NPK inputs. The fact that yield increases are possible using farming practices that restore soils and reduce external NPK inputs confirms the argument that yield increase by soil restoration is a viable alternative to opening up virgin land.

Yield increases through soil restoration are possible only if agro-ecological methods are used to replace external chemical N inputs with organically produced N. The Michigan University researchers calculated that 140 million tonnes of N would become biologically available purely from the planting of leguminous cover crops, without reducing the used cultivation area — this is 58 million tonnes more than the amount of synthetic chemical N fertiliser that was sold in 2003. The argument that additional land is not required is particularly significant, although strongly contested by Bindraban (Bindraban 2009). However, this is a conservative estimate because this ignores all the other agro-ecological practices that are used to biologically transform atmospheric N into agriculturally available N plus other agricultural benefits, namely intercropping, crop rotation, alley cropping with leguminous trees, rotation of livestock with annual crops, and inoculation of soil with free-living N-fixers (Badgley *et al.* 2007: 93).

Like the Michigan University study, the Institute for Social Ecology (ISE) study set out to determine whether organic agriculture could feed the world (Erb *et al.* 2009). What is distinctive about this study is that it is the only one that has attempted to take into account the benefits of a change in diet. In other words, instead of assuming that the Western sugar- and fat-intensive diet will/should become the global diet, they have taken into account the implications for agricultural production of a change in diet. Their conclusion is that unless diets change, it is unlikely that organic farming can feed the world.

The ISE study deploys a completely different methodology to the University of Michigan team because it sets out to estimate demand for food up to 2050, and takes into account the demand for meat and dairy products rather than focusing purely on crops. It also takes into account the FAO projections for the amount of additional land that is required.

In order to build a model, like the Michigan University study, the ISE reviewed the literature on yields from organic production compared to conventional production, and reached conclusions at odds with the Michigan University study. The most significant difference is that because organic farming needs to source biologically produced N to replace external chemical N inputs, the ISE researchers assumed (with no evidence provided) that organic farms require additional land for this purpose. In this regard they are consistent with the evidence advanced by Bindraban (Bindraban 2009). This, coupled with pessimistic assumptions about yield increases per hectare, led the ISE to assume that organic production will result in a 40 per cent reduction in yields in developed countries, and that yield levels in developing countries would remain constant. The ISE must therefore adopt the FAO assumption that more land will be required to meet future demand. By contrast, the Michigan University report argued that no additional land is required to produce N inputs biologically, and that there is no evidence to support the notion that organic yields in developing countries will remain the same or that yields in developed countries would drop as much as 40 per cent. Other reports seem to confirm the Michigan University's more optimistic reading of the evidence (Halberg *et al.* 2005). Nevertheless, the ISE researchers insisted that conservative assumptions help to establish the validity of their findings.

In order to set up their analysis, the ISE team took into account the following factors:

- The global population will be 9.2 billion people by 2050
- They accepted the FAO assumption that 9 per cent more land will be required (120 Mha) to meet future demand, and they also considered FAO's 'massive expansion' scenario, which envisages cultivated land expanding by 19.1 per cent
- As already mentioned, the ISE assumed that organic farming will result in a 40 per cent yield drop in developed countries and no yield change in developing countries — consequently they compared the FAO's intensification of conventional farming option to organic farming, and then an intermediate option was considered, which is a mix of the two with yield drops equal to half the yield drops of the organic option
- As far as animal products are concerned (meat, eggs and milk), it was assumed that there are three options: intensive conventional farming, humane farming (free range), and organic farming (certified organic) — the differences being that humane farming would require 10 per cent more inputs than intensive farming, and that organic farming would require 20 per cent more inputs, and that both humane and organic farming would require 40 per cent more land than intensive farming
- Finally, different diets were considered: a 'Western high meat' diet resulting in everyone needing 3,000 kcal/pp/day; a 'current trend' diet in terms of which all regions attain a level of above 2,700 kcal/pp/day, but the richer regions remain above 3,000 kcal/pp/day due to their high meat diets, while others eat healthier

'low meat' diets (in other words, inequalities in food intake are retained); in the 'less meat' scenario everyone attains the 2,700 kcal/pp/day level by reducing the intake of animal products to 30 per cent of the diet and substituting meat, sugar and oil crops in North American and European diets, with more fruit, vegetables, cereals, roots and pulses, while everyone else improves their kcal/day with similar 'low meat' diets (that is, they catch up, but to a lower level of meat consumption); and finally a 'fair less meat' option, which is what food justice could look like in that animal products are reduced further to 20 per cent of the diet for everyone, an imposed universal caloric level of 2,800 kcal/pp/day for everyone is adopted, which would mean a substantial reduction in both caloric and meat intake by people in developed countries.

These factors are brought together to analyse 72 different scenarios for the year 2050, which are summarised in Table 6.5.[21]

As can be seen from Table 6.5, if the Western diet becomes the norm it will not be possible to feed the world with organic farming methods (including humane or organic livestock systems), even if the cultivated land area is increased by 19.1 per cent! Nor will the intermediate mixed option meet the needs for a Western diet (that is, a mix of conventional and organic farming methods). Interestingly, the only option that is 'probably feasible' if everyone wants a Western diet is conventional HEI farming, but with a massive increase in cultivated land (that is, a 19.1 per cent increase, or well over 200 Mha). The only organic production option that is 'probably feasible' without a massive increase in cultivated land area (120 Mha which is a 9 per cent increase rather than a 19.1 per cent increase) is if everyone adopts a 'fair less meat' diet and the livestock system is either conventional intensive or humane (free range). In other words, this still assumes a 9 per cent increase in cultivated land area. Probably the most realistic scenario is the 'less meat' diet using a mix of farming methods, which will require a 9 per cent increase in cultivated land and it would then also be possible to use a humane/organic livestock system.

As already indicated, the results of the ISE model are overly pessimistic because of the questionable assumptions about organic yields (40 per cent reduction in developed countries, no improvement in developing countries) and uncritical acceptance of the FAO's projected land requirements. It also takes no account of the fact that soil degradation is a key driver of yield losses. The great strength of the ISE analysis, though, is that it factors in diets. It is not difficult, however, to imagine how this model could generate an alternative conclusion, namely that future demand could be met using organic farming methods, without expanding the cultivated land area, if it is assumed that yields in developed countries would drop by only 10 per cent (as suggested by other studies), if yields increased by (conservatively) 50 per cent in developing countries (as suggested by other studies), and the benefits of soil restoration as a means of increasing yields were included in the calculation in order to obviate the need for more land. A

[21] The ISE report also factors in land for bioenergy and the impact of climate change. These are not discussed here due to a shortage of space.

Table 6.5: Feasibility analysis of 72 scenarios

	Crop yields	***FAO intensive***	***FAO intensive***	***Intermediate***	***Intermediate***	***Wholly organic***	***Wholly organic***
	Land use change	*Massive*	*Business as usual*	*Massive*	*Business as usual*	*Massive*	*Business as usual*
DIET	**Livestock system**						
Western high meat	intensive	+/–	–	–	–	–	–
Western high meat	humane	–	–	–	–	–	–
Western high meat	organic	–	–	–	–	–	–
Current trend	intensive	+	+	+	+/–	–	–
Current trend	humane	+	+	+	+/–	–	–
Current trend	organic	+	+/–	+/–	+/–	–	–
Less meat	intensive	+	+	+	+	+/–	–
Less meat	humane	+	+	+	+	+/–	–
Less meat	organic	+	+	+	+	–	–
Fair less meat	intensive	++	+	++	+	+/–	+/–
Fair less meat	humane	++	+	++	+	+/–	+/–
Fair less meat	organic	++	+	++	+	+/–	

(Source: Erb *et al* 2009:23.)

Note: (–) means not feasible, (+/–) means 'probably feasible', (+) means feasible and (++) is very feasible.

variation on this theme would be what is sometimes called the 'middle path', in other words, a mixed system in which agro-ecological principles are applied, but external NPK inputs are used in a limited way when required (in accordance with Altieri's 'burden of proof' principle) — field experience suggests that this can double or triple yields on small farms in developing countries (Halweil 2006).

It is important, however, not to confine the discussion of agro-ecological alternatives to the question of nutrient flows and outputs on cropland. Another set of alternatives lies in different ways of configuring the relationship between forests, agricultural production and population growth. This is relevant because forested areas become important nutrient generators for agricultural production. Lambin and Meyfroidt, for example, have documented in great detail how four developing countries with rapidly growing populations — China, Costa Rica, El Salvador and Vietnam — have managed to simultaneously *increase* forest cover and *increase* agricultural output. Without suggesting that these achievements are due to the application of an integrated 'agro-ecological approach', they do suggest the need for state interventions to shape market dynamics, and (in these four cases) some combination of agricultural intensification, land-use zoning, forest protection, increased reliance on imported wood products, land diversification by smallholders, state intervention in land management, and major capital investments (Lambin & Meyfroidt 2011: 3470).

Conclusion

We opened the chapter by suggesting that the food protests of 2008 marked a key turning point. As Patel and McMichael observed: 'From a world-historical perspective, the food riot has always been about more than food — its appearance has usually signaled significant transitions in political-economic arrangements' (Patel & McMichael 2008: 11). We started off by suggesting that the transition signalled by the food protests is the end of the 'third' and the start of a possible 'fourth' food regime. This echoes many from across the ideological spectrum who have argued since at least 2008 that some fundamental changes to the food/agricultural systems are needed (from Oxfam, FAO, World Bank, UNEP, social movements and academics). The farmer suicides in India are an extreme form of self-defeating protest which signifies how powerless people feel when they face the full consequences of the current crisis.

We have differed from most contributions to the discussion about the solutions to the food crisis by emphasising a key underlying driver of the crisis that has been surprisingly neglected in the literature, namely soil degradation. Indeed, soil degradation is not simply a function of the logic of the 'third food regime' — it is the cumulative outcome of the underlying science and practice that shaped both the 'second' and 'third' food regimes, in particular the Green Revolution technologies and financial arrangements that were so central to the 'second food regime' and persisted into the 'third'. We have, therefore, concluded that unless the 'fourth food regime' factors in a science and practice which comprehends soil ecology, it will fail to generate the kinds of solutions that will support and reinforce the transition to a *sustainable long-term development cycle* (as conceptualised in Chapter 3). We have proposed that the *agro-ecological approach* represents a paradigmatic shift which responds appropriately to the twin challenge of obesity and hunger in an increasingly unfair world. Its incorporation of soil biology, bio-economic regions and an emphasis on smallholder farming is of particular relevance to developing countries facing rural–urban migration, food shortages and unemployment.

We conclude with the prediction that rising food prices over the long term will more than likely trigger more (possibly) sustained, globally connected protest action as increasing numbers of poor people get hungrier. Similarly, it will also lead many to search for local solutions that they can implement themselves, especially if system shocks caused by climate catastrophes, such as heat waves, flooding and drought, continue to escalate. As the entire Cuban nation was forced to do after oil supplies ceased with the collapse of the Soviet Union in 1989, securing agro-ecologically produced food supplies by building the resilience of local food economies may well, in the relatively near future, become a major strategic priority of local communities, as well as local and national governments.

So this is the law of the land, son,
to take, you've got to put back.
And you'll find that your days were full, son,
when it's time to shoulder your pack.

Part III

From Resource Wars to Sustainable Living

Chapter Seven

Resource Wars, Failed States and Blood Consumption: Insights from Sudan

in association with Francis H. Caas

Juba, Sudan, 2008

Peter Pitya is an architect now living in Juba, the pulsating, informal capital of Southern Sudan. He spends much time between meetings, sipping drinks at the bar at Mango Camp, gazing across the strongly flowing White Nile. He is acutely aware that Juba is between wars, tottering uneasily on the thin edge of a fragile peace. He's also painfully aware that Sudan's war is like many others in Africa — he calls it a 'resource war' because it's part of the new scramble for African resources. Until recently, he was one of up to 2 million black African Christian Sudanese who for years have lived in exile, having been displaced by — or having fled from — the 30-year civil war between the North and South, which has ripped so many Sudanese communities apart. After completing school in Juba, he studied in Egypt because he could speak Arabic, and later won a scholarship to do his PhD in Japan — learning Japanese along the way. Like many Sudanese intellectuals, he had to find ways to survive elsewhere because returning home was simply not an option. When the late Dr John Garang, leader of the Sudanese People's Liberation Movement (SPLM), signed the Comprehensive Peace Agreement (CPA) in early 2005, Peter decided to return home — to Southern Sudan — to contribute to the rebuilding of a completely shattered economy and society. Together with a fellow Sudanese architect who had practised in Germany for 24 years, he has set up a consulting practice in Juba — a very different choice to the vast majority of educated returnees who have joined the government of South Sudan (GOSS) or one of the vast number of international aid and NGO agencies that have set up 'camp' in Juba. Peter wants to be innovative and have the freedom to work across institutional boundaries in a context which has an extremely uncertain future but is filled with enormous opportunity.

No one can tell you how many people live in Juba. Estimates range from 200,000 to 500,000, and it changes every day as 'displaced persons' pour in from North Sudan and surrounding countries, some from much further away. They set up their shacks, tents and plastic shelters anywhere they can find a space within a city that is completely unregulated. Even cabinet ministers live in tents in well-guarded camps. Formal buildings for offices and accommodation are simply unavailable. A dominant feature of Juba is the slow-moving 4x4 traffic, carefully negotiating its way over massive potholes and dangerous ravines. Only a few areas have an intermittent supply of electricity, most make do with diesel generators, if they can afford them, while others use kerosene or candles. Water and sanitation infrastructure is practically non-existent, as are street lights. It is a sprawling environment of crumbling, single-storey buildings on large plots, interspersed with traditional mud huts, a few stone houses, and an increasing number of informal settlements. Going out at night is risky. Armed bandits and militia often roam the streets—a problem which gets partly resolved, intermittently, by the armed wing of the SPLM, the Sudanese People's Liberation Army (SPLA) which gets tasked to police the town (for better or for worse). Supplies (including fuel, food, pharmaceuticals and spare parts) are bought at informal markets which, in turn, are supplied mainly by traders trucking goods from Uganda and Kenya. Poor people survive on food brought into town by small farmers who have only recently been able to return to their lands to recommence cultivation.

Juba is a classic, lawless frontier town, contested by an embryonic state (GOSS) trying to assert itself. All sorts rub shoulders in the local market from which they have to buy their supplies for daily existence. There are the hundreds of expatriates in international agencies (most of whom use Juba as the operational base for delivering emergency relief into Darfur); the representatives of the US government who have supported the SPLM (and now GOSS) financially and militarily against the North; traders and hustlers from all over Africa (especially Uganda, Kenya, China and South Africa) looking for deals in one of Africa's resource-rich regions; the 'displaced persons' looking for a place to settle or to reclaim their lost properties; all sorts of adventurers and travellers on faith, charity or development missions; and shady characters on all sides who are there to find out who is arming whom among a myriad of large and small militia with loyalties to different bloodthirsty patrons based in both North and South Sudan. Armed groups that depend on looting local resources include the (Ugandan) Lords' Resistance Army, which still terrorises the communities on both sides of the border with Uganda; ruthless movers in the Arab North who refuse to accept their loss of control of the South's resources; various factions within the South Sudanese liberation movement, loyal to different strong-men; and the so-called 'integrated army' provided for by the CPA (which has its own internal fault lines).

Despite its designation as the official seat of GOSS, Juba is profoundly schizophrenic. Throughout the civil war it was a garrison town for the northern army, and never 'liberated' by the SPLM like most of the other towns in the South. What was left of its local population was effectively starved into submission, and those who did well (or, at least, did not starve) collaborated with the northern forces as government officials and suppliers of various services. The northern army and its Arabic rulers may be gone as formal battalions, but the vacuum is imperfectly filled by GOSS and the vast expatriate community. The motive force of the economy is largely driven by the consumer demands of the so-called 'camps' that are mushrooming across Juba, but in particular along the fast-flowing, wide expanse of the White Nile which links the South to the North. Given the absence of formal houses and hotels, entrepreneurs from Uganda, Kenya and beyond have set up 'camps' comprising large sturdy tents, some of them with built-in toilets and showers, to house the masses of educated returnees, former military commanders who are now government ministers and senior officials, the expatriates, and visiting businesspeople—everyone else who can pay for accommodation, These camps are invariably built around a central area dominated by a rustic bar and restaurant, and a large

TV set blaring broadcasts of the BBC, CNN and a local station. The more established boast mud huts for hire, or prefabricated buildings which rapidly disintegrate in the humidity and tropical rains. There are distinct camps for particular constellations — the UN agencies, GOSS, and relief agencies. Mango Camp, with its idyllic location on the edge of the Nile, is one of the most expensive (US$150 per night), has the best food and bar and is where one can find all sorts of groups meeting late into the night, dining, discussing and shaping Juba's future with their deals, scams and grand plans for reconstruction. Lingering on the edges are the 'ladies of the night', emboldened by the free flow of money, the unlicensed atmosphere and the growing demand for their services.

To the extent that Juba has an identity, it is Africa's latest frontier in the new scramble for African resources, where the whiff of oil and guns combines with the promise of rich pickings. It is also an ideal base from which to co-ordinate resource wars across the fractious East African region. Peter Pitya shares with many, grand dreams of a free Southern Sudan. But these stand in sharp contrast to the dull certainty that awful tragedies perpetrated by various dehumanised, military monsters will persist well into the future, because they are prepared to do the bidding of numerous, increasingly ruthless, resource-hungry global powers.

Introduction

We have argued that the 2007–2010 global economic crisis marks the end of the post-World War II long-term development cycle and the start of a just transition to a potentially more *sustainable long-term development cycle* driven by the deployment phase of the Information Age and the deeper logic of epochal transition to transcend resource depletion and the economic consequences of rising resource prices. However, we also argued that this is by no means inevitable. An *unjust transition* is as likely, informed by the precepts of ecological modernisation and a narrow focus on decarbonisation via large-scale, top-down techno-fixes, co-managed by powerful corporate elites with access to new global funding mechanisms. Another strong likelihood is a series of breakdowns as resource wars and failed states spread — after all, history is replete with cases in which the 'writing on the wall' was ignored (Diamond 2005). Here we go one step further by suggesting that the second and third options may, in fact, be two sides of the same coin: an unjust transition which defends the consumption patterns of the billion or so over-consumers will depend on keeping resource prices down in ways that could exacerbate the current spread of resource wars and failed states. Indeed, the only real alternative to a just transition may well be an accumulation of local resource wars into a global resource war.

As the global economy has expanded and as ever larger quantities of finite primary resources get extracted from the Earth, it is not surprising that resource depletion has become an increasingly significant cause of violent conflict at global and local scales (United Nations Environment Programme 2009). These conflicts have become known as 'resource wars', and if nothing is done to change the way resources are extracted and consumed, 'resource wars' will spread inexorably across an increasingly unsustainable world. At least a quarter of the violent conflicts during the 1990s were about resources, which resulted in the death of 5 million people and the displacement of 20 million more (Renner 2004). It can only be assumed that things have got worse, with resource wars in places such as Iraq and Sudan setting the pace. Furthermore, there is an overlap between resource wars and 'failed states', with as many as 2 billion people living in countries governed by 'failing states' (Ghani & Lockhart 2008).

Political economy of the resource curse

Although large deposits of key resources such as oil would usually be considered a blessing for the development prospects of a country, it often turns out to be a 'resource curse' (Collier 2010; Sachs & Warner 2001). This is particularly the case in countries which suffer from ethnic and religious conflicts, and in which poverty is widespread and governments unstable. Under these circumstances, valuable natural resources such as oil often heighten the danger of civil war, and once violent conflict has erupted it can become endemic and almost impossible to resolve. Furthermore, a dependence on natural resources can make a country more susceptible to civil war when overall growth declines and poverty increases (Ross 2003). It is paradoxical that a 'gift' from nature, such as oil, tends to cause economic distress. Various studies have found that generally speaking, resource-dependent economies grow more slowly than resource-poor ones because it is easier to simply sell extracted resources than invest in innovation and human skills (Ross 2003). Collier's work shows that better governance enhances the potential value of natural resource endowments, but the greater the resource endowment of a country the more likely it will suffer from poor governance. It therefore follows that 'the political systems best suited to harnessing natural assets are those least likely to develop once natural assets have become important in the economy.' (Collier 2010: 1106)

Abundant resource endowments have tended to inhibit the kind of economic diversification that is vital for long-term growth. Resource abundance, such as oil, also reinforces the 'rentier state', which according to Kahl tends 'to be narrowly based, predatory, authoritarian or quasi-democratic and characterised by high degrees of patronage and corruption as well as low degrees of popular legitimacy' (Kahl 2002: 270). All of these aspects were already present in Sudan before oil was discovered in the late 1970s — oil merely exacerbated these problems. The demand for oil and certain strategic minerals is such that 'they are worth controlling and fighting over precisely because they are valued in the global economy' (Dalby 2002).

Spreading resource wars and the related increase in the number of failed states will combine in deadly ways to exacerbate global political instabilities, with increasingly authoritarian responses being the inevitable outcome. Unless unsustainable resource use is seen as the root cause of rising levels of political instability, the wrong solutions will be formulated by powerful political leaders.

Two strategic moments marked the different modalities of the resource wars of the post-Cold War era. In October 1999 authority over US military forces in Central Asia was transferred from Pacific Command to Central Command. This little-noticed event marked a decisive turning point because the Central Asian region (which stretches from the Ural Mountains to China's Western border) had hitherto been a peripheral and insignificant part of the world for military strategists in the Pentagon and North Atlantic Treaty Organisation (NATO). For veteran resource war watcher M.T. Klare, the only reason for this shift in command was to put in place the strategic and military capacities required to manage a highly unstable region in which vast oil and gas reserves had recently been discovered. The Cold War was over; the global war over diminishing resources had moved into a new phase (Klare 2001a).

On 6 April 1994 the President of Rwanda was killed in a plane crash which triggered the horrors of the 100-day genocide that left 800,000 Rwandans (mainly Tutsis) dead. This was, however, merely the spark that ignited a violent reaction to decades of simmering local conflicts over access to a key resource in a society without an energy infrastructure — firewood. As trees disappeared so did the soil nutrients, which exacerbated intense land shortages created by population increases in Africa's most densely populated country. By the 1990s hundreds of localised land conflicts were at breaking point as many people started to run out of food in a country in which ethnic identities had been actively politicised by colonial and post-colonial elites (Mamdani 2002). Settling a land dispute meant eliminating the threat of land claims posed not just by a particular individual, but also his extended family whose rights were entrenched by centuries of common law. Once this technique of settling disputes turned violent, entire families and their communities were killed in a bloody land grab that shocked the world. In a post-Cold War world in which local warlords could no longer access funding from one or other rival superpower to run their murderous little wars, looting the resources of rival groups became an attractive alternative. The Rwandan 'popular genocide' was in reality a resource war, which spun out of control as it connected to deep-seated conflicts over the consequences of depleted soils in a context of high population growth.

These two historical moments — April 1994 and October 1999 — illustrate two things: the central role that resources have come to play in contemporary conflicts, and the wide range of types of conflicts that exist. Although the notion of a 'resource war' emerged from the US security establishment in the early 1980s, it is now used across an ideologically diverse literature (Dangl 2007; Gedicks 1993; Klare 2001b; Le Billon 2001; Le Billon 2005; Renner 2004; Tull 2008). A 'resource war' is, in essence, a violent conflict over access to — and control of — a key (often diminishing) natural resource. Oil is, by far, the most significant of such resources today, but there are others including gas, gold, bauxite, minerals, timber (especially virgin forest that contains valuable hardwoods), fish, gemstones, arable land (for mass monocultures such as cotton or rice production or biofuels), biodiversity (mainly for genetic material and tourism) and water. The contemporary literature depicts the following kinds of resource wars:

1. ***Conventional civil wars or superpower interventions which turn into resource wars.*** Examples include Angola, where oil and diamond revenues financed a civil war after superpower funding dried up after the Cold War; Afghanistan, where the anti-Soviet Mujahideen relied on opium trafficking to partly finance their war

against the Soviets who wanted Afghanistan's natural gas; and Colombia, where coca/cocaine and oil money has fuelled conflicts for years.

2. ***Resource wars which are initiated to capture a particular resource.*** In Sierra Leone it was about control of the diamond fields; in Liberia, Charles Taylor waged war on his citizens so that he could fell the indigenous forests for a profit; in Sudan it was about control of the oil fields that by a freak of geography were located right in the centre of an already divided country; and in the DRC many factions (including the Angolan, Zimbabwean and Rwandan armies) secured rich pickings from many different resources (minerals, gemstones, timber, fuel), which not only funded the DRC wars but also the survival of Robert Mugabe's brutal dictatorship.
3. ***Violent conflicts caused by the resource extraction operations*** of multinational companies allied with powerful political elites and secured via foreign debt that invariably leaves communities impoverished and environments destroyed. Examples include conflicts over oil in the Niger Delta, natural gas and timber in Aceh and West Papua in Indonesia, copper in Papua New Guinea, oil in Ecuador, and water in Bolivia.
4. ***Formal resource wars executed by governments***, in particular the US. Examples include the military conflicts related to Caspian oil and gas resources; the invasion of Iraq; the covert war against Hugo Chavez in Venezuela; military support for the anti-Gaddafi insurgency in Libya; and the disastrous intervention in Somalia to protect oil supplies during the Clinton years.
5. ***State–society conflicts over resource control***, where social movements go up against authoritarian states which are seen to be unjustly using and allocating key strategic resources. Although not strictly speaking 'wars', some of the literature does depict the vast number of such conflicts over land, water, food supplies, seeds, fish, energy and forests across the developing world as 'resource wars' (see Dangl 2007; Gedicks 1993).

There is an obvious connection between resource wars and what has come to be referred to as 'failed states' in the American foreign policy literature (Fund for Peace and Carnegie Endowment 2005; Ghani & Lockhart 2008; Haims *et al.* 2008; Tull 2008). According to the US-based Fund for Peace, failed states are defined in terms of how far they deviate from a Western democratic norm: a competent domestic police force and correctional system; an efficient and functioning civil service or professional bureaucracy; an independent judicial system which functions under the rule of law; a professional and disciplined military accountable to a legitimate civilian government; and a strong executive/legislative leadership capable of national governance. Using this definition, there were between 40 and 60 'failing states' in countries that were home to nearly 2 billion people by 2006 (Ghani & Lockhart 2008). The number of actual 'failed states' has been growing steadily. In terms of the 'Failed States Index', which rates countries on a scale of 1–120 (with 120 meaning total disintegration), a score of 100 and above spells state failure, with terrible consequences for ordinary people and wider regional stability. In 2004 there were 7 countries with scores of 100 or more, increasing to 9 in

2005 and 12 in 2006; 8 out of 12 of these failed states were in Africa, with Sudan in first place with a score of 113.7 (Fund for Peace and Carnegie Endowment 2005). The other African 'failed states' in 2006 were Somalia, Zimbabwe, Chad, Ivory Coast, Democratic Republic of Congo, Guinea and the Central African Republic. The four non-African failed states were Iraq (second after Sudan), Afghanistan, Haiti and Pakistan (Fund for Peace and Carnegie Endowment 2005).

For our purposes, what is important here is that the list of places in which resource wars have occurred overlaps with the list of 'failed states' — Sudan, Afghanistan and Iraq invariably appear at the top of both lists. Significantly, 17 of the 32 weakest states on the Failed States Index in 2008 were African states (Fund for Peace 2008). Nearly 50 per cent of the countries that experienced an end to conflicts since the 1980s (many of them over resources) have reverted to full-scale or partial conflict, thus thwarting efforts to build 'capable states' in these countries (Ghani & Lockhart 2008). However, despite the obvious overlap between conflict over increasingly scarce resources and state failure (Brown 2008), the current concern with 'failed states' has more to do with the fact that they are a security threat and cannot foster environments for foreign investment.

A recent, influential text by two former World Bank officials fails to make the connection between failed states and resource wars, and focuses instead on the need to change the rules of international relations to allow the global governance institutions to take over directly to rebuild state institutions where state failure has become endemic. For them, the problem is political leadership and institutional weakness (Ghani & Lockhart 2008). Despite its eloquence, the authors recommend a solution which ignores the underlying problem and gives Western governments even greater control over these territories in the name of 'state building'. After decades of destroying these states via debt, structural adjustment, neoliberal economic theory and resource extraction at rates well below the value of these resources, this is a cruel recipe which will change nothing.

In his highly influential book, *Plan B 3.0: Mobilizing to Save Civilization,* Lester Brown provides a more appropriate perspective on 'failing states' when he writes:

> *As the stresses from these unresolved problems accumulate, weaker governments are beginning to break down, leading to what are now commonly referred to as failing states. Failing states are an early sign of a failing civilisation. The countries at the top of the lengthening list of failing states are not particularly surprising... And the list grows longer each year, raising a disturbing question: How many failing states will it take before civilisation itself fails? No one knows the answer, but it is a question we must ask.* (Brown 2008)

If Lester Brown is right, surely the solution to failing and failed states is not just to fix them institutionally (invariably using a new generation of remedies concocted in Western policy think tanks and management schools), but rather to find a new way of negotiating the equitable apportionment of the world's remaining resources. Without this, resource wars will spread and so will the number of failed states.

The new scramble for African resources

Two recent volumes by mainly African researchers have raised critical questions about the implications of changed and escalated involvement in Africa by the world's major economic powers (Ampiah & Naidu 2008; Southall & Melber 2009). In his comprehensive introduction to one of these volumes, Southall captures a consensus view when he notes that '[t]he thrust of the new scramble is to systematise the exploitation of Africa's natural resources and markets' (Southall 2009: 20). However, the new scramble is different from the old scramble for Africa: what has changed is that there is a new global configuration of economic and political power; what has not changed is the fact that Africa remains a resource exporter and importer of capital goods and consumables (Southall 2009).

There is a new wave of optimism sweeping across Africa as growth rates climb, consumer spending rises and returns on investment are higher than in most other parts of the world since the onset of the economic recession in 2007. By 2008 Africa's collective GDP was US$1.6 trillion, roughly equal to Brazil's and to Russia's. Real GDP has increased by 4.9 per cent per annum since 2000, more than twice what it was in the 1980s and 1990s. Although these levels of growth are not uniform across all of Africa's sub-regions (see Figure 7.1), at current growth rates, GDP by 2020 is projected to be US$2.6 trillion underpinned by a rapidly urbanising youthful and increasingly educated population, with over 128 million households expected to be moving into the middle class to become vibrant consumer spenders (McKinsey Global Institute 2010). According to the African Development Bank's 2010 *African Economic Outlook Report* released in May 2010 (African Development Bank 2010), the average 6 per cent growth rate for 2006–2008 dropped to 2.5 per cent in 2009. However, the report was optimistic that growth would rebound to 4.5 per cent in 2010 and 5.2 per cent in 2011 due to sound macroeconomic policies, counter-cyclical interventions, sustained aid flows and increased international loans. In reality, it was continued strong demand, despite the economic recession, for primary resources from other fast industrialising Asian countries (in particular China, but also India and Russia) that has been significant in protecting Africa from steep declines in GDP growth rates.

Although the boom in resource prices has clearly been a dominant driver of African economic growth, it would be a mistake to assume that other economic sectors remained stagnant. In reality, growth was spread across a number of sectors with resources reduced to 24 per cent of Africa's total GDP by 2009 (McKinsey Global Institute 2010: 3)

The McKinsey Global Institute has clustered Africa's economies into four distinct clusters (see Figure 7.1). The 'diversified economies' (Egypt, Morocco, South Africa and Tunisia) are Africa's 'growth engines' having significant manufacturing and service industries. These economies are characterised by growth in the service sectors, rapid urbanisation and growth in consumer spending of between 3–5 per cent. The 'oil exporters' have the highest GDP per capita, but they have the least diversified economies. Their key challenge is to ensure that oil wealth is reinvested in education and infrastructure as a basis for more diversified growth. The 'transition economies', such as Ghana, Kenya and Senegal, have lower GDP per capita than the diversified economies

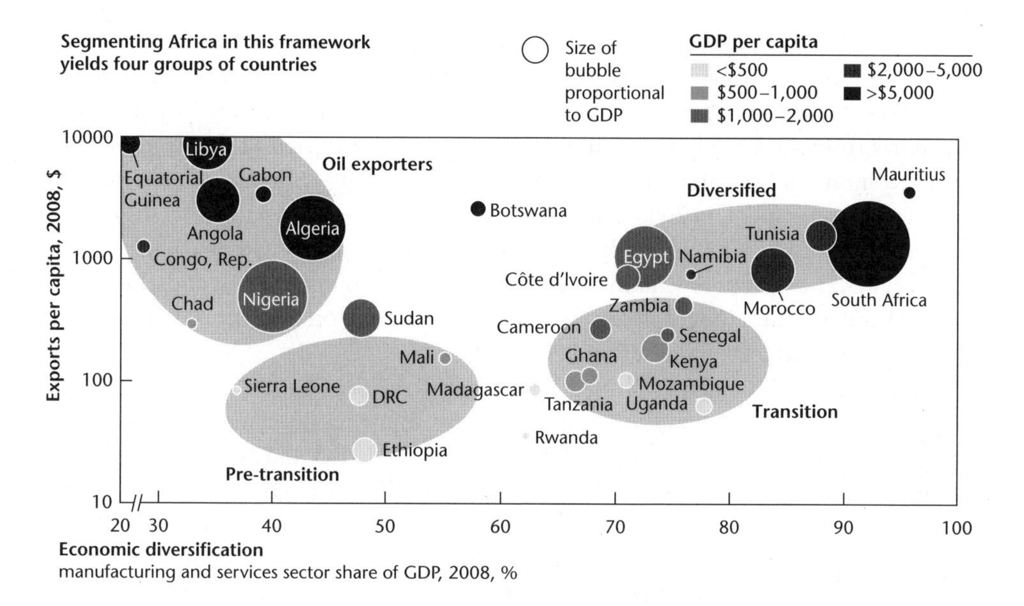

Figure 7.1: Africa's four clusters (compiled by McKinsey Global Institute 2010)

(Source: McKinsey Global Institute, 2010. *Lions on the Move: The Progress and Potential of African Economies.* MGI report, McKinsey & Company.)

Note: We include countries whose 2008 GDP is approximately $10 billion or greater, or whose real GDP growth rate exceeds 7% over 2000–2008. We exclude 22 countries that account for 3% of African GDP in 2008.

and oil exporters, but they are growing steadily as they gradually diversify and benefit from intra-African regional trade. The 'pre-transition economies' are very poor but are growing rapidly, albeit in unstable ways. Much will depend on whether they can get the 'basics' in place, such as stable governments, macroeconomic stabilisation, and reliable food production (McKinsey Global Institute 2010: 5–6).

The optimistic picture painted by the McKinsey report under-emphasises the significance of the fact that primary resources still make up 80 per cent of Africa's exports (which is the highest compared to all other regions). Furthermore, it ignores research by Paul Collier which shows that resource-dependent growth tends to stimulate short-run growth, but undermines long-term growth because there are limited incentives to diversify. For example, after modelling various future scenarios, researchers cited by Collier concluded that the rise in resource prices since 2002 may well result in African output in 20 years' time being 25 per cent lower than if prices had remained at a lower level. In other words, rising resource prices is good for growth in the short term, but not over the long term (Collier 2010: 1105). Future growth and development, therefore, will depend on whether resource rents are in fact reinvested in education/human capital, infrastructure (in particular urban infrastructure) and the effective management of resource exploitation (including ensuring sales at high enough prices).

In 2000, the export of primary natural resources accounted for 86 per cent of all exports from Africa (Mayer & Fajarnes 2005: 8). This was much higher than the rest of the world — the export of primary natural resources accounted for only 31 per cent of all exports from all developing countries in 2000 and 16 per cent of the exports from advanced industrial countries in the same year. According to the UN Conference on Trade and Development, in 2003 many African countries were dependent on the export of a single resource — for example, crude oil (Angola, Congo, Gabon, Nigeria and Equatorial Guinea), copper (Zambia), coffee (Burundi, Ethiopia and Uganda), tobacco (Malawi) and uranium (Niger). Many more were dependent on the export of just two or three primary products (cited in Bond 2006: 60).

In a remarkable 2006 report entitled *Where is the Wealth of Nations?*, the World Bank estimated the 'genuine savings' of all countries by adjusting the national income and savings accounts by deducting the costs of resource depletion and pollution, and then adding investments in education (World Bank 2006). Resource depletion includes the gradual depletion over time of natural assets, which include forests, mineral reserves, and energy resources (such as oil). Echoing the clusters described in the McKinsey report cited earlier, the countries that were the most dependent on exports of primary resources and lowest capital accumulation (measured in terms of 'genuine savings') included some of the largest resource exporters, namely Nigeria, Zambia, Mauritania, Gabon, Congo and South Africa. Indeed, the World Bank report shows that the more dependent an economy is on resource exports, the poorer it becomes over time if the full costs of resource depletion and pollution are taken into account. This, of course, is the end result of trade liberalisation over 20 years and structural adjustment. Contrary to the development strategies pursued by the successful Asian tigers over the same period, African governments were forced to lift protective tariffs, thus killing off local industries that were unable to compete with the prices of imported goods. In the name of increasing trade, the opposite was achieved. According to Christian Aid, '[t]rade liberalisation has cost sub-Saharan Africa $272 billion over the past 20 years. Overall, local producers are selling less than they were before trade was liberalised' (Christian Aid 2005: 3).

Despite increased demand for primary resources caused by Chinese and Indian growth and in line with the general trends (Chapter 2), the real value of Africa's primary resource exports generally declined up until the start of the commodity boom in 2002. This is particularly true for agricultural products (declining from US$15 billion in 1987 to US$13 billion in 2000), but also — according to the World Bank — for non-oil exporting sub-Saharan countries whose terms of trade declined by 119 per cent between 1970 and 1997 (Bond 2006: 60–63).

The global rush for African oil (28 per cent of China's imported oil came from Africa in 2008), as well as minerals and forest products are visible examples of African resources that are extracted for little return. In the biotechnology sector there are mounting examples of global firms which are exploiting the commons with either no — or at best minimal — returns for Africa. This is set to increase as the so-called 'bio-prospectors' comb the African continent for DNA for insertion into all sorts of genetically-modified applications with vast commercial value in global markets. The current examples include a diabetes drug produced from a Kenyan microbe; a Libyan/Ethiopian treatment for diabetes; antibiotics from a Gambian termite hill; an antifungal from a Namibian giraffe; an infection-fighting amoeba from Mauritius; a Congo (Brazzaville) treatment for impotence; vaccines from Egyptian microbes; multipurpose medicinal plants from the Horn of Africa; the South African and Namibian indigenous appetite suppressant, Hoodia; antibiotics from giant West African land snails; drug addiction treatments and multipurpose kombo butter from Central and West Africa; skin whitener from South African and Lesotho aloes; beauty and healing Okoume resin from Central Africa; skin and hair oil from the argan tree in Morocco; skin care from Egyptian 'Pharaoh's Wheat'; skin care from the bambara groundnut; endophytes and improved fescues from Algeria and Morocco; and nematocidal fungi from Burkina Faso (Bond 2006: 87). These bio-resources, and many still to be discovered, will become increasingly valuable in the years ahead as the global biotechnology industry continues to develop at current rates of expansion.

If Africa continues to get poorer as it increases exports of primary resources at discounted prices, it will never build up the financial resources required to invest in the kind of human capital and physical infrastructures that are required for poverty-eradicating development strategies. An obvious question is; what can African governments do to ensure better prices for their exported materials? Unfortunately, all resource-rich countries in the developing world — but especially in Africa — have been pitted against one another within a global free-trade system that is regulated by the rules of the World Trade Organization (WTO). As intended by the World Bank/IMF designers of the system, they are all locked into debt agreements which force them to maximise production output to finance debt repayments, while cut-throat competition in the global market allows buyers to keep prices low. They suffer, therefore, from the consequences of both over-production and low prices. Cartelisation to control both output levels and prices along the lines of OPEC is an obvious solution, but this has not emerged for various complex reasons, not least the influence of powerful players whose interests in cheap resources would be threatened by such a move.

Even the very slight improvements in prices for African resources that were made possible during the growth period before the 2008 crash are now threatened by those who have the power to call the shots. In response to global recessionary conditions the European Union has concluded that '[d]espite recent price falls, raw material prices are still very high from a historical perspective'. In the same statement it responds to this problem in a way that is worth quoting in full:

> *Raw materials are an essential part of both high tech products and every-day consumer products. European industry needs fair access to raw materials both from within and outside the EU. For certain high tech metals, the EU has a high import dependency and access to these raw materials is getting increasingly difficult. Many resource-rich countries are applying protectionist measures that stop or slow down the export of raw materials to Europe in order to help their downstream industries. Many European producers suffer from such practices. On top of this, some emerging countries [Read: China and India] are becoming very active in resource-rich countries,* particularly in

> Africa, *with the aim of securing a privileged access to raw materials. If Europe does not act now, European industry is put at a competitive disadvantage. In response to this challenge, the European Commission launched today a new integrated strategy which sets out targeted measures to secure and improve the access to raw materials for EU industry.* (European Commission 4 November 2008—emphasis added)

To deal with this problem, the European Commission strongly recommends that the commission, member states and industry 'identify and challenge *trade distortion measures* taken by third countries using all available mechanisms and instruments, including WTO negotiations, dispute settlement and the Market Access Partnerships, prioritising those which most undermine open international markets to the disadvantage of the EU' (European Commission 4 November 2008—emphasis added). To enforce this idea of 'market access partnerships', by 2005 the EU had developed no fewer than four Economic Partnership Agreements (EPAs) for sub-Saharan Africa. Critics and many African governments see the EPAs as mechanisms to dump subsidised EU-produced products and to ensure preferential access to EU-based investors in African economies, with special reference to infrastructure development opportunities.

The most significant aspect of the remarkably frank EU statement is the implication that Africa's desire to increase prices to build up its own industries (which, of course, create jobs and reduce poverty) is a practice which 'undermine(s) open international markets' and must, therefore, be resisted by using all the powerful levers available to developed economies, namely the WTO, aid and trade partnerships. 'Open international markets' are seen by the EU as the best means to keep resource prices down. The fact that these resource-rich countries sell their resources at a loss is completely ignored. Nor is the link between low resource prices, resource wars and failed states acknowledged. Although it is all politely articulated in the technocratic language of global diplomacy, this approach reflects very clearly how the global economy is actually managed in the real world of global governance. The direct effect of this approach is the intensification of resource wars and the spread of failing states in the resource-rich countries. So when the EU acts in the interests of the citizens and economies of its member states by pushing down the prices paid for African resources, it is simultaneously promoting an increasingly insecure and unsafe world. Is this, it must be asked, in the best interests of the European Union's citizens and businesses? Does this kind of blood consumption really contribute to world peace?

The EU has always competed with the USA for resources, but it must now also compete with China and India, who are clearly blamed in the EU statement for 'trade distortions'. China, in particular, has become a major economic player in Africa (see Ampiah & Naidu 2008; Campbell 2008). By 2007 China was a greater contributor of economic assistance to Africa than either the USA or Japan. Trade volumes between China and Africa have grown from US$81.7 million in 1979 to $6.84 billion in 1989 to $39.7 billion in 2005 (Campbell 2008). To build on these economic foundations, in November 2006 China organised the China-Africa Forum—or what is generally referred to as the Beijing Summit—of government leaders to consolidate long-term relations between China and Africa. By emphasising the fact that it was not implicated in the slave trade, colonialism or structural adjustment, the Chinese government has convinced Africa that it is a 'friend-in-development', offering a better deal than either the USA or the EU. To back up this commitment, China has agreed to double its economic assistance, increase preferential loans to US$3 billion and preferential buyer's credits to US$2 billion, cancel all debts owed by heavily indebted countries, and set up a US$5 billion investment fund which Chinese companies can access for investments in Africa (Campbell 2008). Nevertheless, China has been prepared to back African governments who have shown no interest in the welfare of their citizens or the protection of the environment — two cases in point being the governments of Sudan and Zimbabwe (supplying both with arms as well as other means).

A recent report from within the US-security establishment identifies southern Africa as strategically important but potentially threatened by resource conflicts and wars. The report deliberates on how the US can secure continued 'access' to key resources such as the platinum group metals, chromium, manganese, cobalt, uranium and the rare earth metals. It refers to shortages of supply, politicisation of mining and, in particular, China's aggressive strategies to secure monopoly control of resource supplies (Burgess 2010).

Resource wars are the outcome of two related processes. Firstly, as the competition for increasingly scarce resources escalates between major global powers, these powers are prepared to intervene in various ways to protect their interests, including militarily. John Perkins, in his bestselling book, *Confessions of an Economic Hit Man,* makes it very clear that these powers maintain a sophisticated legal and clandestine infrastructure to secure their interests by economic and coercive means. Secondly, a context is created for intensified conflicts between local elites within resource-rich countries as they struggle to secure access to resources and position themselves as the key interlocutors in these globalised value chains. Many must extract these resources so cheaply that the costs are carried by exploited populations and degraded environments. Again, they are prepared to use any means necessary, including the organised deployment of conventional military forces and, when necessary, less formal militia when it comes to killing and/or dispossessing large numbers of people (as in Darfur and the DRC).

What most citizens of developed countries do not realise is that the prices they pay for their high-consumption lifestyles are possible only because of the low prices resource-rich countries receive for their exported primary resources. Resource wars and the increasing number of failed states are the logical corollary of this system. 'Blood diamonds' is the term that was coined to refer to the diamonds that were sold into the international market to fund resource wars — maybe we need to extend this idea and start referring to 'blood consumption' if the world's consumers continue to insist on paying discount prices for Africa's primary resources thus reinforcing weak governance and the authoritarian extraction of surpluses.

Sudan, which tops both the 'resource war' and 'failed states' lists, demonstrates in horrifying ways how a resource-rich environment and society can be ransacked, raped and destroyed so that large quantities of valuable primary resources can be sold at discounted prices into global markets.

Sudan's failed state and resource wars

Dar al Harb: Land of war

The word 'Sudan' stems from the Arabic *bilad as-Sudan* or 'land of the blacks'. The term originally applied to the broad belt of sub-Saharan Africa stretching from the Red Sea to the Atlantic Ocean. Up until 2011, the name referred solely to the Republic of Sudan, the largest state in Africa and one of its most troubled and conflict-ridden countries (Petterson 1999).

For most of its modern history, and in particular since independence in 1956, Sudan has been plagued by conflict and civil war. A historic turning point took place in 2011. Contrary to the predictions of most observers, a relatively peaceful referendum took place on 9 January 2011 on whether Southern Sudan should secede from Sudan to become a new independent state. This took place in accordance with the so-called Comprehensive Peace Agreement (CPA) which was signed in January 2005 between the government of Sudan and the Sudan People's Liberation Movement (SPLM), ending a 21-year civil war between North and South. Although flawed in various ways, international pressure and interventions by the African Union succeeded in making sure that the hostile parties to the Sudanese conflict remained committed to the implementation of the CPA, despite continuous, localised, violent conflicts, including the destabilising consequences of the ongoing genocidal violence in Darfur. However, Abiyei remains a flashpoint of violent conflict with many observers predicting that it has the potential for triggering another war.

The various Sudanese conflicts have, over the last five decades, claimed the lives of an estimated 2 million people in direct fighting and related starvation and disease. Some 4 million people have also been displaced either internally or to neighbouring countries. Although the entire country has been affected, the South has been the primary target and has suffered most in terms of the loss of human lives and destruction of infrastructure and resources. This partly explains the northern 'Arab' characterisation and perception of the 'African' South as *dar al harb* or 'land of war' (Goldsmith *et al.* 2002). In contrast, they call their own homeland *dar es islam* or 'land of peace'. This perception is nonetheless rather misleading, as although the southern part of the country has borne the brunt of the conflict, the North has also been impacted, although possibly more indirectly. The economic, social and environmental costs of the civil war, while unevenly distributed, have adversely affected the country as a whole and have been a source of suffering for the majority of the population. The Darfur region, which is administrated by the North, has since 2003 become the latest victim in the country's long history of civil wars. By 2005/06 the conflict in Darfur had claimed the lives of at least 200,000 people and an estimated 2 million Darfurians have been displaced (Crawford-Browne 2007). As in Abiyei, by 2011 the Khartoum government was still engaged in military attacks on the local population, including aerial bombardments.

Division, diversity and marginalisation

Sudan's civil war between the North and South has been the longest conflict in Africa. The first phase which started in 1955 just before formal independence was settled

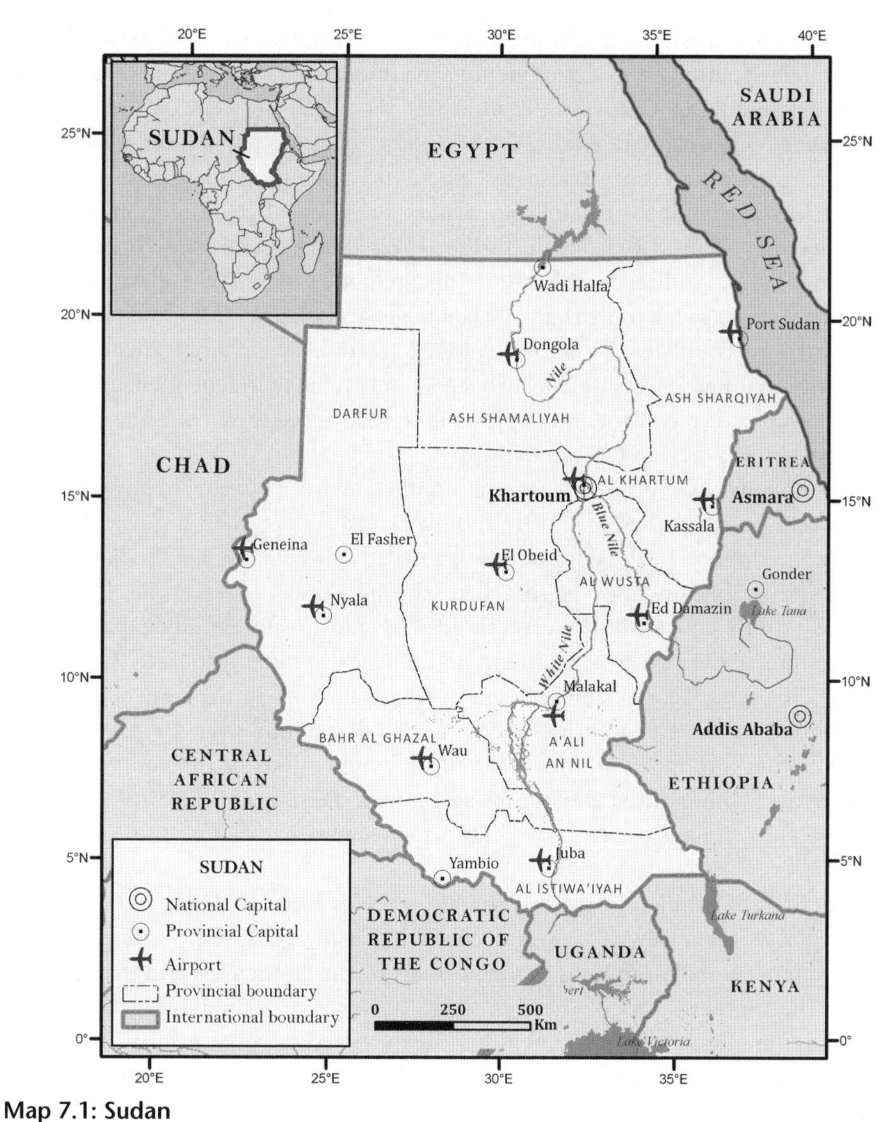

Map 7.1: Sudan

(Source: Centre for Geographical Analysis, Geography and Environmental Studies, University of Stellenbosch, South Africa.)

in 1972. This was followed by 10 years of tentative peace until the second phase was triggered in 1983 and eventually came to an end in early 2005. During the 50 years of independence, both the democratically elected politicians and the military dictators who have alternatively ruled the country from Khartoum, have been 'equally inept at resolving Sudan's basic problems' (Hamdok 2004) and establishing a long-lasting peace.

Conflict in Sudan has often been characterised as a battle between an Arab Muslim North and an African Christian South. While some of the sources of conflict can be traced back to the religious and ethnic differences, it is necessary to look beyond these ideo-cultural designations in order to identify the underlying struggle over the control of key natural resources (Goldsmith *et al.* 2002). About 65 per cent of Sudanese are Africans, while 35 per cent are Arabs. Over 70 per cent of the population is Muslim, a large percentage of whom is of African descent. Of the remaining population, nearly 10 per cent is Christian, while the rest follow traditional religions. Up to 2 million originally Southern Sudanese live in the North, further diversifying the picture (International Crisis Group 2002). Adding to this ethnic and religious diversity, Sudan comprises about 700 tribes who speak more than 300 languages and dialects (Moghraby 2003). As with most African countries, Sudan is a colonial creation which amalgamated people and territories that had never previously been an entity. Since independence Sudanese elites have competed to control and deploy the state to serve their own interests, thus reinforcing what was created by colonialism.

One division, which is relevant to the conflict, exists between a powerful and relatively wealthy centre based in and around the capital and a rather impoverished and marginalised periphery. In the decades that preceded self-rule in 1956, Sudan saw the emergence and establishment of the so-called 'Riverain Arabs', a mercantile class which managed to assume control of the centralised state and to successfully expand large-scale agriculture, while capturing southern and other peripheral resources (Goldsmith *et al.* 2002). The Arab/Islamic rulers based in Khartoum and in the central provinces exert, according to Deng, 'a political and economic hegemony over the marginalised social and cultural groups living in rural and outlying regions of the country' (Deng 1995). Since the signing of the 2005 peace agreement, members of the SPLM were co-opted into a Government of National Unity which somewhat diluted this hegemonic position.

The fault lines of conflict do not just run along a North/South divide but cut across Sudanese society, separating a powerful core from a marginalised periphery which stretches from the Nuba Mountains in the north to Darfur in the west and the southern provinces. The southern part of the country is, however, a particularly extreme case of marginalisation, lack of development and deep-rooted poverty, even in a country like Sudan where human development indicators are already among the lowest in the world.

A tale of two rivers

Covering an area of approximately 2.5 million km^2, Sudan is Africa's largest country, almost the size of India but with a population of only 35 to 40 million (Department for International Development 2004). Sudan's best-known natural and geographical feature is the Nile River. External influences have reached Sudan via this legendary waterway since the time of the ancient Egyptians, right down to its most isolated southern regions (Goldsmith *et al.* 2002). It is also along the Nile that the majority of the country's population and urban centres are concentrated, and where most of its uneven economic development is taking place. The capital Khartoum is located at the confluence of the Blue Nile and the White Nile, a sort of 'permanent way-station between the Arab world and tropical Africa' (Fisk 2005). The Nile River, which could have acted as a unifying factor, has been used by successive northern invaders — the Egyptian Pharaohs, the Mameluks, the Ottomans, and, in the nineteenth century, the British — to gain access to the South's vast natural resources, mainly timber, gold and ivory, as well as slaves. During British rule, the South remained largely inaccessible, despite improved river navigation, and 'the [British] government was able only slowly to bring the vast region and its heterogeneous, non-Arab, non-Muslim population under control' (Holt & Daly 2000).

While there were certainly geographical and natural barriers to expanding colonial authority to the whole of Sudan, the major reason for keeping North and South Sudan separated was political in nature. The colonial government, after having gained control of the South through military action imposed a different system of administration, known as the 'Southern Policy'. The main purpose of this was to try and eradicate all Muslim influence in the area. Christian missionary activities were encouraged and English became the lingua franca in the region. There were even suggestions of federating the South Sudan with Uganda (Suliman 1994). This policy of orchestrated division lasted until 1947 'when London suddenly decided that Sudan's territorial integrity was more important than the separate development which they had so long encouraged' (Fisk 2005). Consequently, the British fused the separately ruled regions and slowly started to devolve most decision-making powers to the northern Arab and Muslim elite. For the people in the South this meant that at the time of independence in 1956, political authority had merely been transferred from one master, the British, to another, the Khartoum-based northern elite. The lingering and simmering animosity that existed between North and South soon caught fire and by 1963, there was fully fledged civil war (International Crisis Group 2002). Ever since then the deeper struggle over resources has been fused with the way in which the protagonists themselves have represented the conflict for reasons that were opportunistically tailored to secure funding support for military action, and rooted in popular perceptions of 'the other' (Jok 2009: 81–115). As Suliman puts it:

> *Few wars are ever fought in the name of their real causes: instead they are fought under old banners and old slogans, based on memories of past conflicts. Most fighters on both sides remain convinced that the war is all about ethnicity.* (Suliman 1994)

During the colonial period, the country was largely shielded from outside economic influence and large parts of the population, particularly those living on the geographical fringes, lived isolated from the outside world. While the North had witnessed some limited and embryonic form of modernisation during British rule, the South was left 'truly underdeveloped' (Petterson 1999). Little has changed since independence and South Sudan is still today 'almost devoid of schools, hospitals and modern infrastructure' (Morse 2005).

At independence, Sudan lacked the major prerequisites for industrialisation, namely capital, technical and scientific expertise, as well as access to markets (Suliman

1994). As a result, the northern-based Sudanese elite that secured control of the post-colonial state, focused on the extraction of natural resources, mainly in the northern regions. The civil war that persisted until the first peace agreement in 1972 effectively excluded the South from the predations of the Khartoum rulers. The exploitation of accessible natural resources in the North was conducted 'in a manner so thoughtless and unscrupulous that it soon endangered the peasant and pastoralist societies of Northern Sudan' (Suliman 1994), including, of course, Darfur. By the late 1970s the Sudanese state was in financial trouble as the onset of a global recession drove down the prices of exported commodities. Like many other African countries, this set Sudan up for a World Bank-imposed structural adjustment programme in the early 1980s. This simply meant cutting back on state expenditure on infrastructure and (what little) human capital development was taking place (health, education), opening up markets for imported products, and lowering prices for exported commodities via intensified competition with other African countries. It was these rather desperate economic conditions that led the Khartoum rulers to look for new opportunities which resulted in 'a new expansion drive to exploit hitherto less accessible resources, mainly in Southern Sudan' (Suliman 1994).

The discovery of oil in 1978 in Bentiu in Southern Sudan created the conditions for a resumption of armed conflict between the North and the South. The final straw came when Khartoum attempted to redraw the administrative boundaries in order to make the oilfields part of the North (International Crisis Group 2002). By 1983 war had resumed between the SPLM and governmental forces, following the unilateral cancellation by the Sudanese government of the Addis Ababa peace agreement, signed in 1972 with the SPLM (International Crisis Group 2002). From this point on, Sudan's conflict was unambiguously an out-and-out resource war.

A harvest of dust

Despite the discovery of oil deposits in the late 1970s and the commencement of large-scale extraction of oil in the early 1990s, the standard of living of the average Sudanese had changed very little since independence (International Crisis Group 2002), and agriculture has remained the basis of Sudan's economy for the vast majority of the population.

During the pre-independence period, the colonial administration promoted the development of large-scale mechanised agricultural schemes, mainly in the country's mid-regions. The intensive exploitation of these areas resulted in extensive soil degradation and the expropriation of traditional farmers who historically inhabited the central regions of Sudan. Agricultural intensification and 'modernisation' was further developed and expanded following independence, 'supported' by the usual gamut of foreign technical advisers and financial institutions (Goldsmith *et al.* 2002), including a significant number of Islamic banks. This move towards large-scale mechanised farming mainly benefited the established elite of large landowners. The rapid extension of cash-crop agriculture dealt a severe blow to small-scale agro-pastoralism. It created a new category of landless and impoverished farmers. In the mid-1990s, the area under mechanised cultivation in the hands of largely absentee farmer-landlords comprised more than 4 million hectares and exceeded the 3.8 million hectares under traditional rain-fed cultivation that supported the livelihoods of nearly 3 million small-scale farmers and their families (Suliman 1994).

The replacement of relatively benign, small-scale methods of exploitation by aggressive and intensive techniques, based on the assumption that natural resources are limitless, gravely degraded the quality of the soils and their ability to sustain adequate agricultural production in the future (Suliman 1994). One example is the Gezira Agricultural Scheme, a large-scale irrigation project started after World War I and officially opened in 1926. This massive agricultural scheme involved building numerous dams and around 10,000 km of canalisation (Moghraby 2003). It was initially supposed to be limited to the irrigation of 300,000 feddans,* but was steadily increased over the years, both by the British colonial rulers and later by the Sudanese government, to eventually cover 2.5 million feddans. The project has had major environmental and societal impacts over the years, including deforestation, salinisation, population displacement, and the spread of water-borne diseases (Moghraby 2003). The Gezira Scheme, although situated in the Northern Sudan, also had right from its commencement a negative effect on the South in that its massive financial costs hardly left any resources for the development of South Sudan and its people (Holt & Daly 2000). The Gezira Scheme and other similar agricultural projects not only proved disastrous from an environmental viewpoint, they also repeatedly failed to fulfil their economic and social development objectives. Gezira, since its inception, concentrated mainly on growing cotton for export purposes. However, in the late 1950s, repeated poor cotton harvests and declining world market prices meant that Sudan was unable to sell most of its cotton stocks, particularly since it insisted on maintaining a fixed minimum price. This resulted in a serious depletion of the country's currency reserves, which were largely dependent on income from cotton sales (Holt & Daly 2000).

Until oil was discovered and exported in the late 1980s, Sudan remained essentially dependent on agricultural products for surplus revenue. Agricultural output and revenue varied greatly from year to year according to external demand and prices, as well as local climatic conditions. In the mid-1970s, the government of Sudan designed and launched a series of ambitious agricultural projects aimed at transforming Sudan into the 'breadbasket' of the Middle East (Suliman 1994). Development projects similar to the Gezira Scheme were embarked upon. Among them was the Rahad Scheme wherein cotton, groundnuts and other crops were cultivated on 300,000 acres of irrigated land, and the Kenana sugar project designed to satisfy Sudanese demand and supply the Middle East region. Construction delays, inattention to existing works, poor maintenance, cost overruns and mismanagement meant that results were mixed. Throughout this period, Sudan's agricultural production declined despite the fact that

* 1 feddan equals 4,200 m^2 or 0.42 hectare.

the area under cultivation had been expanded by 4 million acres. By the mid-1970s sugar cost more to produce in the Kenana project than to import, and by the early 1980s, the country's external debt stood at over US$3 billion (Holt & Daly 2000). As a result, the World Bank stopped providing financial assistance and the International Monetary Fund made emergency loans dependent on the adoption of strict structural adjustment measures. Successive devaluations, the end of subsidies on basic foodstuffs, and a sharp decline in government funding for education and health care, meant that most of the burden of economic decline fell on the poor, particularly in rural areas (Holt & Daly 2000).

All considered, Sudan's massive agricultural development projects created more problems than they solved. They triggered large-scale population movements and environmental deterioration. 'Modernisation' of the agricultural sector — often based on 'advice' from international development organisations such as the World Bank — entailed the expansion of mechanised agricultural practices, which were largely dependent on pesticides and chemical fertilisers, into marginal farming lands, pastures, forests, rangelands and other wildlife areas. Despite vast sums of money invested, agricultural output remained mostly stagnant and the breadbasket dream turned into a nightmare of cyclical droughts and recurring famines. None of the major agricultural projects started between 1975 and 1985 succeeded. In the end, Sudan achieved only a harvest of dust (Petterson 1999).

The creation of scarcity

Sudan's dispersed pastoral and farming communities were most affected by the decades of war, political instability, disastrous and unsustainable agricultural policies, and the gross mismanagement of natural resources. These communities had little opportunity to participate in the decision-making process and were completely under-represented in most federal and local institutions, despite the fact that they formed three-quarters of the total population (United Nations Development Programme 2005). The rapid and disorganised expansion of mechanised, industrial, chemical-intensive agriculture, particularly from the 1960s onwards, from Sudan's central areas towards its peripheral regions, disrupted traditional land tenure arrangements, curtailed transport routes, increased tensions between pastoralists and farmers, and created a large group of landless people. This expansion, combined with increasingly erratic rainfalls and the doubling of the population in less than 25 years, ultimately heightened competition and conflict over resource scarcities that were induced by misconceived policy decisions (United Nations Development Programme 2004). While it was the armed conflicts between northern and southern forces that captured the headlines, low- and high-intensity conflicts over resources continued to take place all over Sudan. The Darfur region and the western areas of Sudan in general have been particularly affected by disruptive agricultural practices which, in turn, led to enhanced competition over natural resources and eventually to conflict (Jok 2009: 115–156). Herein lie the origins of the brutally violent Darfur conflict that began in 2003, killing over 200,000 people by 2006.

In western Sudan, which comprises the Darfur and Kordofan regions, the population is made up of a multitude of different ethnic groups, with some groups specialising in cultivation, while others make a living from cattle rearing or work as camel herders (Manger 2005). However, this division of labour is far from being clear-cut or rigid. Pastoralists often combine their main activity with farming activities during certain periods of the year. Farmers and herders will often have urban-based occupations and cultivators regularly engage in cattle farming using hired herders. These different rural activities form part of the various survival strategies implemented by the people of western Sudan. Some observers have claimed that prior to the arrival of external agents and outside influence, interactions between the many ethnic groups and between pastoralists and farmers were solved rapidly and that conflicts were managed efficiently (Manger 2005). This claim, however, may be somewhat romantic. Clashes over grazing grounds, cattle raiding, trespassing and the burning of crops have existed for centuries, both in Darfur and Kordofan, as well as in many other parts of the country. However, it is equally true that colonial authorities and the subsequent independent governments in Khartoum have intervened in local production systems with profound and often negative consequences (Manger 2005).

In the 1970s, a series of human and natural interventions combined to heighten tensions and trigger conflicts in western Sudan, of which the war in Darfur is the latest illustration. In 1970, the Sudanese government introduced new legislation: the Unregistered Land Act. This Act declared that all land, occupied or unoccupied, belonged to the state and entitlement could no longer be acquired by long use (Suliman 1994). In effect, the Act placed all unregistered land as of 1970, including what was perceived as tribal and communal land, under the ownership of the Sudanese government (Goldsmith *et al.* 2002). A leasehold tenure system was also instituted through which the government could make land available for development projects and other agricultural schemes. The Act enabled the government to distribute 'state land' to its cronies and supporters. In terms of the Act, the government was supposed to be a neutral actor, but instead it became a player in its own right. According to Manger, politicians, leading bureaucrats, army officers and traders obtained access to land resources and schemes by bribing corrupt officials in charge of managing the lease system (Manger 2005). In short, the Act further facilitated an already well-established tendency for land grabbing by the elite. In western Sudan it promoted the rapid expansion of mechanised farming throughout the central plains. This affected the traditional north-south migration routes of pastoralists and herders who travelled between their dry season pastures and their rainy season grazing lands each year. It also pushed traditional farmers onto marginal lands and created a situation of relative over-population in these areas. As a result, more people ended up living in conditions of extreme poverty. This was particularly the case in Darfur, a desolate and marginalised place where most people eked out a living on arid lands (Crawford-Browne 2007).

Adding to the human-created hardship, nature also played havoc on local communities. During the 1970s and 1980s, repeated severe droughts plagued most of the Sahelian regions of Sudan. On the whole, most of the last 30 years have been extremely dry. As a result, more and more pastoralists and farmers moved to urban centres or to those rural areas where agricultural activities could still be practised. The concentration of both people and animals in these areas had many negative environmental consequences, including over-cultivation, over-grazing and deforestation. Small-scale farmers degraded and over-used their land in order to survive, while large-scale landowners over-exploited their resources to maximise profit. The latter had very little incentive to use their land sustainably, since thanks to widespread corruption and the biased land lease system, they could always acquire additional lands to compensate for declining productivity. The same was true for small-scale farmers, who in poorly governed and conflict-vulnerable communities had little incentive to conserve the fertility of their soils or improve long-term productivity. Due to pervasive insecurity, they operated on a short-term basis and more often than not they preferred to simply pack and flee in response to a threat to their lives and livelihoods (De Soysa 2002).

With dwindling resources, competition over the remaining resources increased dramatically. Those tribes, groups or communities with positive links to local or national decision-makers were able to gain access to land assets and resources still worth exploiting. Areas that had previously been regarded as part of the commons were privatised through, for instance, the creation of enclosures or the monopolising of water points, which threatened the livelihoods of herders. These localised pressure points often ended up generating and fuelling conflicts across many different fault lines. Darfur is a particularly good example of simmering low-intensity conflicts that eventually erupted into full-scale armed combat in 2003, which continues to this day (2011). As a region in which almost everyone shares the same religion, it was successive droughts in combination with ruthless land accumulation strategies of government-supported elites and tribes that exacerbated simmering resource conflicts between pastoralists and herders that stretched back over many decades (Goldsmith *et al.* 2002; Jok 2009).

Same actors, similar story

By the 1990s, the situation in Gedarif and Blue Nile States in eastern Sudan, was somewhat a mirror image, with local differences, of what happened in the western parts of the country. In the decades following independence, Gedarif also witnessed the expansion of irrigation-based and rain-fed mechanised agriculture. This form of agricultural development was expanded to the detriment of forests and natural rangelands. It had been a major cause of land degradation through continuous mono-cropping, leading to a decline in soil fertility and productivity. Alongside mechanised farming, small-scale farm holdings were scattered throughout the state, cultivating millet and sorghum combined with sedentary animal husbandry. Pastoralism was also widely practised in all parts of the state, but animal stocks had increased by the 1990s beyond the carrying capacity of the rangelands, thus adding to the pressures on eco-systems. Overgrazing was a major issue, which lead to soil degradation and a decrease in the density of grass and the disappearance of many grass species (Babiker 2005). Deforestation was also taking place at an alarming rate. Trees were cut for wood and charcoal by most people in the state, including the police and the armed forces, both as a means of survival and to supplement low and irregular incomes. Although authorities stipulated in a directive that 10 per cent of the land exploited by agricultural schemes should be planted with trees to enrich vegetation and combat the loss of biological diversity, most scheme owners did not adhere to this directive (Babiker 2005).

Even though population density was relatively low in Gedarif, this state received many immigrants from other parts of Sudan during the 1980s and 1990s, as well as a large number of returning refugees from neighbouring Ethiopia and Eritrea (Babiker 2005). Consequently, Gedarif faced a situation of increasing population pressure combined with rapid environmental degradation and declining agricultural productivity, comparable to that prevailing in Darfur. Similar to the west, conflicts erupted between pastoralists and farmers over ancient pastoral corridors used by nomads. Although local government had reopened some of these routes, they were often not properly designed and did not provide adequate services, such as resting places, water sources and sufficient grazing grounds. They also tended to be too narrow, and as a result herds often ventured into the fields and ended up destroying the crops of pastoralists. Ineffective governance and weak implementation further exacerbated or failed to resolve some of the issues confronting farmers and pastoralists. In 1994, the central government issued a presidential decree which set aside a large area in the South for the sole use of nomads and their cattle. However, the decision was never actually implemented, the reason being that some powerful landowners had already illegally grabbed the land and refused to relinquish it. The influential Farmer's Union, mainly representing large landowners and whose representatives dominated the State Legislative Assembly, also managed to divert some pastoral routes from their original pathways so that they instead passed through the farmlands belonging to small-scale cultivators (Babiker 2005).

Water was also a major source of tension, particularly during the dry season. Farmers often refused to let herders use the water available in their villages or schemes. They tended to fence off water points, which herders believed were communal. Nomads, in turn, used force to gain access, which often resulted in violent confrontations with loss of life. At government-controlled water points, corruption was another major issue. Government water clerks were infamous for their corruption in handling revenues stemming from fees for water use. According to Babiker, the embezzlement of public funds was so widespread 'that nothing was left even for undertaking the routine maintenance of water facilities' (Babiker 2005).

In Blue Nile State, widespread environmental degradation and decreasing agricultural productivity was common by the late 1990s. This was despite the fact that the area received adequate rainfall and had highly fertile clay soils (Babiker 2005). One factor was the waves of displaced refugees from Southern Sudan who escaped the violence by settling in this state. Another was the fact that 'land distribution ... [was] characterised by a clear bias in favour of national and foreign companies at the expense of local communities and the pastoralists' (Babiker 2005). Again, it was political interference combined with weak governance and overall mismanagement that hindered the development of a potentially viable agricultural sector and inflicted severe damage on the environment as land, forest and water resources were over-exploited. Most of the conflicts that have occurred in Blue Nile State have been triggered by multiple ownership claims over the same lands. These conflicting claims grew exponentially over the years because of a dramatic reduction in available arable lands and pasture grounds (Babiker 2005). Most pastoralists and small-scale farmers were repeatedly squeezed into smaller and smaller areas. Not only were the areas decreasing in size but their productivity was also dwindling due to unsustainable agricultural methods. In the years after 2000, competition and conflict over land resources in the Blue Nile State intensified and while still isolated and limited in scale, some observers predicted from as early as the mid-1990s that another Darfur-like conflict was waiting to happen (Babiker 2005).

Alien gods: Controlling water resources

The Khartoum government has always resisted southern separatism because of the presence in the South of vast land and, more significantly, water resources (see also Allan 2001; International Crisis Group 2002). The resumption of violent conflict between the North and South in 1983 was triggered in part by an attempt by the government to capture the water resources of the great *Sudd* wetlands located in the South. The Sudd wetland, which spreads across 5.7 million hectares, is by far the largest wetland in Africa and one of the largest in the world. Due to the fact that it was only registered in 2006 as a so-called 'Ramsar Site' in terms of the 1971 global Convention on Wetlands of International Importance, this unprotected, pristine, ecological paradise and treasure trove was an obvious target for resource-hungry developers looking for water. The key ecological role of the Sudd, however, is to regulate the flow of the Nile River which, in turn, sustains food production in the Nile Valley for the approximately 150 million people who live in the three main countries which share the waters of the Nile (Sudan, Ethiopia and Egypt) (Caas 2004). There are, of course, other countries which depend in various ways on the Nile river system, namely Burundi, DRC, Eritrea, Kenya, Rwanda, Tanzania and Uganda. No other river basin is shared by as many countries and together these countries are home to 40 per cent of Africa's population. Although an agreement between Egypt, Sudan and Ethiopia regulates water use, Egypt extracts the lion's share and makes sure all efforts by the other two to increase extraction are thwarted, including direct military threats. Resource wars will inevitably be waged over the Nile waters in future. The struggle to control the Sudd in the 1980s was just a prelude of things to come (for a discussion of the 'hydro-politics' of this region, see Allan 2001; Collins 1990; Howell *et al.* 1988).

By the early 1970s Sudan's large commercial agriculture schemes that had been established in Northern Sudan were running out of water. Foreign technical advisers and local planners decided to dust off various plans to drain the Sudd, which go back to the heydays of colonialism in the late nineteenth century. However, it was the joint British–Egyptian study between 1946 and 1954 that generated the first serious proposal to build what came to be called the *Jonglei Canal* to divert water away from the Sudd for irrigation. These plans were shelved, however, as attention became focused on the building of the Aswan Dam in Egypt, and the outbreak of civil war in Sudan after Sudanese independence, which made it impossible to contemplate such a large project in Southern Sudan. The idea of the Jonglei Canal was revived in 1972 when the Dutch offered support, complemented later by technical advisers funded by the European Development Fund. Detailed engineering and ecological studies were conducted, mainly by foreign consultants. The proposal was eventually accepted for implementation. The plan was to build a massive canal, 360 km long or twice the length of the Suez Canal. It was designed to be 75 m wide from bank to bank, with a bed-width averaging 28 m and a depth of between 4 and 8 m. This concrete structure was to be driven through untouched African bush and valuable agricultural fields which would cause unmeasured and unimaginable damage. The rationale was simplistic engineering logic, namely that because 50 per cent of the water evaporated once it went into the Sudd, channelling the water away from the Sudd was a more efficient use of the resource (see Howell *et al.* 1988).

The Khartoum government decided to take advantage of the cessation of hostilities between the North and South and launch the Jonglei Canal scheme in 1978. True to form, the Jonglei project was initiated without sufficient consultation with the rural communities who would be most affected, the Dinka, Nuer and Shilluk (Caas 2004). Jonglei means 'alien god' in Dinka, and to most southerners the canal was seen as a foreign enterprise that would benefit mainly Northern Sudan and Egypt while leaving them with reduced and degraded water resources. They feared, with good reason, that this 'alien god' would greatly change their way of life, particularly that of pastoralists whose migrations and grazing system would be disrupted by the canal (Caas 2004). Furthermore, southerners also worried that once drained the Sudd would be utilised to expand mechanised agricultural schemes. While certain promises were made at the beginning of the project to address the needs and concerns of local rural communities, the mounting financial costs of the capital project resulted in the government shelving development projects attached to the main project, such as irrigation farming, cattle centres, social services, bridges over the canal for use by herders, and flood prevention embankments.

Local resource-based disputes triggered by canal construction, coupled with wider political disagreements over the future benefits of oil revenues (after oil was discovered in the South in 1978) resulted in a resumption of the war. The SPLM targeted construction work on the canal, bringing all work to a halt by 1984. By this stage 260 of the 360 km of canal had already been built (Caas 2004). Although the SPLM was in no way interested in the negative ecological consequences of the Jonglei Canal, by preventing its completion they indirectly saved the Sudd, and now that it is a Ramsar Site (as of 2006), it will be very difficult to do anything that could compromise its integrity.

While there have been thus far no attempts to restart work on the Jonglei Canal, the government of Sudan embarked on another potentially disruptive major 'development' project. Work began in the early 2000s on the Merowe/Hamadab Dam located on the

Nile River in Northern Sudan. This dam is currently the largest hydropower project in Africa, and was expected to be completed between 2007 and 2009 at a cost of an estimated US$1.2 billion, mainly financed by Sudan, the China Export and Import Bank and various Arab development funds (Bosshard & Hildyard 2005). China's role is a reflection of the times, in particular when this investment is seen as China's way of entrenching its position with respect to access to the oil reserves.

Most similar dam constructions on the Nile have caused serious environmental damage in the past. There is little reason to believe that this project will be any different. The environmental impact assessments performed so far by companies involved in the project 'have never been properly assessed, and the project has never been certified by the competent Sudanese authorities. On this last score, the project violates Sudanese law' (Bosshard & Hildyard 2005).

According to a preliminary analysis by the International Rivers Network, the Merowe/Hamadab Dam is in breach of the OECD Guidelines on Multinational Enterprises, violates five of the seven Strategic Priorities of the World Commission on Dams (WCD) and contravenes most of the World Bank safeguard policies on natural resources, involuntary resettlements and cultural property (Bosshard & Hildyard 2005). An estimated 50,000 people will or have already been displaced by the project, and more rural communities will be affected downstream of the dam (Bosshard & Hildyard 2005). Furthermore, some Sudanese opposition parties have alleged that the government in Khartoum has simply seized land around the dam without compensation and has handed such land to its supporters in the area (Ajulu *et al.* 2006). Thus far, displaced people have mostly been resettled in inadequate and crowded settlements and have received insufficient compensation for their lost land and houses. According to observers, 'affected people are extremely frustrated about the ongoing process of deception and betrayal', and government authorities have on several occasions used violence to quell protests, resulting in loss of life (Bosshard & Hildyard 2005). Some of the displaced communities are said to be seeking redress through armed insurrection (Ajulu *et al.* 2006).

Apart from the obvious social effects, the Merowe/Hamadab Dam will also most likely have serious environmental impacts. These include sedimentation of the reservoir, invasion by water hyacinths, increased evaporation rates, spread of waterborne diseases, and massive fluctuation of water levels downstream (Bosshard & Hildyard 2005). While nobody is denying the fact that Sudan is in dire need of increased electricity-generating capacity (only 700,000 people have access to the national power grid), most of the investments will go towards large, often unsustainable projects. Of the US$506 million of donor money set aside for the electricity sector, only US$25 million will be dedicated to the development of mini- and micro-hydropower plants and for solar and wind energy (Bosshard & Hildyard 2005). As is often the case with large export-oriented agricultural schemes, most of the electricity produced in Sudan is geared towards urban centres or exported, with little benefit trickling down to the rural poor (Bosshard & Hildyard 2005). Conflict over the benefits of the scheme is seen by all observers as inevitable.

Oil and turmoil

Although land and water resources remain causes of ongoing violent conflicts in Sudan, the conflict over control of oil revenues has effectively resulted in the country being divided into two with complex arrangements to co-manage the oil resources. The Comprehensive Peace Agreement (CPA), signed in 2005, included (at least in theory) a wealth-sharing agreement with respect to oil revenues. In other words, although the discovery of oil in 1978 initially reinforced traditional sources of conflict, the 2005 agreement was based on a realisation by both sides that neither stood to gain from oil revenues if conflict continued. As Salopek put it: 'The rebels control much of the oil country. The government has access to the sea. They need each other to get rich' (Salopek 2003). It took both parties over 25 years to realise this, and in the meantime, ordinary Sudanese paid the price in lost lives, income and opportunities.

Although Sudan started producing oil only in 1999, production by 2007 was at 400,000 barrels per day. Proven reserves have been reported to be 5 billion barrels, compared to Libya which has 41 billion, Nigeria 36 billion and Algeria 12 billion (Energy Information Administration 2008). Sudan, Libya, Nigeria, Algeria and Angola produce 80 per cent of Africa's oil and hold 90 per cent of its reserves (Srinivasan 2008: 61). Consequently, oil's international relevance meant that the Sudanese conflict would eventually acquire a more global dimension with foreign players assuming an increasing role as partisan backers of particular sides, including military support for the Khartoum government from China. However, this globalisation of Sudan's local conflicts also provided international players with leverage to ensure implementation of the CPA.

One of the earliest players in Sudan's oil exploration and exploitation was the US-company Chevron. However, with the resumption of civil war its operations became increasingly difficult to sustain. SPLM combatants repeatedly attacked the company's installations and staff. Three oil workers were killed in one attack, and in 1983 Chevron decided to abandon its oil operations in Sudan (Field 2004), as did the Canadian company, Talisman, and most other Western oil companies. They were later replaced by Chinese, Indian and Malaysian oil businesses which cultivated strong networks in Khartoum (Wescott 2006). This cleared the way for China to become the dominant buyer of Sudan's oil—between 2000 and 2004 China bought 80 per cent of Sudan's oil; this dropped to 40 per cent in 2007 as total production grew. By 2005/06 Sudan was the sixth largest supplier of oil to China and in return received the largest Chinese cash investment in Africa for that period (Srinivasan 2008: 60–61).

The redrawing of Sudan's administrative boundaries in order to exclude southern jurisdiction of oil reserves triggered the second phase of the civil war in 1983. The military regime in Khartoum annexed the oil-rich lands of the South by carving out a new state, ironically called Unity, and by building a refinery in the North instead of the South (see Map 7.2). This brought the fragile peace to an abrupt end. Oil infrastructure, such as pipelines, pumping stations, wellheads and other key elements 'became targets for the rebels from the South, who wanted a share in the country's new mineral wealth, much of which was on lands they had long occupied' (Wescott 2006). The SPLM considered oil installations as legitimate targets and they argued that oil resources belonged to the South, while the regime in Khartoum considered them as a national resource. The Sudanese government also quickly realised that the degree of stability and control it

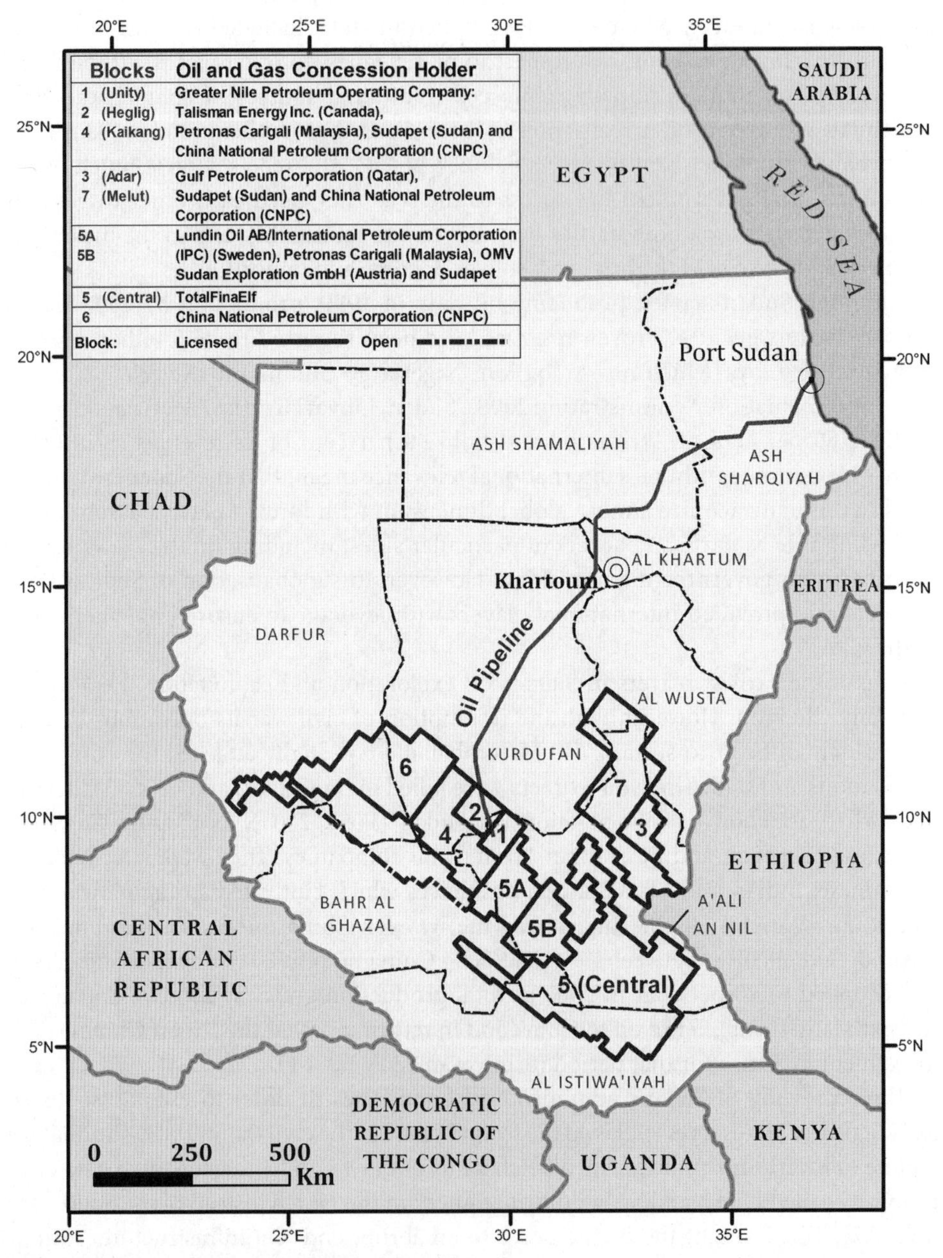

Map 7.2: Oil and gas concession holders in Sudan

(Source: Centre for Geographical Analysis, Geography and Environmental Studies, University of Stellenbosch, South Africa.)

enjoyed in the North depended, at least partially, on its ability to continue exploiting southern resources (Goldsmith *et al.* 2002). Among these, oil soon became its most prized asset and worth waging war for (Jok 2009: 185–238).

The discovery of oil reserves also reignited the South's push for secession and independence from Sudan — a position that was supported by the USA, while China supported the Khartoum regime (Srinivasan 2008). This is a rather familiar occurrence and similar developments have taken place in other parts of Africa, such as in the Biafra region in Nigeria, or the Cabinda enclave in Angola. As Bannon and Collier explain:

> *Where an ethnically different region sees what it considers its resources stolen by a corrupt national elite comfortably ensconced in the capital, the prospect of gaining control over the natural resource revenues ... can be a powerful drive for a secessionist movement.* (Bannon & Collier 2003)

The strategic aim of the secessionist movement in the South was to assert the legitimacy of the territorial integrity and autonomy of the territory, improve bargaining power with the Khartoum government, and profit from the oil business. Here too, oil was considered worthy of a war. Or as a southern fighter put it: 'Now that we know the oil is there, we will fight much longer, if necessary' (Le Billon 2004). Sudanese farmers and rural communities, on the other hand, were far more concerned with the social and environmental impacts of oil exploitation on their daily lives and livelihoods. In its drive to gain complete control of the oil fields, the Khartoum government adopted a scorched earth and starvation policy. Government troops and militias were sent in to appropriate oil rich lands and clear them of their occupants. Some 55,000 people were forced to flee the oil zone and became refugees in their own country (Goldsmith *et al.* 2002). For peasants and pastoralists in the region, it meant being squeezed into smaller areas and having to compete for decreasing resources. The same cycle of environmental degradation, poverty and conflict was again set in motion as had happened during the expansion of mechanised agriculture in an earlier era. What happened in the South in the 1980s and 1990s was basically a rehearsal for similar events a decade later in Darfur, a region also rich in oil, as well as other natural resources such as uranium and gold (Motsi 2006).

It is not surprising that the revenues generated by Sudan's new oil wealth benefited mainly the same Khartoum-based elites. Even if Sudan's macroeconomic situation had improved, 'its people remain impoverished, primarily because oil profits flow to a limited few and are used to fund the war' (International Crisis Group 2002). In 2001, the government of Sudan was spending about US$1 million per day on the war effort, an amount approximating its daily export earnings from oil (Morse 2005). After financing the war against the South, oil revenues after 2003 were used to pay for the war in Darfur. Since the first barrel was produced, oil exploitation has had negative social and environmental consequences for Sudanese society, sustained the central government's appetite for weaponry and generated financial benefits for the usual suspects. Little changed after the SPLM joined the Government of National Unity after the signing of the CPA in 2005, despite the efforts of SPLM leaders.

'When peace breaks out'

One would be forgiven for thinking that Sudan is a desperate cause and a doomed country. Its people have been killed, displaced, starved and impoverished for so many decades that the chances of building a stable, prosperous and peaceful society seem rather remote. Despite a multitude of so-called development initiatives aimed at triggering economic growth, mostly the opposite has happened. The majority of people in Sudan are no better off today than they were at independence in 1956. Sudan's primary reliance on natural resources and its lack of economic diversification puts it in the unenviable category of poor developing and resource-dependent nations that tend to have lower social indicators, are more corrupt, ineffective and authoritarian, and prioritise military spending over social investments (Le Billon 2004). Genocide, ethnic cleansing and crimes against humanity are terms routinely associated with Sudan (Morse 2005). Southern leaders must also share responsibility for aspects of this sad state of affairs. As Salopek puts it '[t]raditionally, the SPLA has mistreated as much as defended Sudan's long marginalised southern people' (Salopek 2003). In short, most actors in the various Sudanese conflicts bear responsibility for the resulting human suffering, recurring humanitarian disasters, lack of social progress, and widespread environmental degradation (Goldsmith *et al.* 2002).

In many ways, the Comprehensive Peace Agreement (CPA) is what Ballentine and Nitzschke term 'a negative peace, where justice and sustainability are deeply compromised and the threat of renewed conflicts remains high' (Ballentine & Nitzschke 2003). Not only is the peace 'negative', it is also far from being comprehensive. Fighting continues in Darfur, where the largest and most expensive humanitarian relief operation in the world has taken place. As many as 300,000 Darfurians may have died in the conflict so far, and about 2.7 million have been displaced.[1] Nevertheless, the referendum prescribed by the CPA did take place in January 2011 with 98 per cent voting in favour of Southern Sudan seceding to become an independent state. Undoubtedly, the major powers played their part in making sure local conflicts did not overwhelm the process, but not so for Darfur where conflict continues unabated and unreported because journalists are kept out of the region.

Although the core argument thus far has been that conflicts over resources have determined the course of Sudanese politics, as Jok has argued it is impossible to ignore the role that race, religion and identity have played (Jok 2009). John Garang, the leader of the southern rebellion who signed the CPA in 2005 but died in a helicopter crash shortly thereafter, dreamed of a unified, democratic Sudan. But the determination of Southern Sudanese to escape Arabic racism (Southerners — who are mostly black Africans — are still often regarded by Northern Arabs as 'slaves'), Islamic rule and systematic discrimination was expressed most clearly in the referendum result. The referendum result has fundamentally weakened the Khartoum government which is

[1] By 2011 Rebecca Hamilton, a veteran journalist who has covered Sudan for years, was still citing the UN estimate of 300,000 deaths in her articles, even though there are higher and lower estimates (Hamilton 2011a).

faced with a powerful Islamic fundamentalist challenge led by Hassan al-Turabi, Bashir's former ally, but now deadly enemy. The Obama Administration makes deals with Bashir (despite his standing indictment by the International Court for war crimes) because the alternative will, in all likelihood, be worse (an abrogation of the CPA and an invasion of the South would follow). At the same time, simmering rebellions amongst the Beja, Dinka, Funj, Nuba and Nuer people in the various provinces have all been encouraged by the referendum. Even Bashir's allies have castigated him for signing the CPA and allowing the referendum to proceed because they all fear the consequences of losing control of the oil resources. The Sudanese Finance Minister was reported to have said in June 2011 that the secession of the South will result in the national budget losing 36.5 per cent of its revenues. With an external debt of US$38 billion and no access to international loans, this is clearly a threat to a corrupt state which depends on cash to service its support networks (Hamilton 2011b).

Ironically, however, it is the geography and infrastructures of oil that may well hold the two sides together and create an economic basis for a two-state solution that may be durable enough to counteract the powerful centrifugal forces leading to total disintegration (Natsios & Abramowitz 2011). As already mentioned, 80 per cent of the oil wells are in the South while the bulk of the infrastructure and pipelines lie in the North. To build an alternative infrastructure through Ethiopia or Kenya will take a decade or more and cost billions of dollars. But with only 5 billion barrels of reserves, at current rates of extraction Sudanese oil could peak before this infrastructure comes on line. So if an acceptable revenue sharing formula can be found, both sides may have an interest in defending separation against the only other alternative, namely violent disintegration in a multiplicity of armed regions with, possibly, a more extreme Islamist party in power in Khartoum. But while a rational case can be made for co-managing the oil resources, a similar common interest is hard to find when it comes to water and, in particular, the future of the Sudd wetland. The same applies to Abiyei, the only region in which no agreement could be reached in the CPA. Once again, at root this is contestation over a key water resource that herders and pastoralists need to access. And nearby, there is oil. Abiyei might not bring the house of cards down, but indications are that it will persist as a flashpoint well into the future.

Conclusion

We began the chapter by suggesting that there is a strong likelihood that concerted efforts will be made by resource-importing economies to reverse the upward trend in global resource prices. Reference was made to the European Union's resource strategy and the EPAs, as well as to emergence of China as the largest investor in Africa which is, of course, aimed at securing long-term supplies of key strategic resources. This provided the context for exploring Collier's dilemma, namely that states get weaker as the value of their endowments grow, while the value of the resource endowment for society increases as states become developmentally accountable (Collier 2010). Sudan clearly confirms this hypothesis. Sudan provides us with a glimpse into a future that Lester Brown contemplates when we asks how many states must fail before civilisation fails. Hopefully the horror of this image is such that it spurs the kind of conscious evolution that is inspired by the vision of a just transition to a fairer and more sustainable world where everyone has sufficient without depending on the misery of others.

Chapter Eight

Transcending Resource- and Energy-Intensive Growth: Lessons from South Africa

*in association with Camaren Peter and Jeremy Wakeford**

We have an opportunity over the decade ahead to shift the structure of our economy towards greater energy efficiency, and more responsible use of our natural resources and relevant resource-based knowledge and expertise. Our economic growth over the next decade and beyond cannot be built on the same principles and technologies, the same energy systems and the same transport modes, that we are familiar with today. (Minister of Finance, Parliament, 20 February 2008)

Introduction

Compared to Sudan, South Africa is not a weak or failed state. Nevertheless, it is a resource-rich, resource-exporting country which has, since the birth of its democracy in 1994, gradually increased its dependence on revenues from primary exports to fund an ambitious developmental and welfare programme aimed at alleviating the high levels of poverty inherited from the apartheid era. The result has been resource- and energy-intensive growth reinforced by neoliberal macroeconomic stabilisation measures. It is now accepted in key government documents that this development trajectory is unsustainable and an alternative must be found. The South African case is instructive because it reveals why decoupling may well be a precondition for future growth and development in countries that are still characterised by high levels of poverty.

In May 2010 the South African government hosted a Green Economy Summit which signalled that there was mounting concern in government policy circles that the resource- and energy-intensive growth path was no longer sustainable. This summit effectively initiated a discursive space for stakeholder engagement about an alternative growth path – a space that the National Planning Commission (NPC) (http://www.npconline.co.za) has kept open with two national workshops in 2011 on the challenge of building a low-carbon economy. Since then a new macroeconomic policy framework, called the New Economic Growth Path (NEGP), has been adopted by government. This identifies *job creation*, *innovation* and the *green economy* as the three pillars of future growth. The NPC's first report concluded as follows:

Given these challenges, there are thus already good reasons to seek to build a new development path that is more inclusive, less dependent on the exploitation of

* This chapter draws from and extends research commissioned by the Development Bank of Southern Africa in 2010 (see, Peter *et al.* 2010).

non-renewable resources and that uses renewable resources more sustainably and strategically. (National Planning Commission 2011: 17)

Although the Integrated Resource Plan (IRP)[1] adopted in 2011 arguably contradicts the 'green' and 'innovation-oriented' aspirations of the NEGP, the Long-Term Mitigation Strategy (LTMS) and the conclusions of the NPC's Diagnostic Overview (for a convincing defence of this view see Trollip & Tyler 2011), the rapid ascendance of a sustainable resource-use perspective within government policy circles between 2008–2011 can be read as a strategic response by key policy actors to the realisation that there are inherent limits to resource- and energy-intensive growth. This raises three questions, which are relevant to many other developing countries that aspire to modernise by pursuing the conventional energy- and resource-intensive growth strategies:

1. What were the main elements of South Africa's post-1994 resource- and energy-intensive growth strategy?
2. What were the conditions that began to undermine the post-1994 resource- and energy-intensive growth strategy?
3. What were the policy responses to these conditions that gradually contributed to the 2010 Green Economy Summit and subsequent policies that have incorporated sustainability thinking?

The approach adopted here differs markedly from the substantial body of critical literature on South Africa's development dilemmas that has — like similar literature on other developing countries experiencing the same pressures to modernise —tended to underestimate the underlying natural resource constraints to conventional conceptions of economic growth (Edigheji 2010; Fine & Rustomjee 1996; Freund 2010; Gelb 2005; Marais 1998; Seekings & Nattrass 2005). Although a handful of analysts have addressed this problem (Bond 2002; Fig 2007; Wakeford 2007), this has not been done in a comprehensive manner which clearly establishes the linkages between macroeconomic policy, resource management and a potential developmental project that substantively reduces poverty and unemployment by focusing on sustainability-oriented innovation as the key driver of a progressive development strategy. In the best overview we have of South Africa's 'development dilemmas' since 1994, Bill Freund calls for a purposive national project of concerted 'modernisation'. What prevents this is, in Freund's view, an inadequate conceptualisation of what this means and as a consequence a poor understanding of economic growth (as mere capital investment) and development (as mere fiscal expenditure). His critique leads him to the conclusion that what is needed is a sustained mobilisation of society by 'human agents of change ... cadres who can use the schools, the media and real local knowledge, coming from or being stationed in every municipality, every significant spatial community, to lead the modernisation drive ...' (Freund 2010: 14). What Freund has not fully realised, of course, is that even if such a pro-modernisation configuration of forces emerged, it would face the reality of rising costs of depleting natural resources and shrinking carbon space (as recognised in the

[1] The IRP is effectively South Africa's national energy policy and strategy.

above cited NPC report). Although Freund and some of his collaborators recognise the need to take into account negative environmental impacts by avoiding the techno-fix approach to mega-projects favoured by those who manage South Africa's 'mineral-energy complex' (MEC) (Freund & Witt 2010), insufficient emphasis is given to the economic implications of resource depletion and, therefore, how dependent future growth and development will be on sustainability-oriented innovations that decouple modernisation from escalating resource exploitation and carbon intensity. Indeed, as argued in Chapter 4, a developmental state which invests in sustainability-oriented innovation as driver of growth and development will succeed in fusing the need to green the economy and to depend more on knowledge and human capabilities than on value generated from resource exploitation.

Resource- and energy-intensive growth

South Africa's democratic Constitution resulted in a thorough review and reform of virtually every facet of policy and practice. The overriding focus of policy and legislative reform was non-racialism and the human rights enshrined in the Constitution. Although poverty was a focus, this was defined in welfarist terms, in other words, basic needs and what the state needs to 'deliver' (for example, housing, education and welfare). Employment creation was seen as a function of economic growth, in turn stimulated by investment. A Bill of Rights formed part of the new Constitution and specifically guaranteed not only individual and social rights (such as the right to housing), but also the right of all South Africans to have the environment protected for the benefit of present and future generations (Section 24 (a)). More pertinently, Section 24 (b) of the Constitution obliges South Africans to 'secure ecologically sustainable development'. How to reconcile the imperatives of economic growth, poverty alleviation and 'ecologically sustainable development' has, unfortunately, not been seen as the central challenge of the post-1994 era. Unfortunately, Section 24(a) of the Constitution has been interpreted narrowly to mean environmental protection resulting in the National Environmental Management Act (Act 107 of 1998) which provides for elaborate environmental controls and environmental impact assessments (EIAs). This dualistic 'environmental impact' approach (Swilling 2007) to development remained largely intact until the adoption by the Cabinet of the National Framework for Sustainable Development (NFSD) in 2008 (Department of Environmental Affairs and Tourism 2007). This was the first policy framework that explicitly used Section 24(b) of the Constitution to suggest that it was time to move beyond the 'environmental impact' paradigm and called into question the prevailing resource- and energy-intensive growth path.

Since 1994 there have been three primary macroeconomic policy frameworks: the Growth, Employment and Redistribution (GEAR) policy introduced in June 1996; the Accelerated and Shared Growth Initiative for South Africa (ASGISA) introduced in 2005; and the New Economic Growth Path (NEGP) introduced in 2011 (which should be read together with the Industrial Policy Action Plan 2 [IPAP2] which was introduced in 2010). The core focus of GEAR and ASGISA was macro-financial stabilisation focusing

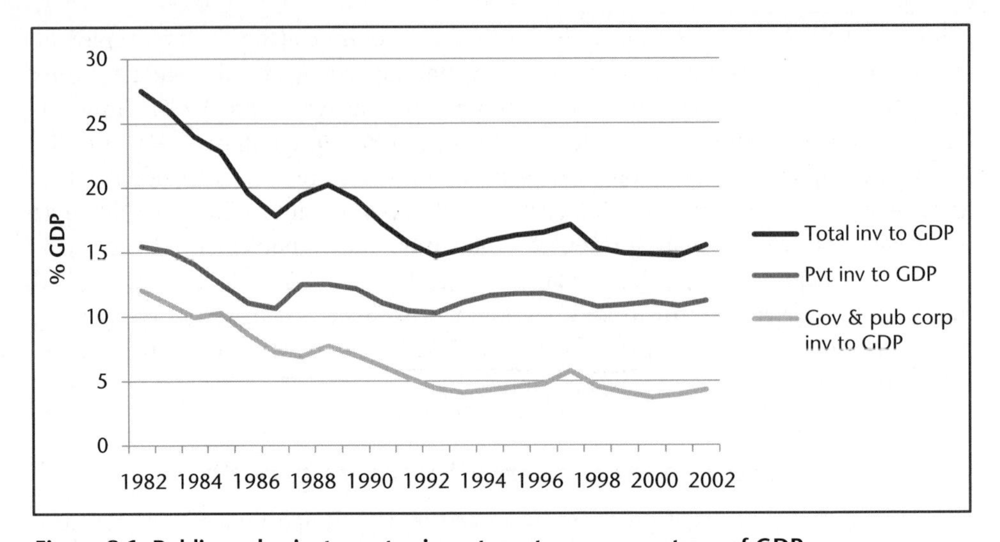

Figure 8.1: Public and private sector investment as a percentage of GDP
(Source: South African Reserve Bank (SARB) *Quarterly Bulletin,* data cited in Quantec 2011) Quantec RSA Standardised Industry Database, www.quantec.co.za)

on inflation targeting, exchange rate stabilisation, deregulation of capital inflows and outflows, deregulation and privatisation of key economic sectors, export-oriented growth and the lowering of trade barriers to consolidate South Africa's position in global markets (for a useful overview see Gelb 2010). The key difference between GEAR and ASGISA is that the latter placed greater emphasis on 'binding constraints' that needed to become the focus of state interventions and on the importance of state-led investments in public infrastructures as a key stimulus of growth as domestic consumption started running out of steam. The rationale of both policies was that economic growth would happen only if the overall decline in investment as a percentage of GDP (see Figure 8.1) could be reversed by substantially increasing the levels of private and public investment (with GEAR more inclined to the former), from both local and foreign sources. Implementing these neoliberal measures to attract private investment, in particular, was regarded as an absolute necessity because South Africa's extremely low savings base was clearly insufficient to finance the levels of investment required to stimulate job-creating growth.[2]

Although private investment picked up a little, total investment remained constrained by the dominance of the so-called 'mineral-energy complex' (MEC). Before discussing these problems in greater depth, it is worth noting that average economic growth levels of 3–5 per cent during the 1994–2007 period were much higher than the decade preceding democratisation in 1994 (see Figure 8.2). The onset of the global recession in 2007/08 clearly brought this relatively long growth period to an end.

Although growth rates for the post-1994 decade were on average higher than the pre-1994 decade, they were, until 2008, nevertheless moderate rather than high, but also just

2 The ratio of gross savings to GDP has declined from a peak of close to 35 per cent in 1980 to a low, but constant, level of around 15 per cent from 1994–2009. However, businesses rather than households account for the bulk of these savings.

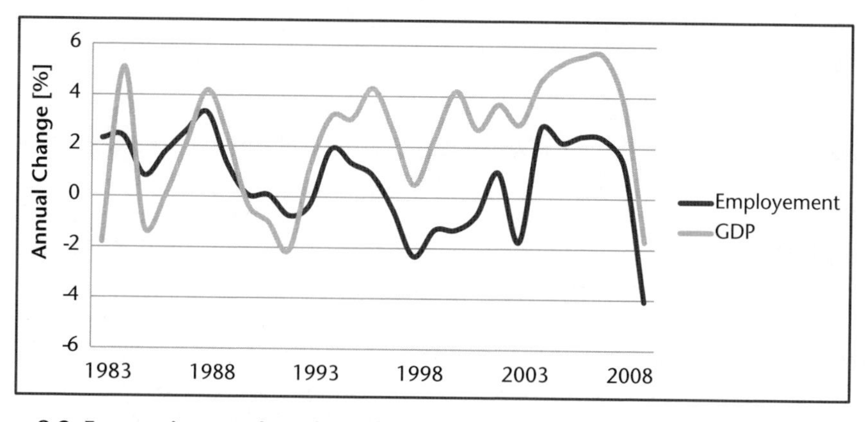

Figure 8.2: Economic growth and employment, 1983–2008
(Source: South African Reserve Bank (SARB) *Quarterly Bulletin,* data cited in Quantec 2010.)

sufficient to support a rapid increase in fiscal spending on key social problems inherited from the apartheid era (especially welfare, education, health and housing) without sliding into a major debt trap. Although there have been many statements in the polemical, academic and policy literature that equate the adoption of neoliberal macroeconomic policies with cutbacks in fiscal expenditure (Marais 2011), the evidence does not support this consensus view. As Seekings and Nattrass argue, South Africa's redistributive measures are, in comparative terms, actually quite impressive (Seekings & Nattrass 2005). Indeed, as Freund insightfully argues (Freund 2010), the result of this achievement is that South Africa's definition of 'development' has now come to be equated with redistribution through fiscal expenditure (or what is referred to in popular political discourse as 'service delivery'). An extensive empirical review of these fiscal trends concluded that real fiscal expenditures have increased rather declined in real terms since 1994 (Swilling *et al.* 2008). Expenditure on social services increased in *real* terms by 57.5 per cent, from actual allocations of R70.2 billion in 1995/06 to R196.6 billion in 2004/05. The result of this trend is that the relative share of social services of consolidated expenditure increased from 45.4 per cent in 1995/96 to 50.9 per cent in 2004/05. Expenditure on economic services increased in real terms by 71.5 per cent, from actual allocations of R16.2 billion in 1995/96 to R49.4 billion in 2004/05. As a result its share of expenditure grew from 10.5 per cent in 1995/96 to over 12 per cent in 2004/05. Furthermore, this was achieved by slightly lowering total expenditure as a percentage of GDP from 30.8 per cent in 1995/96 to 28.6 per cent in 2004/05. Nor did state revenues increase — they remained at around 25 per cent of GDP per annum for most of this period. At the same time, in the six years to 2003/04, personal income tax paid to the South African Revenue Services (SARS) decreased in real terms by 0.9 per cent, while company tax increased in real terms by 12.3 per cent. Achieving this level of fiscal expenditure, while reducing government debt (see Figure 8.3), helped to legitimise the new democratic state during a period that did not result in massive reductions in unemployment or poverty. Business accepted the tax rises as the price to be paid for political stability.

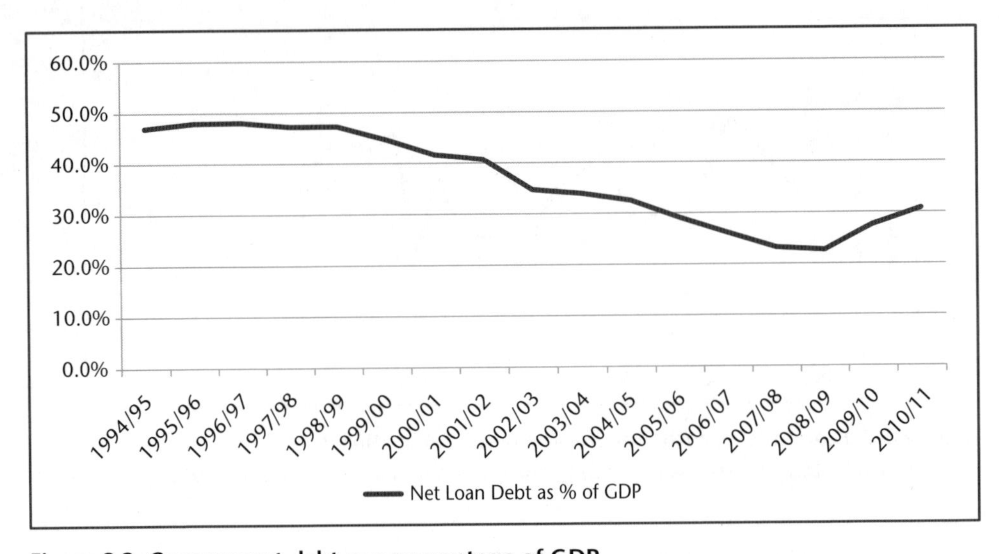

Figure 8.3: Government debt as a percentage of GDP
(Source: Budget Review 2011, National Treasury, Republic of South Africa.)

Despite moderate growth and substantial increases in fiscal expenditure, unemployment and poverty have persisted. According to the National Planning Commission (NPC), using the so-called 'narrow' official definition of unemployment, the number of unemployed people increased from 2 million in 1995 to 4.4 million in 2003. It then declined to 3.9 million in 2007 and increased again to 4.1 million by 2009. The reason for rising unemployment is that employment growth has not kept up with the natural increase in the size of the labour force. Using a R524/month poverty line, 53 per cent lived in poverty in 1995, declining marginally to 48 per cent by 2005. This decline was attributed largely to the impact of social grants, which now benefit more people than the number of people in formal employment. Using the Gini coefficient, South Africa is the most unequal society in the world—according to the NPC, 70 per cent of the wealth accrues to the richest 20 per cent, while the poorest 10 per cent of the society get less than 0.6 per cent.[3]

To find out why South Africa experienced moderate growth plus rising fiscal expenditure without substantial reductions in poverty and unemployment, it is important to note that South African economic growth has been driven by a combination of expanded domestic consumption (primarily via the expanding multi-racial middle class) financed by rising levels of household debt (securitised against residential properties) and exports of primary resources, which entrenched the hegemony of the so-called 'mineral-energy complex'. Unfortunately, resource-based manufacturing has tended to grow faster than the other manufacturing sectors (see Figure 8.4) in response to a vigorous strategy to lower import tariffs and liberalise the capital markets (thus favouring investments in liquid assets rather than long-term fixed investments).

3 The figures cited in this paragraph are drawn from the National Planning Commission (Republic of South Africa: National Planning Commission 2010).

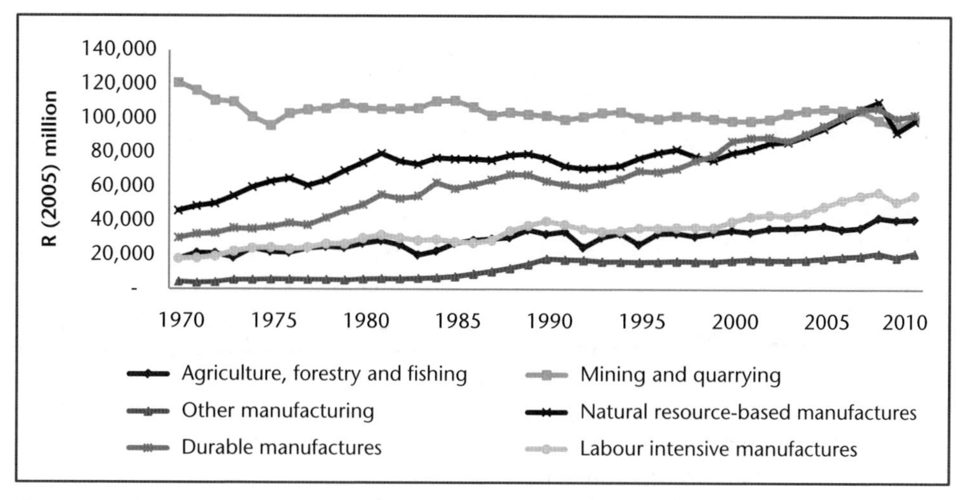

Figure 8.4: Sector value-added, 1970–2007, R million (2005 prices)
(Source: Statistics South Africa, data cited in Quantec 2011)

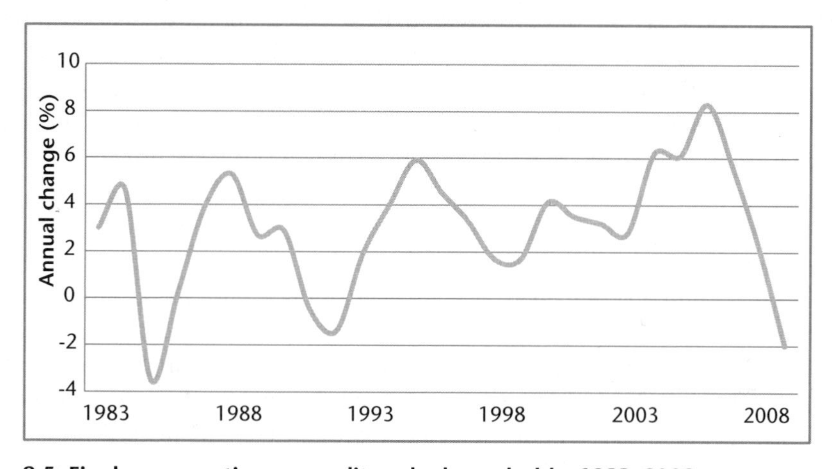

Figure 8.5: Final consumption expenditure by households, 1983–2008
(Source: South African Reserve Bank (SARB) *Quarterly Bulletin*, data cited in Quantec 2010)

Figures 8.5 and 8.6 reveal the rise in consumption spending and the relationship between household disposable income and debt.

The growth in real consumption expenditure (that is, final real demand) is shown in Figure 8.5, but when read against rising debt levels and the declining contribution of manufacturing relative to mining and natural resource industries as reflected in Figure 8.6, it is clear that debt-financed consumption has been the driver of consumer demand for an increasing quantity of imported products, while growth has shifted to an *increasing* reliance on the extraction and export of natural capital. By 2010 minerals comprised 60 per cent of total exports. The balance of payments pressures this created was at first mitigated by the beneficial impacts of rising commodity prices. But with

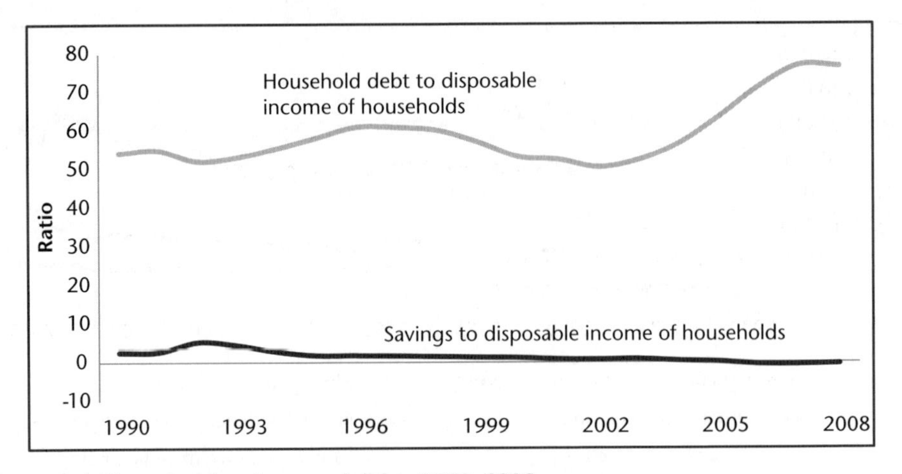

Figure 8.6: Household savings and debt, 1990–2008
(Source: South African Reserve Bank (SARB) *Quarterly Bulletin*, data cited in Quantec 2010.)

the global economic crisis, both easy credit to drive consumption and high commodity prices came to an end, although it is clear that commodity prices did start to recover from mid-2009 as China and a few other industrialising, developing countries continued to grow at much higher rates than the global average. South Africa's admission to the so-called BRIC club[4] in 2010 has reinforced South Africa's role as a key provider of primary resources into other fast industrialising, developing countries (in particular, China and India).

The basic structure of the South African economy has its roots in what the NPC calls 'natural resource colonialism' (National Planning Commission 2011: 16). This basic structure was consolidated by a political deal between English-speaking mining interests and the Afrikaans-speaking business elites who gained influence after the electoral victory of the National Party in 1948, and led in turn to the formal introduction of the apartheid policy that reinforced the pre-existing colonial structure of the society. The result was the forging of a 'minerals-energy-complex' (MEC), which consolidated an accumulation regime based on state-owned, cheap energy production (via the state-owned entity called Eskom), cheap labour, the incorporation of Afrikaner business elites into the mining sector and the rationalisation of finance houses, tightly bound to energy and mining capital (Fine & Rustomjee 1996). Their predilections for grand mega-projects (such as power stations, dams, highway infrastructures, huge mines, giant steel factories) spawned a body of expertise that is responsible for the regime 'lock-in' that suppresses innovation today. Ever since then, the MEC has exercised great control over the nature of growth in the economy, not purely through production, but through its networked linkages to the manufacturing and industrial sector, deeply entrenched knowledge

[4] BRIC is an acronym for Brazil, Russia, India and China and refers to the formal alliance of these countries in international relations.

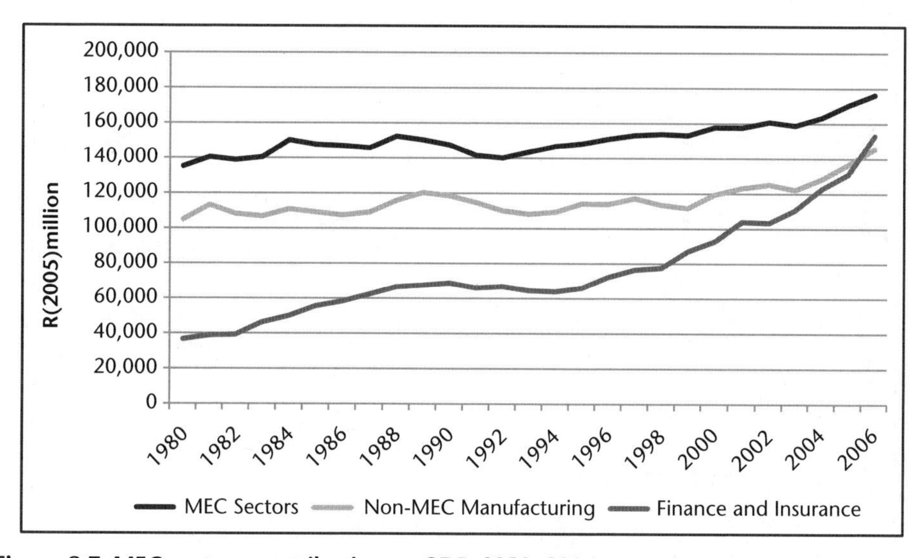

Figure 8.7: MEC sectors contribution to GDP, 1980–2006

(Source: Roberts, S. and Z. Rustomjee (2009). Industrial policy under democracy: Apartheid's grown up infant industries? Iscor and Sasol. *Transformation*, 71: 50–74.)

infrastructures which link industry and the universities, and especially through the extensive control of investment flows that mining finance houses enjoyed until the 1980s (Fine & Rustomjee 1996; Mohamed 2010). According to Mohamed (2010), with the exception of a few sectors, such as automobiles and components manufacturing, manufacturing after 1994 has remained dominated by sectors with strong links to the MEC. As Figure 8.7 shows, the manufacturing sectors that have grown maintained strong links with the MEC, while the non-MEC-related sectors have not grown as fast (Fine & Rustomjee 1996; Mohamed 2010). Renowned Harvard economist Dani Rodrik developed a similar argument when he pointed towards a shrinking non-minerals trade sector, particularly export-orientated manufacturing, as an explanation for low growth and persistently high levels of unemployment (Rodrik 2006).

This nexus between energy and minerals is clearly evident in the fact that 39.2 per cent of all electricity consumed in 2008 was attributed to the following economic activities: mining, traction, basic chemicals/refined fuels, non-metallic mineral processing, basic iron and steel production, ferro-alloys and non-ferrous metals (Rustomjee 2011). The provision of cheap electricity through coal-fired power stations, originally intended to enable the large-scale extraction of raw materials, continues to underwrite the shape of the economy significantly today (Tyler & Winkler 2009). Indeed, Tyler and Winkler state that the 'minerals energy complex is so central to the economy that it is likely to take decades to change dramatically' (Tyler & Winkler 2009: 3).

A key consequence of the MEC-centred structure of the South African economy has been a long-term consensus that cheap energy is this economy's strategic competitive advantage. The result is that it has become the most carbon-intensive economy in the

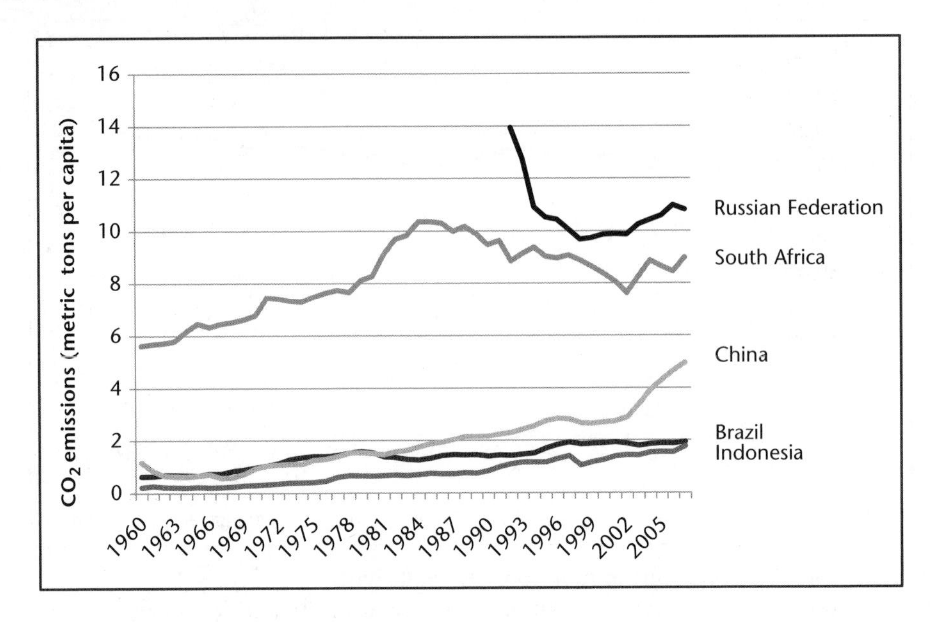

Figure 8.8: Carbon intensity of selected developing countries (CO_2/t/cap)
(Source: World Bank Indicators, CO_2 emissions)

world compared to other major developing economies, with the notable exception of the Russian Federation. For the last two decades CO_2 per capita hovered between 8 and 10 tonnes, a figure twice as high as in China (despite recent carbon intensification), 4–5 times higher than Brazil, Indonesia and India (World Bank 2010), and similar to the United Kingdom and Germany (Figure 8.8).

South Africa's carbon-intensive economy stems from the fact that it is also one of the most energy-intensive economies. Figure 8.9 reveals that while energy resource productivity improved in most of the other fast-industrialising, developing countries, energy productivity in South Africa has generally stagnated. This is most likely a logical consequence of the strategy to keep energy prices low, thus incentivising inefficient use of the resource. Instead of gradually increasing prices over a long period, sudden large increases were introduced in 2009, which had negative economic consequences.

As far as resource extraction is concerned, Figure 8.10 reveals the significance of ore and fossil fuel extraction as a percentage of total domestic extraction.

At the same time, coal extraction has increased to fuel the coal-based electricity-generation industry, which supplies extremely cheap electricity to South Africa's economy. The policy of keeping the prices of coal and minerals as low as possible has constrained diversification of the economy into more knowledge-intensive sectors and encouraged high levels of operational inefficiency. There is, however, evidence that this has started to change. Figure 8.11 reveals improvements in the material efficiency (DMC/GDP) of the South African economy. This spontaneous, albeit limited, improvement in resource efficiency suggests that this may be practically viable as a long-term policy objective.

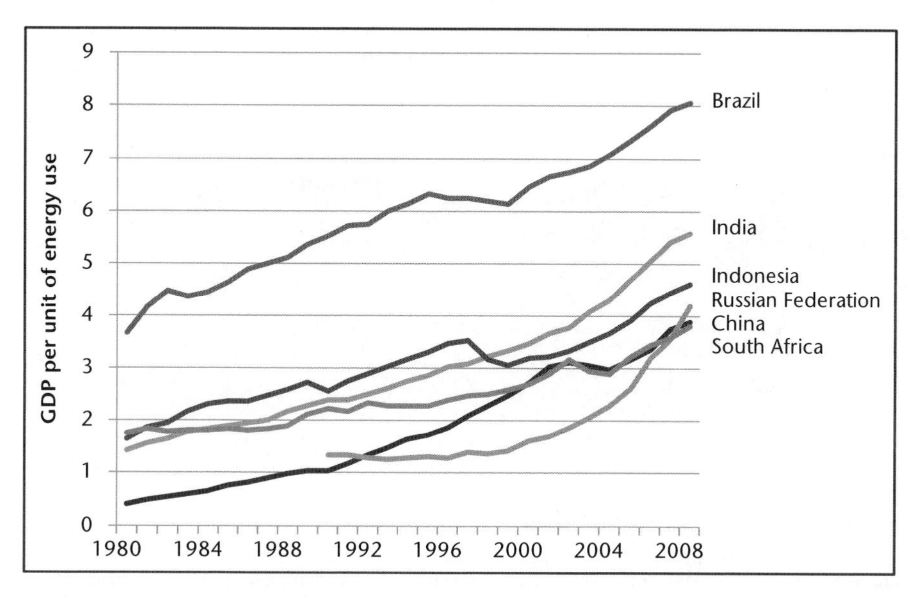

Figure 8.9: Energy intensity of selected developing countries (GDP per unit of energy use)

(Source: International Energy Agency (IEA Statistics © OECD/IEA, http://www.iea.org/stats/index.asp) and World Bank Indicators, GDP per unit of energy use, http://data.worldbank.org/indicator/EG.GDP.PUSE.KO.PP)

Note: GDP per unit of energy use is the PPP GDP per kilogram of oil equivalent of energy use. PPP GDP is gross domestic product converted to current international dollars using purchasing power parity rates. An international dollar has the same purchasing power over GDP as a US dollar has in the United States.

Figure 8.12 reveals that exports of primary resources have continued to grow during the post-1994 period, with coal exports making the largest contribution to this increase. Although the share in tonnes of primary resource exports as a percentage of total exports declined after 1994, their share of total revenue generated from all exports has increased significantly to 60 per cent of total revenues from exports by 2010 as commodity prices have escalated on the global market due to rising demand from rapidly industrialising countries such as China and India. To enhance this dependence on revenues from primary resource exports, in the late 1990s the government decided to sell the state-owned steel manufacturer (Iscor) to the global steel giant, Mittal Steel, with a back-to-back agreement that Mittal could buy South African iron ore for cost plus 3 per cent forever (or, to be precise, 25 years). This is why Mittal could testify at the Competition Tribunal hearings in 2005 that its South African operations are its most profitable, worldwide.

In short, South Africa is a good example of an economy caught up in the financialisation of a globalised economy, with commodity exports and debt-driven consumption as the key drivers of growth. This has undermined manufacturing, reinforced by the lowering of tariff barriers and the rise of cheap imports from Asia. The unsustainability of this growth strategy is recognised partially by the

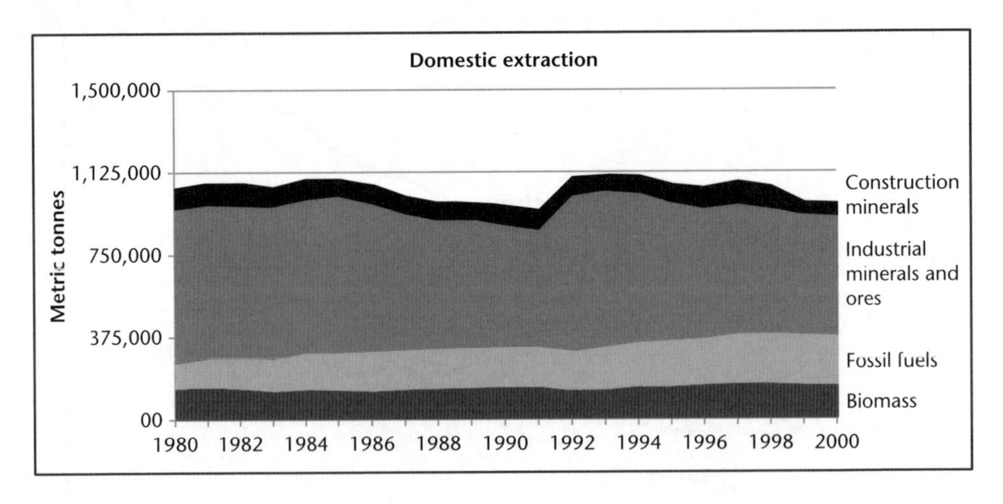

Figure 8.10: Domestic extraction (DE) in metric tonnes, 1980–2000

(Source: United Nations Environment Programme (2011) Decoupling natural resource use and environmental impacts from economic growth. A report of the Working Group on Decoupling to the International Resource Panel. Fischer-Kowalski, M., Swilling, M., Von Weizsäcker, E.U., Ren, Y., Moriguchi, Crane, W., Krausmann, F., Eisenmenger, N., Giljum, S., Hennicke, P., Romero Lankao, P. & Siriban Manalang, A.)

government and key stakeholders, and various interventions began to be considered by a wide range of government departments and state institutions. However, policy discussion can easily bounce off entrenched socio-technical regimes with long-term capital investments in fixed infrastructures shored up by generations of technical and professional expertise that manage to monopolise the definition of the problems South Africa faces and the solutions that the country needs.

During 2010 government consolidated what it has called the Green Economy policy. It hosted the Green Economy Summit in May 2010, which provided a forum for many key ministers and the President to announce commitments to inter-sectoral co-operation. In February 2010, the Cabinet approved a Department of Environmental Affairs document entitled *Proposals on Green Jobs: A South African Transition.* This was followed up by an inconclusive process to develop a National Green Economy Strategy. Two key state-owned financial institutions — the Industrial Development Corporation (IDC) and the Development Bank of Southern Africa (DBSA) — worked closely with the Department of Environmental Affairs to develop detailed financial plans for implementing the Green Economy Strategy. In the meantime, the Gauteng government (which presides over South Africa's industrial and financial heartland) adopted what it called a Developmental Green Economy Strategy, which focused heavily on decoupling by targeting investments in renewable energy, water efficiency, recycling of solid and liquid wastes, moving people into public transport and massively increasing locally produced food to

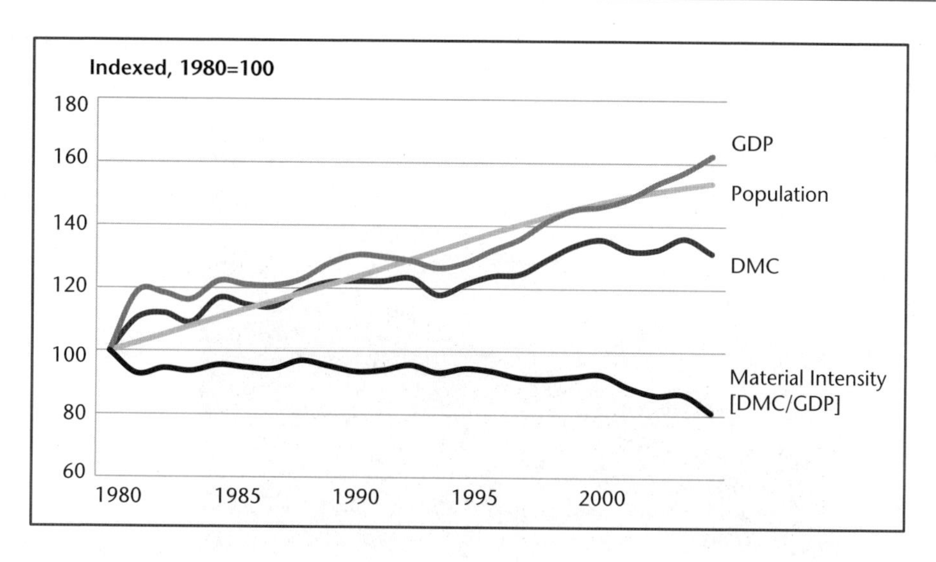

Figure 8.11: Material efficiency, 1980–2000

(Source: United Nations Environment Programme (2011) Decoupling natural resource use and environmental impacts from economic growth. A report of the Working Group on Decoupling to the International Resource Panel. Fischer-Kowalski, M., Swilling, M., Von Weizsäcker, E.U., Ren, Y., Moriguchi, Crane, W., Krausmann, F., Eisenmenger, N., Giljum, S., Hennicke, P., Romero Lankao, P., Siriban Manalang, A.)

Note: Domestic material consumption (DMC) = Domestic extraction minus exports plus imports.

improve food security and create jobs. This was the general context for the adoption of the NEGP and IPAP during 2010/11, both key economic policy documents that addressed the need for South Africa to find a less resource- and energy-intensive growth path.

This raises a key question: given the inherent limits of debt-financed, consumption-driven growth, and given the dependence on primary resource extraction for domestic use and exports, what explains the rising interest in a Green Economy? The answer, it will be argued, lies in the fact that resource depletion had implications for prices, profits and employment. Interestingly, as in many countries, these signals are interpreted in purely economic terms, giving rise to ad hoc policy shifts which often fail to recognise the underlying resource-depletion drivers. If these signals are seen for what they are — indicators of resource-depletion thresholds — this would make it possible to generate a comprehensive response. It will be suggested that the Green Economy initiative within South Africa is an example of a response which could lay the basis for a more long-term sustainable growth and development pathway, which is less resource- and energy-intensive and driven more by sustainability-oriented innovations. This is not to say that the policy-makers who are involved see the significance of this clearly enough, but at least the policy discussion has begun. Much, however, depends on whether the capacity for innovation can be upscaled to deal with these challenges.

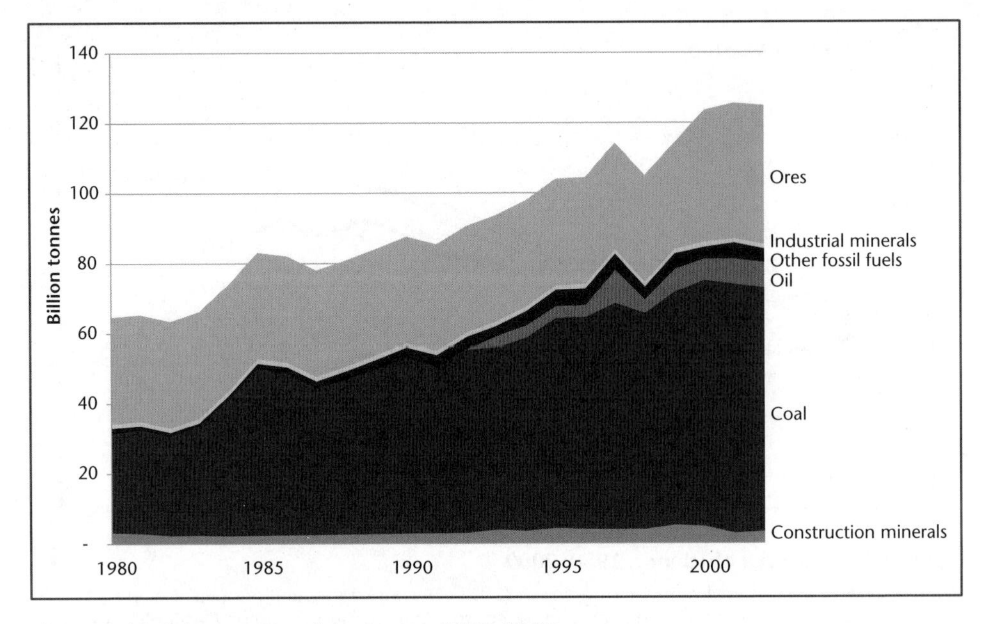

Figure 8.12: Primary material exports, 1980–2000

(Source: United Nations Environment Programme (2011) Decoupling natural resource use and environmental impacts from economic growth. A Report of the Working Group on Decoupling to the International Resource Panel. Fischer-Kowalski, M., Swilling, M., Von Weizsäcker, E.U., Ren, Y., Moriguchi, Crane, W., Krausmann, F., Eisenmenger, N., Giljum, S., Hennicke, P., Romero Lankao, P., Siriban Manalang, A.)

Innovation drivers

Climate change[5]

Using global climate models, significant changes to the South African climate over, the next 50 years were predicted, with drastic impacts on national water availability, food and biomass production capacity, incidence of disease and the country's unique biodiversity:

- Continental warming of between 1 °C and 3 °C
- Broad reductions of approximately 5 to 10 per cent of current rainfall
- Increased summer rainfall in the north-east and south-west, but reduced duration of summer rains in the north-east
- Nominal increases in rainfall in the north-east during winter season
- Increased daily maximum temperatures in summer and autumn in the western half of the country
- Extension of the summer season characteristics.

CO_2 is South Africa's most significant greenhouse gas (GHG), contributing more than 80 per cent of its total GHG emissions. The main source of CO_2 emissions is the

[5] Based on the work done by researchers for the Department of Environmental Affairs (Scenario Building Team 2007).

energy sector, which generated 89.7 per cent of total CO_2 in 1990 and 91.1 per cent in 1994. These high emission levels relate to the high energy intensity of the South African economy, which depends on large-scale primary extraction and processing, particularly in the mining and minerals beneficiation industries. Although still a developing economy, its energy-intensive nature and its dependence on coal-driven energy sources results in an extremely high carbon emission level per unit of GDP compared to the rest of the world.

In July 2008, the Cabinet adopted a document generally known as the Long-Term Mitigation Scenario (LTMS) which was commissioned by the Department of Environmental Affairs and Tourism and compiled mainly by a group of University of Cape Town researchers. This report produced two primary scenarios, namely the *Growth without Constraints* scenario and the *Required by Science* scenario. The first models long-term implications of current economic policy, and concludes that emissions will grow from 440 megatonnes of CO_2-eq in 2004 to 1,600 megatonnes of CO_2-eq by 2050. This would involve increasing fuel consumption by 500 per cent, building 7 new coal-fired power plants or 68 integrated gassification plants, constructing 9 conventional nuclear and 12 Pebble Bed Modular Reactor (PBMR) plants, and introducing 5 new oil refineries. Needless to say, renewable energy would play a negligible role in this scenario. The *Required by Science* scenario envisages radical interventions to position South Africa in a post-carbon world. The result would be a 30–40 per cent reduction of CO_2-eq emissions by 2050 from 2004 levels. The scenario views this ambitious programme of decoupling as necessary, but admits it cannot be reliably costed as the required technologies must still mature. The LTMS document was adopted by the South African Cabinet in July 2008, with a commitment to the *Required by Science* scenario as the preferred option. At the Copenhagen Climate Change Summit, in December 2009, South Africa pledged to reduce its carbon emissions by 34 per cent by 2020, and by 42 per cent by 2025 compared to 'business-as-usual' (Trollip & Tyler 2011). Although 'business-as-usual' was not defined, thus making this pledge somewhat vacuous, this pledge did nevertheless have major implications for economic and development policy. However, there is limited evidence that these implications have been registered, in particular after the Copenhagen Summit in 2009, which the Minister of Energy interpreted as giving developing countries such as South Africa a mandate to build more coal-fired power stations (despite the fact that South Africa's CO_2 emissions profile is the same as that of the UK). The NEGP and the IPAP2 did recognise the importance of the carbon issue, but there was no reference to the drastic structural changes required to realise the goals of the LTMS.

Although researchers, civil society groups and the Department of Environmental Affairs have actively championed the climate change cause for at least a decade, the National Treasury has quietly introduced a set of low-level carbon taxes as part of a long-term programme of gradual increases. However, since the NPC took up the issue in 2011, there is evidence that there is a growing realisation that South Africa's carbon-intensive growth path could well become a liability in the near future (Trollip & Tyler

2011). The problem, though, is that Eskom controls and manages the socio-technical regime that dominates the sector. There is little evidence that Eskom understands the global economics and ecology of the transitions underway elsewhere in the world.

Energy

By 2009, South Africa had an installed capacity of 42 GW and 27,000 km of transmission lines. Of the 253,798 GWhr of electricity generated in 2006, 235,548 GWhr (93 per cent) was produced from coal. For the 20 years to 2009 Eskom supplied electricity at the second lowest tariff in the world (5 US cents/kWhr compared to Canada at 6.5, Australia at 6.5 and USA at 9) (Heun, Van Niekerk, Swilling, Meyer, Brent and Fluri 2010). This was possible because of over-investments in generation capacity during the 1980s, an abundance of cheap coal, and under-priced water provided by state-subsidised infrastructures. Despite repeated warnings that demand would outstrip supply during the late 1990s/early 2000s, investments in additional capacity did not take place, resulting in an alarming reduction in the size of the reserve margin (see Figure 8.13).

Since 1996, Eskom has been squeezed between two titanic economic dynamics that ran in diametrically opposed directions. The first has been the commitment to export-led growth. For this to work, South African raw materials and manufactured goods needed to be cheap enough to break into international markets. Given the strength of the trade unions, forcing down wages was not an option, and so resource prices needed to be kept low, in particular energy prices. The second dynamic, however, was government's determination to deracialise the business elite by using state-owned enterprises to leverage so-called 'black empowerment' deals to build privately owned

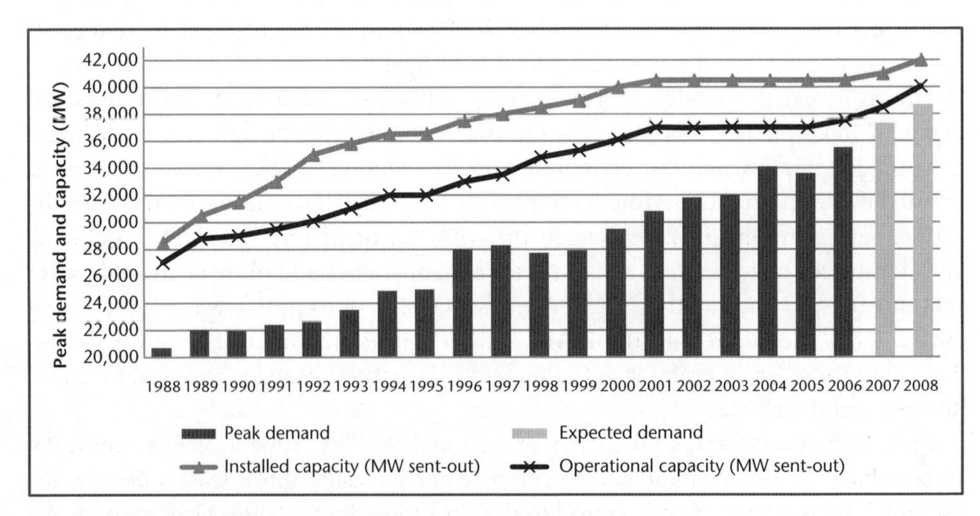

Figure 8.13: South Africa's electricity reserve margins

(Source: adapted from Interventions to address electricity shortages (2008) http://www.info.gov.za/view/DownloadFileAction?id=77837)

Note: Reserve margin equals 25% in 2002, 20% in 2004, 16% in 2006 and 8–10% in 2008.

power stations. For this to work, energy prices needed to be high enough to make this profitable. But this contradicted what the National Treasury thought was good for exports. This paradox resulted in a policy paralysis which culminated in rolling blackouts that cost the economy US$7.1 billion (1.4 per cent of GDP) in 2007/08, price rise shocks (25 per cent increases in 2010, 2011 and 2012) and finally a decision in 2010 to take a World Bank loan to build the third largest coal-fired power station in the world.

By the end of the 1990s China and South Africa had no significant investments in renewable energy. By the turn of the millennium, China had decided to invest heavily in subsidised production of renewable energy systems, while South Africa decided to invest in the unproven Pebble Bed Modular Reactor (PBMR) nuclear technology. By 2010 South Africa had spent US$1.4 billion on this programme with no significant results, leading eventually to its closure in 2010. In the meantime, China became a world leader in solar energy production, including global exports to South Africa and elsewhere. To be sure, the South African government did promote energy efficiency and a feed-in tariff was promulgated, but Eskom's control of implementation meant little had been achieved by 2010 (Heun, Van Niekerk, Swilling, Meyer, Brent and Fluri 2010).

In 2011 the South African government adopted the *Integrated Resource Plan for Electricity, 2010–2030* (commonly referred to as the IRP) (Department of Energy 2011). Although marketed as a strategy to realise the *Required by Science* targets formulated by the LTMS, the IRP leaves unchallenged Eskom's business model—centralised production via a few large plants, dominance of non-renewable energy, grid-based delivery with no storage, and limited roles for independent power producers. The IRP envisages an investment of nearly R1 trillion to increase output from 260 TWh in 2010 to 454 TWh per annum by 2030. This demand forecast is based on the assumption that demand will increase by 10 TWh/y for the period 2010–2030 which is *double* the annual increase in demand in the period leading up to 2010. Nevertheless, this forecasted growth in additional generation capacity is slightly lower than forecasted GDP growth, which means a very small measure of decoupling is in fact anticipated. The forecasted demand will be met by increasing coal-fired generation by 6.3 per cent, nuclear by 9.6 per cent, hydro by 2.6 per cent (mainly imported), gas by 2.4 per cent, diesel-based peak generation by 3.9 per cent, and renewables by 17.8 per cent (more or less evenly divided between wind and solar). This means that renewables will comprise 42 per cent of the new installed capacity that will be built by 2030. The end result will be that by 2030 coal will generate 65 per cent of the 454 TWh that will be generated (that is, down from the current 90 per cent), nuclear 20 per cent, hydro 5 per cent, gas 1 per cent, diesel less than 0.1 per cent and renewables 9 per cent (calculated from Department of Energy 2011).

In a paper commissioned by the NPC, it was demonstrated that the IRP is not in fact aligned with the LTMS, NEGP and the decarbonisation pledge made in Copenhagen in December 2009 (Trollip & Tyler 2011). The main problem with the IRP is that it is based on a demand forecast compiled by Eskom which was, in turn, based on confidential data that was never independently verified. As the NPC-commissioned paper notes, this demand forecast 'doesn't incorporate structural transformation of electricity demand sectors to a low-carbon economy' (Trollip & Tyler 2011: 3). Nor does this demand forecast include a significant role for increased energy efficiency.

The IRP's strategy cannot be regarded as an ambitious programme appropriate for the challenges of our times. Investments in coal-fired generation will remain dominant by 2030, thus crowding out the new, more economically generative technologies, and the targets of the LTMS will not be realised by increasing investments in coal-fired power at the rate that is envisaged (Trollip & Tyler 2011). More worrying is that the financial implications of peak coal and peak uranium production have been ignored (Dittmar 2011; Hartnady 2010; Patzek & Croft 2010; Rutledge 2011). Why the IRP has under-emphasised solar power, which is the only energy source that is projected to drop in price over the long term, can only be explained in terms of the technological 'lock-in' that is reproduced by Eskom's business model and knowledge base. The only reasonably significant counter-knowledge lies within civil society, the universities and certain business networks, but these are largely ignored by government and Eskom. Even the NPC's concerns about carbon intensity have not been translated into the IRP's targets. Non-renewables, by contrast, are extremely vulnerable to rising resource prices as global reserves are depleted (unless, of course, South Africa wants to follow the Chinese example of restricting the export of strategic resources, but this might not affect prices if the mining industry remains privately owned). The IRP is, therefore, potentially a threat to long-term macroeconomic stability. As the NPC-commissioned report concluded:

> *[T]he IRP is not aligned with supporting transition to a low-carbon economy on the demand side and poses a serious risk that post-2030 there will be significant stranded assets.* (Trollip & Tyler 2011: 3)

Oil

After being relatively stable and cheap for nearly two decades from the mid-1980s, the international price of oil followed a rising trend from around 2003, reaching an all-time high of US$147 per barrel in July 2008. Subsequently, the oil price fell dramatically in the wake of the global financial and economic crisis, but since 2009 it rose steadily ending up at US$120 per barrel by April 2011, which is high in inflation-adjusted, historical terms. A fifth of South Africa's primary energy supply is derived from oil (Department of Minerals and Energy 2006), of which 70 per cent is imported, and 30 per cent produced domestically. National crude oil reserves were 15 million barrels as of January 2010 (Energy Information Administration 2009). The transport sector consumes approximately 75 per cent of petroleum fuels in South Africa. Absolute production of petroleum products rose between 1998 and 2009, but the petroleum dependency of GDP has decreased, indicating some level of decoupling (South African Reserve Bank 2010). This is expected to continue which, in turn, eases pressure on the domestic oil refinery industry, which by 2010 was at maximum capacity.

As demonstrated in Chapter 2, the era of cheap oil is over, and oil prices are expected to be more volatile in an era of peak oil production. With a projected decline due to peak oil, which could lie between 2 and 5 per cent per annum (Hirsch 2008), the future volatility of oil prices is expected to be high with major implications for foreign exchange rates and, therefore, the price of fuel in South Africa.

Total liquid fuel supply disruption cost the South African economy an estimated R1 billion per day in 2005 (Department of Minerals and Energy 2007). Future supply shocks should not be ruled out, thus reinforcing the need to reduce dependence on oil imports. Measures to protect current and possible future crude oil reserves should be supported by resource rents paid by the major suppliers from a strategic perspective, ensuring resource asset security. The transport sector, which is the largest consumer at 75 per cent, is arguably the first and most obvious point of intervention. During the course of 2010, the Department of Minerals and Energy called for proposals for the development of the liquid fuels 'roadmap' which will address the challenge of oil supplies. The contract was awarded in 2011 and represents the first really significant opportunity to tackle South Africa's dependence on imported oil.

Minerals

South Africa produces a wide range of minerals, including coal, gold, platinum group metals (PGMs), iron ore, diamonds, copper, nickel, manganese, chromium, uranium and several others. In addition, South Africa is one the four largest producers of rock phosphate, which is an essential mineral for agriculture. Despite the historical significance of minerals, the share of the mining and quarrying sector in the economy as a whole declined from a high of 21 per cent in 1970 to less than 6 per cent in 2009. Nevertheless, mineral products still account for over half of all export earnings and are a significant source of tax revenues (Government Communication and Information System 2010). Production volumes for most of the principal minerals have grown reasonably steadily over the past three decades, the major exception being gold, the output of which has followed a declining trend (see Figure 8.14).

Due to the decline in gold, total mineral production has been almost flat for the period. South Africa boasts over 80 per cent of the world's platinum group metal (PGM) reserves (Government Communication and Information System 2010) and is by far the world's top producer of these. Iron ore is exported via the port of Saldanha Bay and is also used domestically to produce steel. Coal is of particular importance to the South African economy because of its role as the principal energy source. Coal provides over 70 per cent of South Africa's primary energy supply, supports over 90 per cent of electricity generation, and provides feedstock for nearly a quarter of the nation's liquid fuels via Sasol's coal-to-liquid process. Coal is also used in steel production and converted by Sasol into petrochemical products. Furthermore, a quarter of the annual coal output is exported, generating significant foreign exchange earnings. Uranium, another strategically important mineral, has historically been produced in South Africa mainly as a by-product of gold or copper mining, although several dedicated uranium mines are currently under development (World Nuclear Association 2010). Phosphate rock reserves are concentrated in only a few countries and South Africa is

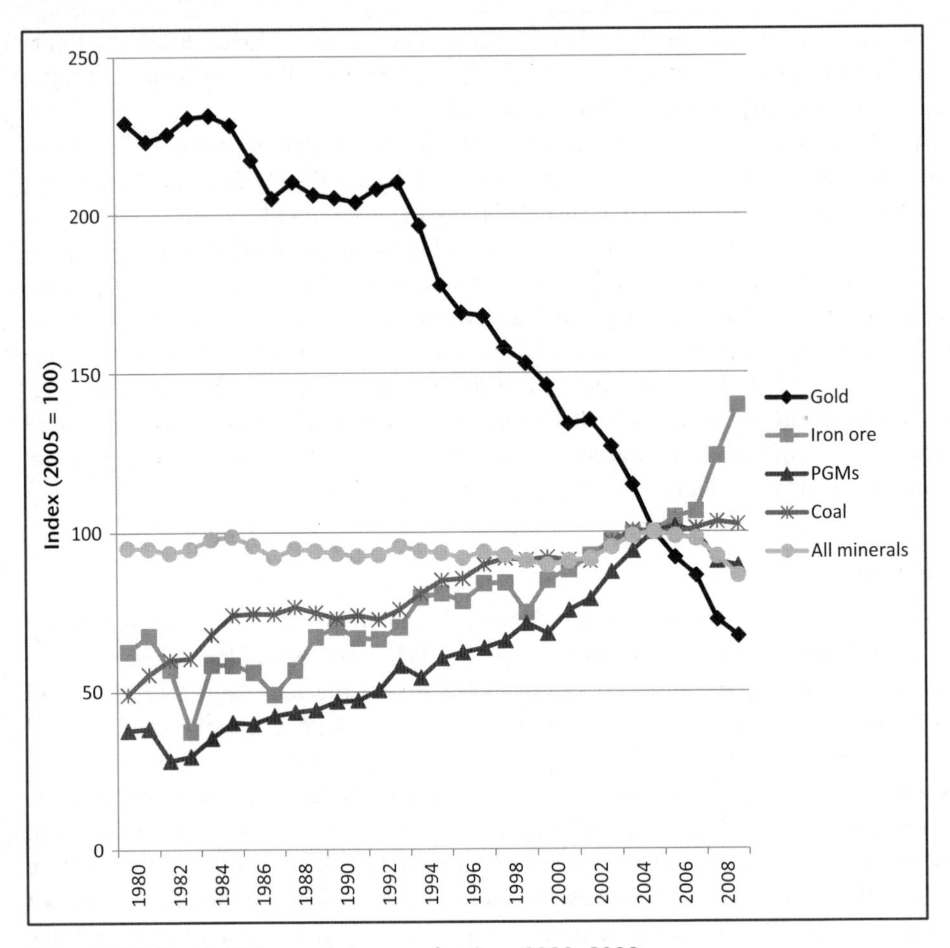

Figure 8.14: South African mining production, 1980–2009
(Source: Statistics South Africa (2010) Statistical Release P2041: Mining Production and Sales.)

in the fortunate position of possessing the world's third largest reserves (9 per cent of the world total) after Morocco/Western Sahara and China. South African production in 2009 was 2.3 million tonnes (United States Geological Survey 2010). Foreign demand for South Africa's export minerals can be expected to continue to grow, especially from BRIC countries.

Demand for gold will likely rise in a global economy plagued by economic crises, and mineral commodities such as iron ore are underpinned by strong economic growth in emerging economies, notably China and India. Domestic consumption of coal increased by an average of 2.5 per cent per annum between 1980 and 2009 (BP 2010), and is set to grow further. Eskom is in the process of returning to service three coal-fired power stations and is constructing two new large coal power stations (Medupi and Kusile). The combined consumption of these five power plants could raise Eskom's annual coal consumption by 50 million tonnes (from 121.6 Mt in 2009). If Sasol were

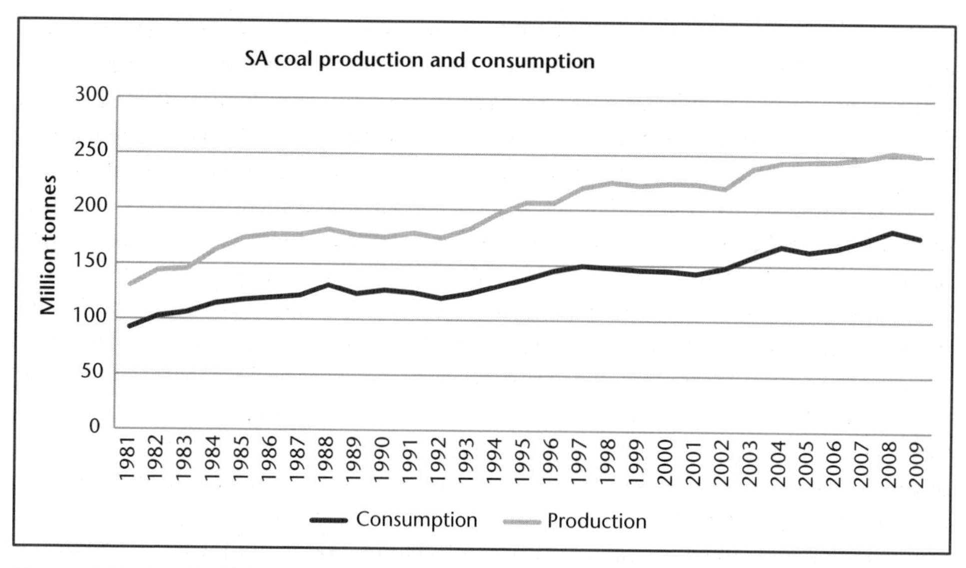

Figure 8.15: South African coal production and consumption, 1980-2008
(Source: BP. (2010). Statistical Review of World Energy 2010. Available from: http://www.bp.com/liveassets/bp_internet/globalbp/globalbp_uk_english/reports_and_publications/statistical_energy_review_2008/STAGING/local_assets/2010_downloads/statistical_review_of_world_energy_full_report_2010.)

to build the mooted new coal-to-liquid plant in the Waterberg area (Mafutha), it would require about another 20 Mt of coal a year (a 50 per cent increase on Sasol's current consumption). Thus the domestic demand for coal could rise by some 70 million tonnes or by 60 per cent over the next decade.

The prices of mineral commodities are determined in international markets. In the case of coal and gold, although South Africa is an important producer, it is a price taker on global export markets. For PGMs, however, South Africa's dominance of world production means that domestic production levels can influence world prices, as was evident, for example, during the 2008 electricity crisis, which curbed production. In recent years, international coal prices have generally tracked the world price of crude oil (International Monetary Fund 2010). The price of gold has recently risen to a nominal record, although in inflation-adjusted terms the metal is still cheaper than it was during the spike of 1980. The platinum price rose substantially in the 2000s, along with many other commodities, but the collapse in demand (mainly as a result of lower demand for automobiles, which use platinum as a catalytic converter) resulted in a sharp — although temporary — drop from 2008 to 2009. In general, robust demand growth from China, India and other emerging economies is likely to continue to support mineral prices over the medium to long term, although further volatility is a distinct possibility. Global phosphorus production from mined phosphate rock reserves could reach a peak and begin to decline from the 2030s onwards, which is likely to lead to the continuation of much higher prices for this essential mineral (Cordell *et al.* 2009).

While demand for and prices of mineral commodities look set to remain strong, there is growing evidence of supply constraints for at least some of South Africa's key minerals. South Africa's annual gold output reached a peak in 1970 at approximately 1,000 tonnes and has been on a downward trend ever since (Hartnady 2009). This is despite massive capital investments, technological improvements, and the recent upward trend in the gold price. The cause is clear: more than 100 years of exploitation and consequently declining ore grades. After being the world's foremost producer of gold in the twentieth century, in recent years South Africa has slipped to fourth position. According to the government, the country's gold reserves stand at 36,000 tonnes (Government Communicationa and Information System 2010), representing some 40 per cent of the global total. However, using mathematical techniques developed by Deffeyes (2001) and historical production data, Hartnady (2009) estimates the ultimate recoverable resources (URR) of gold over the *entire historical period* to be less than 54,000 tonnes, with just under 3,000 tonnes of remaining reserves after subtracting cumulative production through to 2007. This implies that the Witwatersrand goldfields are approximately 95 per cent depleted and production is likely to fall permanently below 100 tonnes per annum within the next 10 years (Hartnady 2009: 329).

The production of uranium associated with gold mining can be expected to continue to decline as gold reserves dwindle. However, there appears to be significant scope for new uranium mining. According to the World Nuclear Association (World Nuclear Association 2010), South Africa had 435,000 tonnes of known recoverable sources of uranium, representing 8 per cent of the world total as of 2007. However, a recent review predicts that global uranium production will peak by 2015, which will obviously push up prices and undermine long-term strategies for nuclear generation (Dittmar 2011).

South Africa's phosphate rock reserves are estimated at 1.5 billion tonnes (United States Geological Survey 2010), which would last 650 years at the 2009 rate of production (2.3 million tonnes per annum). However, phosphate prices have rocketed in recent years (see Chapter 6) which again raises questions for South Africa: will phosphate prices float upwards with market trends or, as in China, be regulated to ensure domestic food security?

The case of coal reserves and potential future production is even more controversial than that of gold. The official government figure for reserves was revised sharply downward from nearly 50 billion tonnes in 2003 to approximately 28 billion tonnes in 2007 (Hartnady 2010). However, recent peer-reviewed research casts doubt on even this latter figure. Rutledge (2011) estimates that remaining reserves of coal in southern Africa may be as low as 10 billion tonnes. Similarly, using the Hubbert Linearisation technique, Mohr and Evans (2009) estimate South Africa's ultimately recoverable reserves (URR) of coal at 18 Gt (including 8 Gt of cumulative, historical production which means there is around 10 billion tonnes left in the ground), which yields a forecast of peak annual production in 2012. Using similar techniques and historical production data, Hartnady (Hartnady 2010) estimates that coal output is likely to peak by 2020. Patzek and Croft (2010), recognising that over time the quality and energy content of mined coal deteriorates, estimate that South Africa's coal production from existing mines, when measured in energy units, peaked in 2007. They further contend that future new mines are unlikely to reverse the trend given the long lead times required for their development, and the fact that the most accessible and high quality coal reserves tend to be mined earlier on. Hartnady (2010) points out various problems with the relatively undeveloped Waterberg coal field, including geological and water constraints which means the cost of coal from this last large reserve will be higher than anticipated.

Although other mineral deposits in South Africa (for example, PGMs) are less depleted than gold (and possibly coal) reserves, it is merely a matter of time before they too run into production constraints as a result of their non-renewable, finite nature. In addition to the physical resource limits of the various minerals themselves, two other factors could potentially constrain future mineral production, namely constraints on the country's supplies of energy and water—both of which are used extensively by mining. The electricity crisis in 2008 clearly demonstrated the energy-intensiveness of mining, and enforced production cuts at a number of mines. Eskom has warned that electricity supply will be under severe pressure for at least the period 2011 to 2016, which could put a brake on growth. Furthermore, as mineral ore grades decline over time, the amount of energy required to produce a unit of mineral output will rise (Bardi 2008). Water scarcity could limit future production of both coal and PGMs (Glaister & Mudd 2010).

In addition to resource scarcity and depletion, another critical sustainability issue facing the mining sector is environmental degradation. This involves land degradation, but also pollution of water resources with toxic metals and acid mine drainage, which are reaching critical proportions in areas surrounding Johannesburg (Turton 2008). The Department of Mineral Resources recently acknowledged that at current rates, it would take 2,900 years to rehabilitate South Africa's derelict and abandoned mines alone (*Business Day*, 2010).

To conclude this discussion of minerals it might be worth noting the warnings by Rutledge and Hartnady that historically, estimates of recoverable reserves tend to remain overly optimistic while actual extraction rates are rising, but rapidly reduce or even collapse when depletion (often previously denied) results in declining levels of extraction (Hartnady 2010; Rutledge 2011).

Water and sanitation[6]

With an average annual rainfall of 497 mm South Africa is a dry country, and 98 per cent of available water resources have already been allocated (Department of Water Affairs and Forestry 2004). This means that 'South Africa simply has no more surplus water and all future economic development (and thus social wellbeing) will be constrained by this one fundamental fact that few have as yet grasped' (Turton 2008: 3). This is why

[6] This section relies on the following documents: Ashton & Turton 2008; Department of Water Affairs and Forestry 2006; Republic of South Africa, Department of Water Affairs and Forestry 2004; Republic of South Africa, Department of Water Affairs and Forestry 2002; Turton 2008, plus additional research by Turton commissioned by the Sustainability Institute in 2010.

water prices in the sub-Saharan region are expected to rise as much as 40 per cent in the medium and long term (Muller 2007).

Cheap water (and energy) has historically been the backbone of macroeconomic growth in South Africa, which has involved externalising costs on social-ecological systems (Adler *et al.* 2007). The total national water resource, including mean annual runoff and groundwater is 59.5 Bcm in non-drought years (WWR 2005: Middleton & Bailey 2008). Full supply capacity of dams is 31.721 Bcm, translating into a surface capture of 64.4 per cent (Middleton & Bailey 2008). Demand pressure for economic growth is greatest on the Limpopo and Orange River basin systems, which sustain 70 per cent of the total national economy. South Africa also captures the major portion of the water shared with Lesotho, Botswana and Namibia, Swaziland and Mozambique in trans-boundary river catchments.

However, there is evidence that critical water availability and eco-system reserve limits have been breached. Eighty-two per cent of freshwater eco-systems are officially classified as threatened, 50 per cent of wetlands have been destroyed and 36 per cent of fish are threatened (Driver *et al.* 2005). Surface capture of 64.4 per cent is above the 60 per cent margin, above which ecological problems can be expected (O'Keeffe *et al.* 1992; Rabie & Day 1992). In both the high and low water-use scenarios evaluated in the National Water Reconciliation Strategy (Department of Water Affairs and Forestry 2004: see Tables 2 and 3), shortfalls are projected, that is, 2,044 Mm^3/yr by 2025 in the high-use scenario and by 234 Mm^3/yr by 2025 in the base scenario, respectively.

The country therefore has no further 'dilution capacity' when it comes to absorbing effluents in its water bodies. The Johannesburg–Pretoria complex — South Africa's most significant urban-economic conurbation — is located on a high-altitude watershed, which means that outflows of waste water pollute the downstream water resources on which Gauteng depends for its water supplies. The result is that after China, South Africa's national water resources contain some of the highest toxin levels, in particular mycrocystin, for which no solution currently exists. Cyanobacteria blooms, caused by end-of-pipe NPK loads, threaten national water security. Inter-basin water transfers have degraded the ecological integrity of aquatic systems, and radionuclides, heavy metals and sulphates from mining activities have polluted valuable water resources. In short, the combination of low average rainfall, over-exploitation and re-engineered spatial flows have led South Africa to an imminent water crisis in quantity as well as quality (Turton 2008).

The Water Crowding Index (WCI), used by the Department of Water Affairs and Forestry to assess population needs and water availability in different catchments, shows that South Africa is projected to be over twice the limit associated with social cohesion and stability by 2025 across all major South African river basins (Department of Water Affairs and Forestry 2004). This will be exacerbated by climate change. Climate change projections all point to greater variability in temperature, precipitation and wind. While precipitation may be projected to increase in the eastern escarpment of the country, it will decrease in the western half.

The National Water Act is the key policy framework governing water in South Africa, and in a sense, establishes the basis on which the resource rentals for water are established.

The mining sector is not subject to this policy. If it were, this would transform the mining sector into the largest investor in pollution reduction using new technologies, including technologies that make phosphate removal viable, as well as the reversal of endocrine disruption via chemical removal and extensive mine-water treatment. The biggest challenge might well be future technologies which could productively reuse acid mine drainage by decanting it as feedstock for various productive purposes. A policy on resource rents to force mining into water management needs to take into account the large amounts of mining-related waste that is currently classified as 'spoil' and thus as a future resource. This is the single, largest waste stream in the country (which has very large remedial costs for biodiversity restoration and agricultural systems) and remains invisible to waste management legislation (Adler *et al.* 2007; Godfrey *et al.* 2007; Oelofse, Hobbs, Rascher & Cobbing 2007; Oelofse 2008a & 2008b).

There is scope for major water saving in two sectors — urban and domestic use, and the agricultural sector. Recycling urban waste water is an urgent priority. For example, between 40 per cent and 50 per cent of all water piped into households is used to flush toilets. Yet it is technically possible to flush toilets from on-site grey-water resources (in particular for large middle-class homes), or via neighbourhood-level closed loop systems which recycle treated water back into households. Rainwater harvesting and grey-water supplies for irrigation also have potential. The second major water-saving priority is in agriculture, especially in combination with organic farming methods that simultaneously rebuild the biological capacity of soils, sequester carbon and protect moisture levels in the top layers of the soil.

The government is aware of these severe water-supply constraints. Since her 2007 Budget Speech, the Minister of Water Affairs and Forestry has dedicated considerable space to her water-efficiency campaign, with apparent emphasis on regulations and tighter controls. But unless more immediate and drastic action is taken, economic growth will soon be undermined by water shortages and related dysfunctionalities (such as salinisation of aquifers). The research results are clear: available physical extra capacity in 2000 was at most 1.7 per cent higher than existing requirements, while growth in water demand could be as much as 25 per cent higher than available yield by 2025. Even if demand increases by only 1 per cent per annum, by 2014 the economy will already be facing severe shortages on a number of fronts. By 2019, water shortages will have pulled the economy into a downward spiral of low growth and growing socio-economic inequalities, with associated mini-'resource wars' over water supplies.

Solid waste[7]

Solid waste includes all municipal, mining and industrial waste. As of 2005, the solid waste system managed the disposal of 20 Mt[8] of municipal solid waste (MSW), 450 Mt of mining-related wastes and 30 Mt of power-station ashes. MSW quantities are

[7] Based on Von Blottnitz 2005.
[8] Mt =1 million metric tonnes or 1 billion kg.

growing faster than the economy in many cities.[9] The typical daily average of 2 kg per person is higher than that in many European cities. Both the quantity and nature of solid waste differs considerably across the socio-economic spectrum. A person in an informal settlement generates on average 0.16 kg/day, whereas a person in an affluent area will dispose of over 2 kg/day. Food and green waste makes up 35 per cent of waste in affluent households, compared with 20 per cent for poor households. In Cape Town 60 per cent of industrial waste is recycled, compared to only 6.5 per cent of residential and commercial waste (among the lowest in the world). There is no reason to believe that the situation is much different in other South African cities.

While many countries have moved away from 'disposal-to-landfill' as the primary means of solid waste management, in South Africa the bulk of MSW is disposed of in landfill sites spread across the country. Although national costs have not been calculated, they are probably similar to those in Cape Town, where the cost of managing landfills — and related dumping — doubled between 2000 and 2004. Recent research revealed that whereas landfill costs in 2005/06 were R54 per tonne, they had increased to R192 per tonne by 2007/08 and are projected to increase to R235 per tonne by 2013 (De Wit 2009).

Growth in the minerals and coal-based energy sector leads directly to increased industrial wastes with limited productive recycling and reuse — a clear example of the way unsustainable resource use is coupled to growth and poverty reduction (Adler *et al.* 2007; Oelofse, Hobbs, Rascher & Cobbing 2007). Yet technologies and processes for decoupling waste from growth and poverty eradication are simple, low cost, job creating and extensively used throughout the world.

Waste recycling represents one of the most available, immediate, tangible and low-cost investments in decoupling. It saves on capital costs, creates jobs, and forces the middle classes to take greater responsibility for the resources they throw away. It is also normally a highly competitive sector, with sophisticated value chains with respect to resources such as used engine oil, used vegetable oils, a wide range of plastics, building rubble, organic matter for composting, glass, cans and paper. Even e-waste is becoming a resource input now. Numerous studies confirm that recycling has very positive economic benefits with respect to job creation, manufacturing, technology and innovation. Furthermore, waste recycling also has significant export potential (Botha 2007).

The National Integrated Waste Management Act adopted by Parliament in 2009 will force every local government to prepare an Integrated Waste Management Plan with defined targets for recycling, thus paving the way for a recycling revolution in South African cities. The stage is now set to move South Africa decisively into a post-disposal approach with respect to MSW, with a special focus on middle- and high-income consumers. The Mineral and Petroleum Resources Development Act (2002) makes specific provision for waste management and pollution control in the mining sector. This Act, together with the emerging MSW approach, provides the basis for the emergence

[9] For example, in Cape Town MSW grew by 7 per cent per annum up until 2008, after which it has started to decline.

of a vast, decentralised network of market-driven and community-based recycling businesses. In addition, the National Cleaner Production Strategy is being beefed up, establishing incentives and legal requirements to stimulate cleaner production systems (CPS) in the business sector — particularly mining and construction — with a special focus on investments in recycling enterprises.

Soils[10]

South Africa falls within the so-called 'third major soil region' typical in mid-latitudes on both sides of the Equator. The result is that South Africa is dominated by very shallow sandy soils with severe inherent limitations for agriculture. Only 13 per cent of the land is arable and just 3 per cent high-potential land. The result is overexploitation and the use of inappropriate farming methods, as South Africa tries to exceed its soils' capacity to meet growing food requirements. All this has resulted in far-reaching, nationwide soil degradation. Although no detailed study has been done, it is estimated that 5 million of South Africa's 14 million hectares of arable land are degraded.

Water erosion remains the biggest problem, causing the loss of an estimated 25 per cent of the nation's topsoil in the past century (although erosion rates may be declining, not because the problems are being resolved, but because there is less and less to erode away). Other factors include: wind erosion affecting 25 per cent of soils; soil compaction due to intensive mechanised agriculture; soil crusting caused by overhead irrigation systems; acidification of more than 5 million hectares of arable land, caused by poor farming practices, particularly incorrect fertiliser and inadequate lime applications; soil fertility degradation resulting from annual net losses of the three main plant nutrients (nitrogen, phosphorous and potassium); soil pollution caused by various human practices; and urbanisation often spreading across high-value arable land on the outskirts of cities.

Once land is degraded, there is little potential for recovery. Areas in which degradation is limited must be prioritised so that efforts can be focused on prevention via appropriate farming practices. Reversing the above trends will require locally trained soil scientists who recognise that South Africa's soil conditions are unique (because they are 'third major soil region' soils) and therefore we cannot copy solutions generated in countries with a different soil profile. Location-specific technical solutions are required as blanket solutions have proven unworkable. Locally-trained soil scientists must work together with local leader farmers via horizontal learning practices. This has worked in India, Cuba and many other places in the developing world and is urgently required in South Africa.

Biodiversity[11]

South Africa is recognised globally as the third most biologically diverse country in the world, yet this diversity is one of the most threatened on the planet (see Table 8.1).

[10] Based on Laker 2005.
[11] Based on Driver *et al.* 2005.

Table 8.1: Key threats to South Africa's eco-systems

	Officially classified as threatened	*Main issues and causes*
Terrestrial eco-systems	34%	• Degradation of habitat • Invasion by alien species
Freshwater eco-systems	82%	• Pollution
• Wetlands destroyed • Fish threatened	50% 36%	• Over-abstraction of water • Poor water course condition
Marine eco-systems • Estuaries endangered	65% 62%	• Climate change • Unsustainable marine harvesting • Seabed destruction by trawling • Coastline urbanisation • Marine pollution

Significantly, this concerns not just the prevalence of plant and animal species, but also critical eco-systems which provide vital services to human society.

Although South Africa has invested enormous public, private and community resources into the expansion of protected areas, conservation areas and reserves, in future innovative partnerships will be required to ensure that the burden for all this is not carried entirely by the fiscus. To this end, the Protected Areas Act offers a unique opportunity. It provides for any land, including private or communal, to be declared a formal protected area, co-managed by the landowner(s) or any suitable person or organisation. This means that formal protected area status is not limited to state-owned land, and that government agencies are not the only organisations that can manage protected areas, opening the way for a range of innovative arrangements not previously possible. A related challenge is to make the links between protected area development, sustainable tourism, and benefits to surrounding communities who should be key stakeholders in protected areas.

The National Environmental Management Act provides for a comprehensive regulatory framework for the protection of key environmental resources. The core instrument used to give effect to this Act is the Environmental Impact Assessment (EIA). Although development projects must be subjected to an EIA, the focus is on costs of pollution and environmental impacts, and not resource inputs and prices. This does not provide a sufficient basis for decoupling over the long run.

Food[12]

Environmental degradation, high dependency on fossil fuels and marginalisation of small farmers have exacerbated the problem of food insecurity. Fifty-one per cent of South African households experience hunger; 28 per cent are at risk of hunger; in urban areas 70 per cent of poor urban households reported conditions of significant and severe food insecurity; and 50 per cent of women and 30 per cent of young men are obese, and

[12] This section is derived from the contributions by Candice Kelly and Gareth Haysom to the green economy research commissioned by the Sustainability Institute (Peter et al. 2010).

follow unbalanced diets, high in saturated fats and sugars, leading to chronic disease among the poor.

The main causes of a limited supply of affordable food include a shortage of high-fertility soils, the reduction in the number and scale of agricultural subsidies since 1994, and the failure of land reform to successfully generate a substantial class of black commercial farmers. Affordable and accessible water will be affected by current and projected climate change patterns as soil moisture levels decline and rainfall patterns change — agriculture currently consumes over 70 per cent of the national water supply, yet contributes only 4 per cent of GDP and employs only 8.5 per cent of the workforce. Energy and oil price increases and related price volatility will have negative impacts on farmers, who struggle to prevent profit margins from shrinking in order to survive. Pressures exerted by global and local retailers have artificially lowered food prices for farmers, leading to the global downward trend in farming, which is now globally one of the professions with the highest suicide rates in both small-scale and commercial-scale farmers. More food is sold to South Africans (just over 50 per cent) through commercial supermarket chains than any other country in the world (Reardon *et al.* 2003). This gives these retail chains extraordinary market power to determine prices.

Policy responses

The evidence clearly shows that a powerful set of diverse global, national and local pressures for change affect a wide range of systems and sub-systems: climate regulation, the energy regime, oil supplies, minerals extraction and use, water use, soils and the myriad eco-systems that make up South Africa's rich biodiversity. The solid waste and food supply regimes are more complex emergent outcomes of a range of interconnected sub-systems, which are, in turn, subject to change (from transportation and landfill space for solid waste, to soils, climate and water supplies for food production). Because each of these so-called 'sectors' are understood via specialised professionalised 'disciplines', they are rarely brought together in this way to provide a broad generalised understanding of the serious pressures experienced by a wide range of strategic socio-technical regimes on which the wider system of economic production and consumption is dependent.

If macroeconomic policy-making ignores the resource-depletion effects and environmental impacts of current modes of production and consumption, there will be little understanding of the reasons why many of the long-term growth, development and employment targets will not be attained. More significantly, investments in strengthening adaptive capacities to manage the transition will be neglected. However, as in many similar countries in the developed and developing world, what is interesting is that there is clear evidence of an emerging trend in South Africa's national policy discourse which is calling for more responsible use of natural resources, albeit in relatively uninformed ways for certain resources. Growing numbers of policy statements acknowledge that the country's economic growth and development path is too resource intensive and that this needs to change (see in particular Chapter 12.2 of the IPAP2). Unfortunately, this way of thinking is by no means a dominant paradigm in policy-making circles.

Section 24(b) of South Africa's new Constitution provides the point of departure for the National Framework for Sustainable Development (NFSD) adopted by the Cabinet in June 2008 (Department of Environmental Affairs and Tourism 2007). During the course of 2009, the Cabinet mandated the Department of Environmental Affairs to transform this policy framework into a more binding 'strategy', hence the commencement of the complex process of formulating the *National Strategy for Sustainable Development* (originally planned for adoption in 2010, but still under discussion by early 2011). However, key macroeconomic policy documents made no reference to Section 24(b) of the Constitution, the LTMS, the NFSD or the mountain of evidence contained in the government's 2007 *State of Environment Outlook* (Department of Environmental Affairs and Tourism 2007).

National Framework for Sustainable Development (NFSD)

The NFSD was adopted by Cabinet in June 2008. In sharp contrast to macroeconomic policy, it explicitly acknowledged the growing stress on environmental systems and natural resources from economic growth and development strategies, and mapped out a vision and the following five 'pathways' to a more sustainable future:

1. Enhancing systems for integrated planning and implementation
2. Sustaining our eco-systems and using resources sustainably
3. Investing in sustainable economic development and infrastructure
4. Creating sustainable human settlements
5. Responding appropriately to emerging human development, economic and environmental challenges.

The NFSD committed South Africa to a long-term programme of resource and impact decoupling. The government resolved in 2009 that the NFSD should be converted into a full-blown National *Strategy* for Sustainable Development by the end of 2010, including specific targets, commitments and budget allocations under the above headings.

Ad hoc policy shifts

Because resource constraints and negative environmental impacts are material realities, they show up in all sorts of unanticipated ways. Although some of these unanticipated consequences are reflected in prices which, in turn, loop back into production costs and consumption prices, others are deeper and more long-term with consequences that get reflected in less economic indicators, such as quality of life, the wealth of future generations, the burden of disease, and youth under- or unemployment. What is interesting is that there has been a wide range of policy shifts over the last few years which reveal that government and stakeholders are responding in ad hoc sectoral ways to these underlying resource constraints and environmental impacts without necessarily developing the institutional capacities and knowledge sets to manage a more holistic, nationwide transition. Examples include the following:

- The passing of the National Integrated Waste Management Act in 2009, which provided for the introduction of the globally recognised '3R' framework (reduce, recycle and reuse).
- The National Environmental Management Air Quality Act of 2004 was a response, albeit lacking in several respects, to the declining quality of the air that South Africans breathe, in particular in urban areas, resulting in rising health costs to treat respiratory diseases.
- The introduction by the Department of Agriculture of a 'land-care' programme aimed at finding ways to rehabilitate soils, although virtually no funds are allocated for this task—contrast this to Cuba and China. Cuba made soil rehabilitation via organic farming methods a top priority after the fall of the Soviet Union in 1990 (Wright 2008), while China decided to incentivise organic farming methods as a key element of its rural development programme. Unfortunately, South Africa's 2009 Rural Development Strategy makes no mention of the need to rehabilitate South Africa's soils, even though it assumes that millions of people will make a living cultivating these soils.
- The National Water Resource Strategy has been a focus of the Department of Water Affairs (DWA) for a number of years, but a greater sense of crisis is evident in the work that is going into the Water for Growth and Development strategy, which acknowledges that South Africa has a limited water supply which will negatively affect future growth and development strategies.
- A Renewable Energy Feed-In tariff was introduced in 2009, which made it possible, in theory, for investors to establish renewable energy plants that will feed energy into the national grid in return for payments from Eskom. In practice, however, little progress has been made due to policy and regulatory uncertainties and Eskom's refusal to buy renewable energy at higher prices than coal-based energy.
- In 2009, the National Department of Human Settlements introduced a national programme to bring sustainable resource-use criteria into the design of the settlement projects and houses that it subsidises across the country, with special reference to issues such as densities, orientation of the buildings, roof overhangs and insulation, installation of solar water heaters, and sustainable use of water and waste resources.
- The National Treasury's Carbon Tax Discussion Document released in 2010 makes it clear that the National Treasury wants to use taxation to reduce carbon intensity and incentivise green economy investments—carbon taxes have been included in national budgets since 2007 and a commitment to gradually increase carbon taxes in order to manage the transition to a low-carbon economy via a slow and gradual decrease in the carbon intensity of the economy.
- The Department of Science and Technology adopted what has come to be called the Global Change Grand Challenge policy framework which provides for a 10-year investment in sustainability science, which will develop the skills sets and knowledge base for designing and implementing more sustainable production and consumption systems.
- There has been a big shift within the transport sector in favour of large-scale investments in public transportation systems, which will, hopefully, gradually lead

to much more aggressive measures to retard the growth of private car consumption and use.

- The significant moves emanating from the Cabinet's Economic Cluster resulted in the formulation of a 'green economy' approach, including the Green Jobs and Industries for South Africa initiative (known as Green-JISA) and the 'green economy' component of the NEGP. Much now will depend on the capacity of the Ministry of Economic Development to lead a sustained effort to ensure that the 'green economy' is not reduced to a 'sector' but is seen as a macroeconomic policy perspective that is translated into specific investments. Although this is a brand new ministry with limited internal capacity, the key institutional resources available to it are the state-owned financial institutions (including the Industrial Development Corporation and Development Bank of Southern Africa) that fall within its policy mandate.
- Section 12.2 of the much praised Industrial Policy Action Plan, released in February 2010 by the Minister of Trade and Industry, refers specifically to the need for 'green and energy-saving industries' because rising energy costs 'render our historical capital and energy-intensive resource processing based industrial path unviable in the future'.
- The slow and painful process of formulating and adopting a National Climate Change Policy Framework was largely driven by the need to position South Africa as a player in the global climate negotiations. Unfortunately, this policy is caught between the Department of Energy, which effectively operates in accordance with a script written by Eskom, and the Department of Environmental Affairs, which is home to the those involved in global climate change negotiations, who are mandated to commit South Africa to the *Required by Science* scenario defined by the LTMS.
- One positive thing that the Department of Energy did in December 2009, was to release the Solar Hot Water Framework which promised to ensure the delivery of 1 million solar water heaters. These devices could eliminate up to 50 per cent of household electricity, generate many jobs all along the value chain and attract large amounts of carbon finance (Mokheseng 2010). Unfortunately, constantly changing subsidy levels, regulatory uncertainty and consumer ignorance combine to undermine this potential.
- As far as key eco-system services are concerned, significant progress has been made, including the passing of the National Environmental Management Biodiversity Act (NEMBA), the National Environmental Management Protected Areas Act (NEMPAA) in 2004, and the National Environmental Management: Integrated Coastal Management Act in early 2009.
- Connections between sustainable resource use and livelihood creation are recognised in programmes such as Working for Water, Working for Wetlands, Working for Land, Working for Energy, Land Care, Coast Care and the Integrated Sustainable Rural Development programmes. Largely funded by the state, these programmes create employment and address key environmental challenges (such

as alien tree removal, fire breaks and coastal conservation). Working for Water is a major success story combining capacity to manage a well-established knowledge network rooted in practice and well-funded research institutions and NGOs.

All these initiatives can be traced in one way or another to resource constraints and negative environmental impacts which affect growth and development to a lesser or greater extent. These pressures affect a wide range of regimes with varying capacities to respond.[13] Given South Africa's energy and carbon-intensive economy, the energy regime is strategically central. Although it is a sector with high-level capacity to manage change it has weak links to external knowledge sets — a situation which makes it possible for Eskom to control the transition on terms that will not threaten existing vested interests. By contrast, the biodiversity sector has significant adaptive capacity, but is heavily dependent on an external knowledge network, which means a more fundamental transition may well be possible in this sector. Waste recycling will clearly run into problems due to weak adaptive capacity at municipal level and weak links to external knowledge resources (which are, in turn, limited to only a few consultants with expertise and no major university-based research programme). Green jobs might be an example of a transition driven by limited state capacity to coordinate actions, thus opening up space for knowledge sets drawn from the private sector, NGOs and the state-owned financial institutions.

Decoupling: Opportunities for action

Perhaps the most significant prospect for a purposive transition in South Africa is the massive injection of public and private investment funds to drive a vast multi-year infrastructure investment programme that will cost between R800 billion and R1 trillion over the period 2007–2014. A cornerstone of the government's long-term growth strategy, this national programme offers a unique opportunity to advance towards a more sustainable future. There is no doubt that public investment in infrastructure is a powerful way to ensure that growth sets up the conditions for meaningful poverty reduction. But there are two key questions.

The first is whether these investments address the challenges discussed above. There are some obvious positive investments, such as in public transport, upgrading of the rail infrastructure, and sustainable approaches to housing and waste. These are already government priorities. There are also some obvious gaps, for example, investments in soil rehabilitation, sustainable water and sanitation, air quality and renewable energy on scale.

The second question is less about *what* is being built, and more about *how* it will be built. There is an enormous opportunity to design and build low-carbon infrastructures and buildings that could contribute significantly to resource decoupling. Furthermore, the way infrastructures and buildings are developed on scale could be the single biggest catalyst to drive a long-term commitment to sustainable resource use which, in turn,

[13] For a significant attempt to theorise the relationship between transition and governance, see Smith *et al.* (2005).

frees up resources for poverty eradication. Finally, doing things in new ways opens up a wide range of new value chains that could be exploited by new entrants into the sector with major employment-creation opportunities — the best example being the potential of solar water heating.

Conclusion

In a resource-depleted world at a time of rising resource prices, South Africa will undoubtedly exploit its natural resources to generate private and public revenues. The same is true for many other resource-rich developing countries. Now that the BRIC countries have joined the traditional developed economies in the global scramble for resources, rising resource prices translates into rising state revenues that can, in theory, be redeployed to address the poverty challenge, build infrastructure and invest in non-extractive job-creating economic activities in the secondary and tertiary sectors. This is, in essence, what economic development means today. In practice, as argued by Collier and more generally by the institutional economists, the conversion of resource rents into development depends on the political commitments of the government, the institutional accountability of policy elites and the capacity of state agencies to implement coherent programmes (Collier 2010; Evans 2005; Rodrik *et al.* 2004).

The South African case suggests that there are two problems with this mainstream approach. The first is that a resource-extractive growth path builds up a powerful 'mineral–energy complex' which develops the institutional, financial and knowledge capacity to successfully reproduce itself at the expense of other more labour-intensive and/or innovation-driven sectors. Using highly technocratic language to justify their strategies and the 'too big to fail' logic to intimidate decision-makers, the policy networks that represent the MEC are able to justify massive expenditures and subsidies that effectively 'crowd out' alternatives which are often significant job creators. Combine this with state-supported black empowerment deals in the mining sector, and the result is spreading rent-seeking behaviour that weakens governance capabilities. These are not the conditions that stimulate innovation-driven investments in new job-creating enterprises.

The second is the contradictory consequences of resource depletion, especially when the use of a particular resource hits a peak of some kind (whether this is for a non-renewable resource such as minerals or a renewable resource like water). This can be contradictory because prices rise as the peak approaches, creating an illusion of long-term prosperity in the case of minerals (which encourages over-optimistic projections about reserves), but the opposite in the case of a public resource such as water. In theory, it should be possible to respond to this by planning and plotting a transition to a mode of production and consumption that is less resource and energy intensive (as has happened in some other countries). In practice, the dominance of the 'mineral–energy complex' makes even this minimalist position very difficult to achieve, never mind the more ambitious restorative strategies that are actually required.

Although we discussed a wide range of natural resource constraints to South Africa's current resource- and energy-intensive growth path, we identified three that are of particular significance: water resources, peak coal and limits to carbon space. These three alone should be driving a high-priority concerted effort to map out alternatives, including raising big question marks about the wisdom of investment in more coal-fired and nuclear power stations. Unfortunately, the IRP falls far short of this ambition.

Finally, we have also suggested rather optimistically that there is a double movement afoot. As South Africa becomes increasingly unsustainable by massively escalating its dependence on carbon- and water-intensive, coal-fired energy generation and nuclear power, quite a number of key policy documents do recognise that the unsustainable use of resources is a key obstacle to future growth and development. The LTMS, NEGP, IPAP2 and the Diagnostic Overview of the NPC have all recognised in one way or another that something is wrong with current trajectories. These documents, together with the range of sectoral responses, clearly reflect the fact that key stakeholders have started to respond to some underlying realities that have hitherto been consciously filtered out as inconvenient truths.

The double movement referred to here reflects a dim realisation that the nature of the developmental state needs to be reconceptualised. As argued in Chapter 4, development economics on its own — even with an institutional orientation — is simply not fit for purpose any more. If the insights from ecological economics are incorporated, it becomes possible to conceptualise a developmental state which promotes sustainability-oriented innovations that create jobs by working with, rather than against, natural systems. Decoupling growth and development from escalating resource use in this way would, in turn, create the basis for a new over-arching national strategy that positions South Africa as a potential leader of the next — hopefully more sustainable — long-term development cycle. The alternative is a recipe for sluggish growth through intensified resource exploitation and debt-funded consumption that works for the elites but fails to create the job opportunities and material basis for poverty eradication. The long-term outcome would be a 'rentier state' wracked by resource conflicts as elites fight over the remaining scraps.

Chapter Nine

Decoupling, Urbanism and Transition in Cape Town

Introduction

We argue that the transition to the next long-term development cycle will be shaped by two dynamics: the modes of governance that emerge in response to the global crisis (Chapter 4), and the restructuring of urban space (Chapter 5). In this chapter we explore these themes by taking a closer look at how they play out in one particular city, namely Cape Town. The struggles for an inclusive urbanism after democratisation in 1994 initially ignored key resource constraints, but now there is a need to find alternatives that combine a more sustainable use of resources and practically address the poverty challenges. In this chapter we suggest practical strategies for reconfiguring urban infrastructures to decouple growth and development from escalating resource use.[1]

Since democratisation successive political coalitions have competed for the political support of the previously disenfranchised black majority by attempting to extend services infrastructure to nearly every household in Cape Town. The challenge has been to reconcile differentiated levels of service with a commitment to inclusive urbanism in order to realise the vision of a post-apartheid non-racial city. This inclusive urbanism agenda was executed via the City of Cape Town's[2] large, institutionally mature, technical services bureaucracies who were committed to using tried and tested conventional technologies, coupled with institutional innovations which allowed for a limited role for market-based delivery mechanisms. However, for a range of community organisations active in informal settlements, inclusive urbanism was not enough because top-down delivery did not necessarily translate into bottom-up empowerment.

Over time, however, the limits to inclusive urbanism have become increasingly apparent: fiscal constraints have stimulated projects that resemble splintered urbanism, and ecological constraints have resulted in an increasing interest in green urbanism to make inclusive, splintered and even slum-type urbanisms more sustainable. Cape Town's

---

[1] This chapter is based on an extensive three-year research project on Cape Town's urban infrastructures, with special reference to energy, waste, water and sanitation. Called the Integrated Resources Management for Urban Development project, the results were published in reports commissioned by the Sustainability Institute and supervised by the authors (see De Wit et al. 2008; Sustainability Institute & E-Systems 2009a; Sustainability Institute & E-Systems 2009b; Sustainability Institute & E-Systems 2009c). Funding for the preliminary phase of the research was provided by the Isandla Institute (main results published in Swilling 2006) and subsequently by United Nations Development Programme (UNDP) and DANIDA. Some sections of this chapter are drawn from a published paper co-authored by Mark Swilling and Martin de Wit (Swilling & De Wit 2010).

[2] City of Cape Town is the formal name of the municipality that governs Cape Town (henceforth referred to just as CCT). The CCT is a strong metropolitan government structure.

property industry responded by setting up the Green Building Council (GBC), which modified Australia's Green Star Rating Tool for application to commercial buildings in South African conditions. Inspired by the 'minimising damage' ethos, the GBC is at the vanguard of South Africa's nascent green urbanism movement, with significant backing from the property and financial services sectors.

We will explore these dynamics and the patterns of relations that fostered them, and conclude by considering the prospects for a more liveable urbanism and the coalitions of actors that are — or could be — inspired by this alternative.

Context

Like all post-apartheid cities, Cape Town has faced the twin challenge of overcoming the spatial divisions created during the colonial/apartheid era and addressing the endemic poverty that these divisions reproduced for over three centuries. This marked the start of 10 years of almost continuous institutional and organisational change. By 1994 Cape Town was governed by 61 municipalities and managed by 17 separate racially based administrations. In 1995–1996, the first democratic, local government elections took place in integrated municipal areas. Initially, the 61 former racially segregated municipalities were amalgamated into seven local government authorities, with a weak over-arching metropolitan level of governance. In 2000 the Municipal Structures Act was passed by the national legislature (*Government Gazette* No. 19614, 1998) which provided for a new system of metropolitan government for South Africa. This led to the establishment of the 'so-called' Cape Town Unicity, a single-tiered form of strong metropolitan government. This new structure made possible a single, integrated, metropolitan tax base in order to create the fiscal resources to address the inequitable access to services, characteristic of all apartheid cities. This process of municipal restructuring and its impact on service provision has been well documented (Jaglin 2008; Jaglin 2004; Khosa 2000; McDonald & Smith 2002; Parnell *et al.* 2002; Parnell & Pieterse 2007; Watson 2002). However, because the focus of this literature has been on governance, spatial planning, service-delivery systems, inequalities and economic policy, it has not provided an adequate basis for analysing the ecological limits to inclusive urbanism (Crane & Swilling 2008).

Privatisation/corporatisation of state-owned enterprises, the formulation of public–private partnerships, greater emphasis on service delivery on a cost-recovery basis and the implementation of performance management systems were part of the process of institutional reform, albeit only partially implemented in the Cape Town context (Smith & Hanson 2003; Smith 2004b; Watson 2002; Wilkinson 2004). Despite arguments that this has resulted in neoliberal governance approaches and, by implication, splintered urbanism in Cape Town (Miraftab 2004; Smith & Hanson 2003; Smith 2004b), the fact of the matter was that by 2008 all municipal services in Cape Town were still being delivered by the CCT's integrated municipal administration, and despite trends that could have 'splintered' Cape Town's post-apartheid urbanism, the vast majority of poor households were connected to publicly managed, networked infrastructures. Yet it is also true that a half-baked 'new public management' discourse was incorporated, giving rise to an awkward hybrid: publicly managed service delivery with an inclusive bias, wrapped in the language of commodification, markets and cost recovery (which did not reflect the real complexities of actual service delivery).

By 2005 there were nearly 850,000 households in Cape Town (Haskins 2006) with a population of approximately 3.25 million people (Romanovsky 2006). Population growth rates peaked at about 2.5 per cent per annum in 2000/01, but have declined since then, down to 1.24 per cent by 2009 (Romanovsky 2006). Table 9.1 presents the class structure of these households with just over 50 per cent classifiable as poor and working class, 16 per cent comprise the wealthy elite, with a relatively small middle class comprising 31 per cent of households.

For the period 2000–2005, economic growth as measured in gross geographic product (GGP) averaged 4.7 per cent per annum for the Western Cape provincial economy as a

Table 9.1: Household class structure in Cape Town

Cluster group	*Percentage of suburbs*	*No of households*	*Percentage of total households*
Elite suburbs	14	54,630	6.4
Upper middle class	19	68,129	8
Sub-total	**33**	**122,759**	**14.5**
Middle suburbia	20	77,380	9.1
Inner city	1.5	17,564	2
Semi/skilled labour pool	9.5	42,404	5
New bonded areas	13.5	101,638	12
Sub-total	**44.5**	**238,986**	**28**
Traditional townships	4.5	80,980	9.5
Dense run-down high-rise	13	170,752	20
Urban and working poor	2	26,108	3
Below the poverty line (mainly formal townships, with some shacks)	3	111,770	13
Sub-total	**22.5**	**389,610**	**46**
Mainly stand-alone shack settlements, with & without tenure, approx half 'unserviced'	Not calculated	94,766	11
Total	**100**	**846,121**	**100**

(Source: calculated from Haskins 2006; & Swilling 2006)

whole, of which the economy of Cape Town makes up 80 per cent. These growth rates, which were higher than the national average, gave rise to the optimistic pre-financial crisis 2007 estimate by the Western Cape Provincial Government that growth was expected to remain at around 5 per cent per annum up to at least 2010 (Western Cape Provincial Government 2007). Unfortunately, these fairly impressive economic growth rates did not translate into similar rates of employment creation. The reason for this is that for the period 2000–2005 this growth was driven mainly by growth in the services sector (excluding government) at 6.1 per cent per annum. Employment in this sector, which makes up two-thirds of the Cape Town economy, tends to absorb higher-level skills, not the in-migrating unskilled or semi-skilled blue-collar workers — although the growth of the security industry, which is counted as part of the services sector, does suggest employment growth for un- and semi-skilled workers. The sector in which unskilled and semi-skilled workers did find new employment opportunities in the period 2000–2005 was in construction, which grew at 8 per cent per annum, but comprises only a relatively small proportion of the total Cape Town economy. By contrast, the manufacturing industry in the Western Cape — which makes up half of the non-services sector of the Cape Town economy, was on the decline and grew at a much slower rate of 3.1 per cent per annum. However, this figure masks sharply deteriorating export growth in the period 2003–2005, a sharp drop in business confidence from 2006, mainly due to skills shortages in the sector, and the rapid lowering of tariffs by government, resulting in the closure of many textile factories which could not compete with cheap Chinese imports.

The good news is that although employment and unemployment statistics are not considered to be very reliable, there was some optimism before the 2008 recession that overall unemployment in the city was on the retreat. According to the annual Labour Force Survey, unemployment (in the age group 15–64 and excluding discouraged job seekers) was measured at 21 per cent in September 2005, compared to 15 per cent in September 2006. It is interesting to note that this estimated drop in unemployment took place at the same time as the economy was not significantly increasing its growth effort, possibly signalling that positive structural shifts rather than growth itself was responsible for creating new employment. But with the onset of the recession in 2007/08, these gains were reversed. Absolute figures on unemployment are still substantial, amounting to almost 225,000 unemployed people in the city in 2007/08, compared to 212,000 in 2000 and 175,000 in 1995 (City of Cape Town 2006).

Cape Town's formal economy is dominated by a large number of small-to medium-sized businesses concentrated in the services sector (Western Cape Provincial Government 2005) which absorb relatively skilled labour, with a declining manufacturing sector and growing employment levels in the construction sector. Government expenditure is a relatively small proportion of the local economy, which diminishes its leverage. But when added together expenditures on energy, waste, water and sanitation do make up a significant portion of the economy. As elsewhere in the world, the decline of manufacturing undermines the backbone of inclusive urbanism, namely the employed (usually unionised) working class with stable long-term jobs and nuclear families, who are willing and able to pay city rates and taxes. This was the same group of people who were the most powerful and well-organised components of the struggle for democracy and from whose ranks many city councillors were drawn after 1994. Ironically, their vision for an inclusive urbanism for Cape Town was out of kilter with a local economy that was adjusting rapidly to global economic trends, which favoured high-skilled service sectors, the importation of cheap manufactured goods (mainly from China) and the proliferation of ultra-cheap, informal sector services and the casualisation of labour (for similar trends in Johannesburg, see Beall *et al.* 2000). The only real countervailing employment creators for blue-collar workers from the mid-1990s onwards were the growth of the tourism, security and construction sectors.

Despite the many political changeovers in Cape Town's municipal government since 1994, a constant theme of successive administrations has been the need to address the service backlogs in the poorer areas of the city. This has had major implications for capital and operating expenditures in the energy, waste, water and sanitation (EWWS) sectors, which together accounted for nearly half of the annual expenditure of the CCT up until 2007/08. R9.3 billion, or 47 per cent of the CCT's 2007/08 budget, was spent on capital and operations to extend and maintain the networked urban infrastructures that deliver Cape Town's EWWS services. This equated to 8 per cent of the GGP of the Cape Town metropolitan economy at that time (De Wit *et al.* 2008). One key outcome has been that over 90 per cent of all households are connected to urban infrastructures, but this does not mean they all have houses (which are financed and delivered by national and provincial government). Indeed, there are between 150,000 and 200,000 shacks in Cape Town, of which about 100,000 are serviced.

It is therefore not surprising that expenditures in just these three sectors (that is, EWWS) have been at the centre of Cape Town's political dynamics. They also directly affect the local economic growth potential and natural resource base. Our previous published research has demonstrated that Cape Town's urban system is coming up against serious resource limits and scarcities that cannot be transcended if resources are managed in the old ways (Crane & Swilling 2008; Swilling 2006). This includes the resources conducted via vast networked infrastructures, such as electricity, water, sanitation, solid waste, transportation and affordable food supplies. However, there are also a range of eco-system services that are under threat, including biodiversity (the fragile fynbos biome), the region's rivers which are the most polluted in the country, rapidly depleting fisheries, degraded agricultural soils and deteriorating air quality. There is now a consensus that Cape Town and the wider Western Cape region face severe climate change-related threats which require a co-ordinated government response (Western Cape Provincial Government 2008). Although originally ignored by the post-1994 policy-makers, the constraints imposed on Cape Town's spatially segmented inclusive urbanism by the realities of ecological degradation are slowly making themselves felt. The result is that Cape Town has more policy frameworks that include commitments to sustainable development than any other South African city (City of Cape Town 2008; Clark, Dexter & Parnell 2007; Western Cape Provincial Government 2008).

Cape Town's inclusive urbanism

The integration of a large number of smaller EWWS departments into vertically integrated metropolitan bureaucracies after 2000, with district-level satellite offices, was seen by strategists as a necessary precondition for the delivery of uniform services to all households. Like so-called 'large technological systems' (LTS) in other contexts (Guy *et al.* 2001; Summerton 1994), Cape Town's EWWS systems were highly complex networked infrastructures managed by a specific set of vertically integrated institutions that were required to operate in accordance with (at times contradictory, but ever-changing) national, provincial and local government regulations (Jaglin 2008). Timothy Moss observed that since the late nineteenth century networked infrastructures:

> *... under the direction of their* system-builders, *... have generally followed the familiar pattern of development of LTS into large-scale, centralised and hierarchical systems built around a dominant technology (such as mains sewers) and comprising several sub-systems (such as storm-water collection).* (Moss 2001: 44 — emphasis added)

As discussed in Chapter 5, these vast networked infrastructures regulate the way natural resources are sourced and processed, and after use, disposed of or reused. These systems, in turn, set the physical context for everyday urban life from the intimacy of the morning shower, the daily trip to work, the ever-present telephone call, to the taken-for-granted ubiquity of electricity at a flick of a switch. The system-builders develop, over time, a reliable set of routines and technologies to do this, which they systematically defend against any competing system. Indeed, the better they are at their jobs the more risk averse they become. This is because their success as *system-builders* depends on reducing risks to a minimum and breaking everything up into small, repeatable actions that can be programmed, controlled and monitored in accordance with detailed, documented, operating procedures. Alternatives are regarded with deep suspicion and vigorously opposed, mainly because they inevitably mean raising the risk levels as unknowns are allowed to enter the system, thus undermining their rewards and promotions. The result is that these LTS (or networked infrastructures)

> *... develop a momentum of their own. Early decisions on the type of technology chosen or the way utilities are regulated limit the options for future development... Once established, the homogenous, standardised networks become stable forces which continue to reinforce themselves internally and sustain other systems,* such as urban development, *externally.* (Moss 2001: 44–45 — emphasis added)

The longer the system has been around, the more inflexible and committed the system-builders are to a particular technology for sourcing, processing and disposing of resources. These systems do, however, change in response to 'reverse salients' (Moss 2001: 45) which are basically existing system conditions that either slowly dissipate (for example, abundant water supplies), temporarily break down (for instance, electricity blackouts) or cease to exist altogether (such as space for more landfills or cheap oil); or, alternatively, new system conditions emerge that force change (such as the institutional amalgamations of the kind that took place in Cape Town after 1994, or regulations to limit CO_2, enforce recycling or redistributive measures). Reverse salients (which are, in complexity language, 'negative feedback loops') can lead system-builders to incorporate alternative technologies (for instance, solar water heaters) or expand delivery systems to meet the needs of new users (for example, Cape Town's commitment to extend services to all), or it can lead to a straight fight to the finish — what Moss tellingly calls the 'battle of the systems' as old ways come up against their economic, ecological and technical limits without the system-builders acknowledging that change is needed (Moss 2001). Given the enormous knowledge-power that these system-builders can mobilise, they can very often ensure that new problems arising from 'reverse salients' are constructed in ways which generate solutions that do not subvert the prevailing techno-institutional path dependencies (as the saying goes — 'if the only solution is a hammer, every problem will be a nail'). But when they fail to retain control of the discourse, significant system change has a chance. These are usually brief moments of opportunity — often driven by some kind of underlying crisis or imposed demand on the system — which need to be noticed and grabbed with alacrity by reformers within and outside these (often ossified) bureaucracies.

The ecology of actors that set Cape Town's post-1994 agenda in motion included the newly elected councillors (many of whom represented under-serviced constituencies which had never voted before); a set of senior officials with entrenched positions by virtue of their long-standing roles as system-builders; a new layer of top managers, many of whom were political appointments that reflected the value commitments of the newly elected councillors; a network of highly influential professionals in private consulting practices (engineering, architecture and town planning in particular); a network of developers with significant landholdings and capital for property investments; a small group of (at most four) major financial institutions who were not only funders of the property development industry but also significant shareholders in major property investments; and a small but influential set of professionals located in NGOs and universities who had hitherto exerted limited formal influence by virtue of their alliance with the democratic movement. Social movements such as the SA Homeless Federation, the Congress of South African Trade Unions (Cosatu) and various local civic organisations were also actors, often supported by NGO-based professionals. Land invasions by the homeless were the kinds of direct actions that shifted the balance of power in favour of the poor, but these were rare. All these actors took the key natural resource actors for granted — bulk water resources, energy supplies, landfill spaces, and the resource carriers such as large-scale networks of pipes and cables, and sewage treatment works. However, these actors were all committed in one way or another to the extension of services to the under/un-serviced areas of Cape Town on an equitable and uniform basis (City of Cape Town 2003). Indeed, as in South Africa as a whole, so-called 'service delivery' became synonymous with development.

Needless to say, the commitment to a vision of *inclusive urbanism*, as expressed in city-wide planning frameworks for Cape Town, does not mean that this was actually achieved in practice. Although the literature has clearly demonstrated the vast gap between vision and reality when it comes to integration of the city (due to the logic of the property market and rights, limits to planning capacity and powers, lack of political will, consequences of neoliberal public management approaches and the marginalising effects of housing subsidies) (see Haferburg & Obenbrugge 2003; Jaglin 2008; Jaglin 2004; McDonald & Smith 2002; Miraftab 2004; Robins 2006; Ross 2005; Smit 2006; Smith & Hanson 2003; Smith 2004a; Swilling & De Wit 2010; Swilling 2006; Watson 2002), what cannot be denied is that the EWWS administrations did extend services to nearly every household in Cape Town. However, the EWWS administrations achieved this by working within existing spatial relations to connect nearly everyone into the networked infrastructures of the city. The result was uniquely South African: *spatially segmented inclusive urbanism*. This more limited conception of inclusive urbanism within a spatially fragmented landscape is very different to the more idealised vision found in all policy documents of a post-apartheid city, which envisages a city that has managed to somehow overcome the stark geographical separation of (overwhelmingly poor) black areas and (increasingly multi-racial) middle-class and elite areas. Although richer suburbs slowly deracialised, racial segregation remained a reflection of the city's class divisions.

Since 2000, tariff policy has benefited the poor significantly.[3] Water tariffs have included a progressive five-step structure which resulted in large consumers paying more per litre than small consumers, and a 7 per cent surtax levied on businesses to cross-subsidise the 6,000 litres of free water which needed to be provided to all households in line with national government policy. As for sanitation, the first 4,200 litres were provided free of charge. Solid waste was standardised via a 240-litre bin and the service was provided free of charge for properties valued at less than R88,000 and heavily subsidised for houses worth between R88,000 and R160,000. A progressive approach to electricity tariffs has been much more difficult because Eskom supplies approximately one-third of Cape Town's households directly. Nevertheless, all consumers on the Domestic 2 tariff, who consumed less than 450 kWh per month on average, got 50 kWh free electricity per month. After Eskom refused to apply this to the areas they served, the city had to step in to pay Eskom for this benefit to the poorer households.

As far as rates are concerned, the 2007/08 budget provided for properties valued at less than R88,000 (about 85,000 property owners) to pay zero rates, get free refuse collection, and a basket of free services. Properties valued at R199,000 received a R20 per month discount on their rates, and households that earned less than R1,740 per month, listed on the City's Indigent Register, received a 100 per cent rates rebate and a R20 per month subsidy of their services account. Finally, since 2003/04, tariff increases have taken into account the affordability levels of poor households — the result being average increases at above inflation, but with much lower increases for poor households (and in some cases even decreases) compensated for by much higher increases for richer households.

[3] The data referred to in the following paragraphs is drawn primarily from Jaglin (2008), but also from Swilling and De Wit (2010).

Table 9.2: Key elements of the Progressive Equity Model in Cape Town

Water	First 6,000 litres free
Sanitation	First 4,800 litres free
Electricity	First 50 kWh free
Solid waste	Free service on properties valued <R88,000 and heavily subsidised properties R88,000–R160,000
Rates	Properties <R88,000 zero rates Properties <R199,000 20% discount Households earning < R1,740/month receive 100% rebate and R20 subsidy on services account
Tariff increases	Since 2003/04, average increases at above inflation, but with much lower increases for poor households (and in some cases even decreases) compensated for by much higher increases for richer households

(Source: Jaglin 2008)

To complement the progressive aims of rates and tariff policies, the general approach to services from the mid-1990s onwards was that the levels and standards applied in the former white areas must be applied to all areas. This had major implications for capital budgets, reinforced by increasingly large inter-governmental transfers. However, it is one thing to extend infrastructure using capital budgets and transfers, it is a completely different matter to make sure that operating budgets are expanded accordingly in order to maintain and repair these infrastructures into the future, and that provision is made in capital budgets for refurbishment and upgrade.

While tariff/rates policy and capital expenditure policy aimed to achieve equity via cross-subsidies, the energy and water/sanitation departments have become increasingly strident since 2004 about the need for so-called 'economic tariffs' for each service. By this they meant a 'corporatised' cost-recovery model that would allow each sector to define its own costs of operations (maintenance and repairs) and capital expenditure so that revenues from the sale of their services could be ploughed back into their sectors, rather than used to cross-subsidise the rates account and corporate services. The underlying reason for this response from these two departments is that the progressive equity model pursued via tariff/rates policy and capital expenditures came to be financed by surpluses creamed off from the sale of electricity and water services. Whereas 10–11 per cent of expenditure by the electricity department was transferred to the rates account in the 1980s and early 1990s, this had increased to 15 per cent by for the financial year 2004/05. Similarly, the water sector has contributed significantly: the total contribution as a percentage of expenditure increased from less than 11 per cent in 2001 to nearly 19 per cent by 2004/05 (Jaglin 2008). The water and electricity departments have argued since at least 2002 that these contributions to the rates account and to corporate services undermine their capacity to finance essential upgrades and repairs. However, the fact of the matter is that without these cross-subsidies, a more inclusive urbanism would have been impossible.

Cape Town's socio-metabolic flows

An early estimate for the period 1996–1998 was that Cape Town consumed 365 million tonnes of 'raw materials' per annum, and disposed of 208 million tonnes of wastes into local water, air and land sinks (Gasson 2002: 5). For the purposes of this chapter, the most significant annual resource flows through Cape Town's urban systems were as follows:[4]

- ***Electricity***: approximately 10 billion kWh of grid-supplied electricity for the year 2006/07, which equals nearly 3,500 kWh/cap/yr.
- ***Oil***: over 2 billion litres of crude oil for the year 2006/07, or 666 l/cap/yr.
- ***CO_2***: nearly 20 million tonnes emitted for the year 2006/07 from all energy users, including transport, which is nearly 7 t/cap/yr (which is lower than the national average of 9.8 t/cap/yr but equivalent to per capita emissions in Italy, France, Spain, Poland and Malaysia).
- ***Water***: 247 million kilolitres of water per annum in 2006, or 82 kl/cap/yr which is higher than the global average of 57 kl/cap/yr (and also higher than the European average).
- ***Sewage***: 200 million kilolitres of sewage for the year 2005/06, or 67 kl/cap/yr.
- ***Solid waste***: 2.9 million tonnes of solid waste in 2007/08 (which is over 2 kg/cap/day, higher than the European average), of which about 0.4 million tonnes was recycled.
- ***Building materials***: by the early 2000s, 6 million tonnes per annum of building materials entered Cape Town for conversion into buildings and infrastructure, with an output of about 1 million tonnes of builder's rubble, much of which is recycled and reused as inputs (estimates range from 35 per cent to 75 per cent).

Without these inter-connected flows, Cape Town would grind to a halt. The same is true for any other city. However, if one assumes that locational advantages for investors are determined by the 'cost of doing business', the relative efficiency of Cape Town's infrastructure, compared to other cities, could mean that the 'cost of doing business' in Cape Town may be higher, in particular as resource prices rise. This, in turn, will put downward pressures on wages and retard the potential expansion of the tax base to finance service delivery with obvious negative consequences for the urban poor.

[4] This summary data is derived from the results of a three-year research project managed by Mark Swilling (see Sustainability Institute 2007a; Sustainability Institute 2007b; Sustainability Institute 2007c). Supplementary data has been drawn from Gasson (2002) and Hansen (2010).

Food as a resource flow?

Although the focus in this chapter is the socio-metabolic flows of energy, waste, water and sewage, it has been estimated that the food chain (which includes all the inputs, energy and materials required to produce and move food from 'farm to fork') is responsible for about a third of Cape Town's ecological footprint. Gasson estimated that Cape Town consumed 1.3 million tonnes of food per annum during the period 1996–1998 (Gasson 2002: 6).

Each of these flows (excluding CO_2—at least for now—and building rubble) is managed by a specific set of highly regulated administrations which operate vast networked infrastructures built around a well-tested set of technologies. An integrated 'resource management' authority does not exist. Each 'sector' runs according to its own logic on the assumption that the underlying resource base will always be there at an affordable price.

Cape Town's energy system

Cape Town's major energy sources are shown in Table 9.3.[5] Read together, this means that Cape Town is more dependent for its energy supply on imported oil than on grid-supplied electricity. This, in turn, is because it is dominated by road-based mobility, which is entirely dependent on oil. This also means that whenever the oil price increases, Cape Town transfers large quantities of extra cash out of the local economy with no added benefit. In 2006/07 just over 2 billion litres of crude oil was required by the privately owned CALREF oil refinery, located in Cape Town, to supply the city with its petrol and diesel requirements at a cost to consumers of R900 million for that year. This was, however, when oil was US$60 per barrel—by 2008 the oil price had doubled, meaning that an extra R1 billion or so was transferred out of the Cape Town

Table 9.3: Energy use profile, 2006

Energy use by source		*Energy use per sector*		
	Cape Town		*Cape Town*	*Western Cape*
Electricity	29%	Transport	47%	34%
Petrol/diesel	46%	Industry/ Commerce	38%	48%
Other oil-based products	17%	Households	14%	9%
Coal	7%	Agriculture	0%	5%
Wood	1%	Other	1%	4%

[5] Most of the data cited in this section are derived from the Sustainability Institute report on Cape Town's energy system, supplemented by the updating of this information by Spencer (2010; 2009).

economy, contributing to reduced investment and the consequent reversal of the rising employment levels of the previous few years.

Approximately 72.5 per cent of all electricity supply in Cape Town is provided by the CCT via traditional or pre-paid meters, while 27.5 per cent is supplied directly by Eskom,[6] mainly via pre-paid meters. This means there are two electricity supply administrations that service Cape Town, with little co-operation between the two. Eskom likes to believe that it is just another private company selling commodities, not a key pillar of the public service with a developmental mandate. For 15 years now it has waged a concerted campaign to remove electricity provision from local government — a splintered urbanism logic informed by a neoliberal vision that the post-apartheid government has done little to challenge (for an account sympathetic to Eskom, see Pickering 2008). Unfortunately for Eskom, local government's right to deliver electricity is entrenched in the Constitution.

Cape Town's electricity supply system comprises the following infrastructures:

- The bulk of the electricity supply comes from coal-fired power stations to the north of the country carried into Cape Town via a 400 kV line which can deliver a maximum of 2,600 MW (with no serious plans in place by anyone to expand this).
- The nuclear power plant at Koeberg (1,800 MW) and the gas turbines at Acacia (171 MW) provide additional supply from units located within the Cape Town functional region. Two gas turbines at Roggebaai and Athlone (40 MW each) are used only during emergencies.
- Hydro power is generated from the surrounding region at the Palmiet (400 MW) and Steenbras (160 MW) pumped storage schemes. Surplus energy at night is used to pump water up a hill which then drives a turbine during the day when electricity is expensive.
- The Darling Wind Farm on the West Coast has been operational since 2008 and supplies electricity into the grid from four 1.3 MW turbines (5.2 MW).
- By 2007 there were an estimated 10,000 solar water heaters (SWH) in Cape Town, representing 4.2 MW of renewable energy.

If the last three are combined, 10 per cent of Cape Town's installed capacity is derived from a combination of energy-efficiency (pumped storage) and renewable sources (569.4 MW).

Although CO_2 emissions are associated with all forms of electricity production (including renewable energy because of the energy embodied within the physical materials of the infrastructure), electricity derived from coal is by far the most serious. Of the 19.4 million tonnes of CO_2 emitted by energy use in Cape Town in 2006/07, 11.4 million tonnes was emitted from Eskom's supply of electricity derived from coal and

[6] Eskom is the name of the large public utility that is responsible for managing the generation and distribution of all electricity in South Africa. It is now a company with the state as its shareholder. Its core business is coal-fired power generation and it defines its mission purely as the provision of electricity as a commodity to its 'customers'.

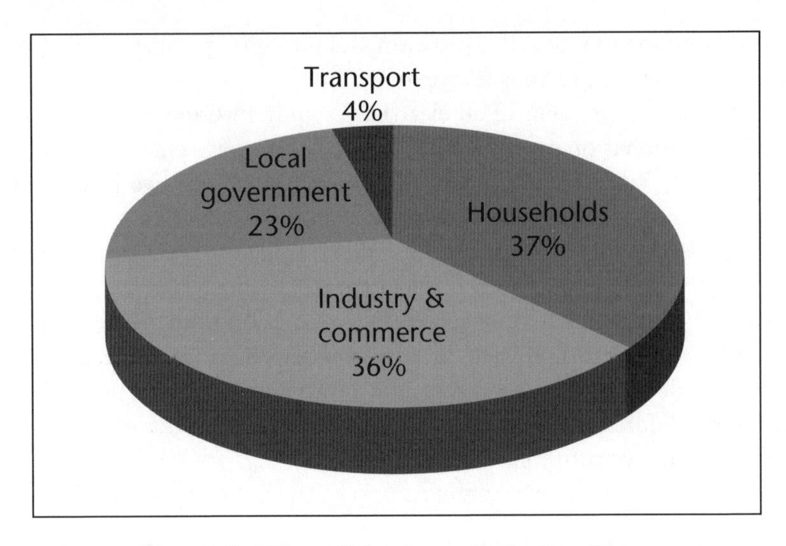

Figure 9.1: CO_2 emission by sector in Cape Town

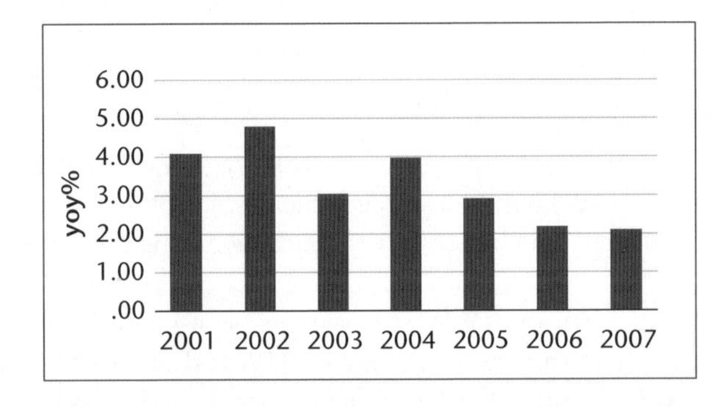

Figure 9.2: Growth in electricity use (year-on-year per cent)

nuclear. The Western Cape Provincial Government has already formulated a climate change response strategy which effectively promotes Cape Town as a future low-carbon city (Western Cape Department of Environmental Affairs and Development Planning 2008).

Given the lead times for building large coal-fired or nuclear power stations, Cape Town's future growth and development will be severely constrained by endemic limits to the expansion of conventional energy sources in the short-term (5–10 years), and possibly even the medium-term (7–15 years).

The good news is that there is evidence of relative decoupling since 2004 between economic growth and the growth in demand for electricity. As Figure 9.2 demonstrates, the latter has been declining since 2004, reinforced by the onset of rolling blackouts since 2006.

The Western Cape government identified the following combination of energy efficiency and renewable energy potential:

Table 9.4: Energy efficiency and renewable energy potential

Wind	3,000 MW
Ocean	1,000 MW
Solar—PV	247 MW
Hydro	15 MW
Solar thermal	1,400 MW
Pumped storage	1,800 MW
Total	**7,452 MW**

(Calculated from source: Western Cape Provincial Government 2008. *A Climate Change Strategy and Action Plan for the Western Cape.* Cape Town: Western Cape Department of Environmental Affairs and Planning.)

As Table 9.4 makes clear, there is more potential here than the current installed capacity (+5,000 MW) that services Cape Town now.[7] However, the existence of two major electricity suppliers in Cape Town will more than likely make it impossible to develop a coherent actionable plan to implement such a vision.

To understand why Eskom has emerged as a direct supplier one needs to go back to strategic decisions made by Eskom when it was 'corporatised' during the last years of the apartheid regime (early 1990s). Since then, Eskom has relentlessly pursued the idea that electricity generation and distribution should be a self-contained business, including the notion that electricity is nothing more than a commodity that must be delivered on a cost-recovery basis. Motivated by this vision (which changed along the way from full-scale 'privatisation' to 'corporatisation'),[8] and reinforced by the ANC government after 1994, despite the splintering effects at the urban level, for nearly 20 years Eskom has actively tried to find ways of centralising generation and distribution, including trying to remove electricity provision from local government. The reason for this is that local governments have always used profits from the sale of electricity to cross-subsidise other services. Indeed, after 1994 progress towards a more inclusive urbanism would have been unviable without these cross-subsidies. In Cape Town's case, for example, approximately R300 million was transferred from the Electricity Department to the rates account to cover general expenses in 2007/08. For Eskom, using these surpluses for anything other than the 'electricity business' contradicted the technical efficiency and cost-recovery principles so valued by the private sector management culture that had established itself within Eskom.

Fortunately, there were strong believers in local governance who managed, at the time of the writing of the Constitution before 1994, to insert a clause which defined

[7] Since the publication of this report further research has revealed the potential of Concentrated Solar Power plants, which also have the advantage of storage capacity (see Heun, Van Niekerk, Swilling, Meyer, Brent & Fluri 2010).

[8] Or what Smith has called 'first and second generation neoliberalism' (Smith 2004a).

'electricity reticulation' as a local government responsibility. Without this simple clause, there is little doubt that Eskom, supported by national government acting on the advice of influential consultants who focused narrowly on technical efficiencies, would have stripped local government of the surpluses that have been used to cross-subsidise the post-apartheid vision of inclusive urbanism. So to counteract the Constitution, Eskom has subverted local government control of electricity distribution by stealth. This was done by making the approval of additional supply for new developments conditional on direct supply by Eskom rather than via the municipality—which is why 40 per cent of Cape Town is supplied by Eskom. It is no coincidence that the areas supplied by Eskom are overwhelmingly newly developed areas in which many of the new upwardly mobile, largely black citizens live (now comprising 40 per cent of total supply). Furthermore, as Figure 9.3 reveals, Eskom's sales volumes are lower than CCT's, but its profits are greater. The CCT also compensates Eskom for the free basic electricity supply, which means Cape Town's ratepayers cross-subsidise Eskom's profit margins. This confirms that when service delivery agencies cease to be controlled by elected bodies, they are able to implement with impunity profit-generating cost-recovery systems that have splintering effects.

Eskom's long-term aim, however, was to establish Regional Electricity Distributors (REDs) to remove electricity provision entirely from local government. On 17 April 2001 the South African Cabinet approved a documented entitled *Blueprint of the EDI Restructuring Committee (EDIRC),* more commonly known just as the *Blueprint.*[9] Inspired by crude neoliberal conceptions of governance (and ignoring the fact that the outcome would be splintered urbanism), the *Blueprint* laid out a plan to set up REDs that would take over from both local government and Eskom the rights to supply electricity

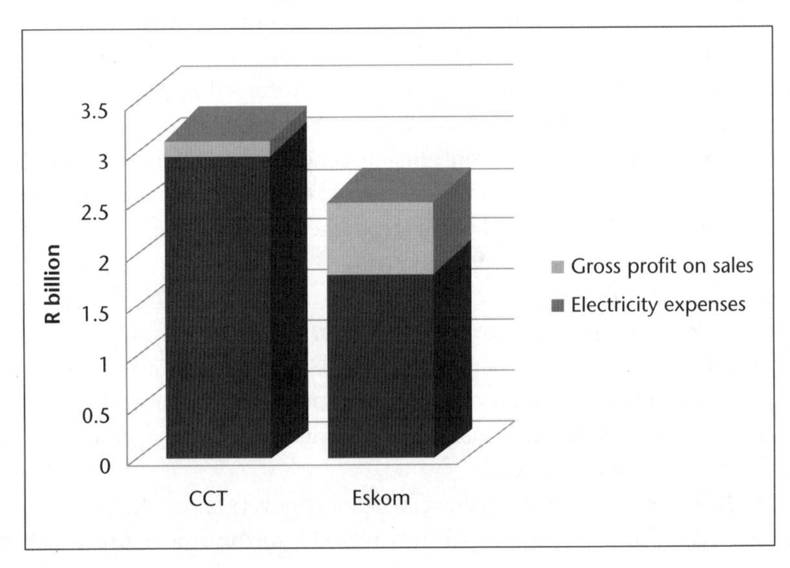

Figure 9.3: Electricity cost and profit
(Source: Spencer 2010)

[9] EDI = Electricity Distribution Industry

to end users. In September 2005 the Cabinet decided to establish a RED for each of the six metropolitan areas. The electricity assets of each metropolitan government, plus Eskom's distribution assets in these cities, would be transferred to the REDs, which would then be required to operate as 'businesses'. A giant national RED would then be established for the rest of the country (Pickering 2008). Although vehemently denied by Eskom and national government, the negative financial implications of this decision for local government were clear. Cape Town was the site of the battle to establish the first RED — the last round (which ended in December 2006) going undoubtedly to Cape Town, with major national implications.

Although the September 2005 decision was clearly a significant step towards removing electricity from local government, major local government players were won over when government agreed that REDs would be constituted as so-called 'municipal entities' in terms of the Municipal Systems Act.[10] Although this meant full-scale 'corporatisation', at least it allowed local government to extract surpluses to cross-subsidise other services. At a meeting on 28 August 2006 between the Mayor of Cape Town, Mayoral Committee members,[11] RED1 (which had been constituted by then) and national government representatives, it was agreed that the CCT would support RED1 and the transfer of CCT and Eskom assets to RED1 on condition RED1 remained a 'municipal entity' (that is, a company owned by CCT). This would have effectively created an integrated energy delivery system under CCT control. On 25 October 2006, Cabinet decided that the REDs would be constituted as 'public entities' in terms of the Public Finance Management Act and be accountable to the Minister of Minerals and Energy. The CCT interpreted this as a direct threat to the autonomy of local government and contradicted the 28 August 2006 agreement. It would also, of course, threaten the flow of surpluses from electricity sales, which were needed to cross-subsidise a cash-strapped metropolitan government. The CCT immediately responded by announcing that all contractual agreements with RED1 would be terminated as from 31 December 2006 and that no asset transfers would take place. The CCT tabled this decision at a public hearing of the National Energy Regulator of South Africa (NERSA) on 7 December 2006, which was opposed by RED1 and national government. NERSA ruled in favour of the CCT on 15 December 2006. RED1 (as 'public entity') was effectively dead and with it the biggest threat to an inclusive urbanism agenda Cape Town as a city had ever faced. This, however, was a battle won; the war was far from over — not least because Eskom remains determined to retain control of 40 per cent of Cape Town's energy supply. This stubborn attachment to a politically unacceptable model will prevent the establishment of an integrated, sustainable resource management approach to Cape Town's long-term energy needs. Ultimately, it is the urban poor who will suffer the consequences as resource prices skyrocket.

[10] Municipal entities are legally constituted as stand-alone companies but owned by local governments. They are required to operate in terms of the Municipal Finance Management Act.

[11] The Mayoral Committee was, at that stage, controlled by the Democratic Alliance, the official opposition to the African National Congress-controlled national government.

Cape Town's water and sanitation system

Cape Town's integrated water and sanitation system is a vast networked infrastructure with an estimated replacement cost of R17.5 billion (as of 2003) and an operating budget of R2.6 billion (2006/07). It has evolved into a *large technical system* in the classic incremental manner over nearly two centuries, with technological shifts tracking international trends. During the colonial and apartheid eras the mission was racially limited, inclusive urbanism. Since 1994 universal access has been its overriding goal.

Figure 9.4 is a graphic representation of the resource flows which this networked infrastructure makes possible, from rainfall catchment, to dams, water purification works, consumers, waste water treatment works, and then back into the natural system.

Over the past 15 years the administration of the Cape Town water and sanitation system has been reorganised several times as local governance has been de-racialised and restructured to meet the needs of historically disenfranchised citizens, many of whom did not benefit from water and sanitation services before 1994. In the process, many skilled personnel were lost, together with big chunks of the institutional memory. As will be explained later, after 1994 the justifiable focus of all parties who won control of the CCT was on capital expenditures to meet the needs of the poor. However, this was often done by cutting budgets for operations and maintenance.

In 2007 the CCT considered a crucial report, the 2007 *Water Services Development Plan* (WSDP), which warned that R750 million was needed for capital expenditure to keep up with demand, R300 million more than was allocated in the 2006/07 budget. Furthermore, the report argued, the budgets allocated to the Water and Sanitation Department over the years had been shrinking and the infrastructure was starting to break down due to inadequate budgets for replacement and upgrade, and for operations and maintenance. It was this report that triggered a policy change, reflected in substantial

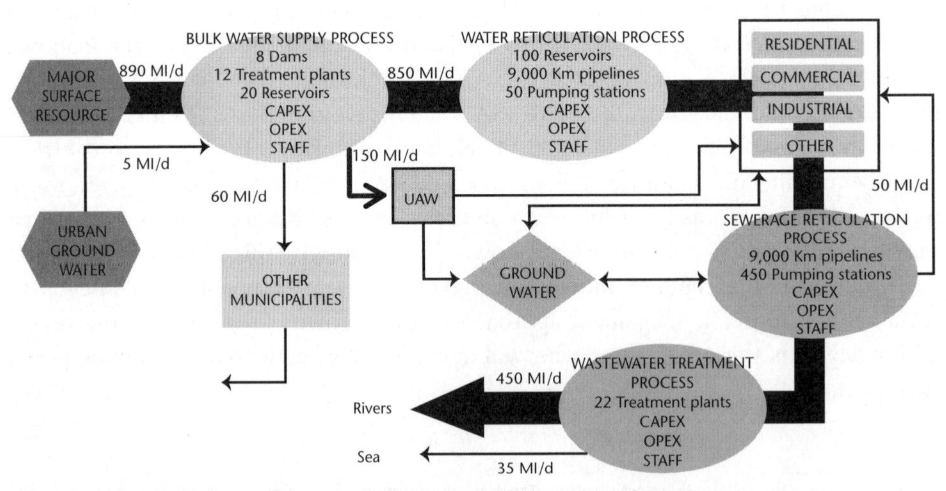

Figure 9.4: Resource flows in Cape Town's water and sanitation system

(Source: CCT Department of Water and Sanitation Services quoted in Sustainability Institute & E-Systems 2009b: 74)

budget increases for water and sanitation in the 2007/08 budget — a trend that seems set to continue.

Cape Town depends almost entirely on surface-level storage of winter rainfalls for its water supply. As a result it must rely on a massive water-catchment area that incorporates the Berg, Breede, Gouritz and Olifants-Doorn Water Management Areas (WMAs) (see Map 9.1). The Berg WMA, which supplies the Cape Town region, has 77 per cent of the Western Cape population, contributes 86 per cent to the Western Cape's GDP, and may therefore experience significant increases in water demand. The main water-use drivers in the Berg WMA are urbanisation, industry, agriculture and tourism. Agriculture is the largest user of water in the Western Cape (Department of Environmental Affairs and Development Planning 2007). Given that the bulk of this water is used for the cultivation of wheat, citrus and wine which are mainly for export, this means that the Western Cape is an exporter of water, or what in the literature is called 'virtual water' (Hoekstra & Chapagain 2007).

Demand for water in the Western Cape is fast outstripping supply. The Western Cape's rivers are the most polluted in South Africa (Western Cape Department of Environmental Affairs and Development Planning 2007), and climate change is already resulting in less frequent but more heavily concentrated rainfalls, leading to less absorption and therefore more run-off and related erosion (Western Cape Provincial Government 2008).

It has been estimated that annual unrestricted demand for all uses in Cape Town itself and surrounding towns that form part of the Berg WMA is 510 million m^3 whereas maximum bulk supply is only 475 million m^3, with the new Franschhoek Dam intended to increase supply by at most 18 per cent at a cost of R1.5 billion (Shand 2005).

Groundwater accounts for only 1.5 per cent of total yield, and is believed to represent an under-exploited resource. A recent assessment by the Department of Water Affairs and Forestry (DWAF) suggests there are between 100–200 million m^3 of groundwater in the Berg and Breede WMAs available for use in these areas, after taking account of current use, and these are conservative estimates (Department of Water Affairs and Forestry 2005). Sustainable harvesting of aquifers is a real option, including the Table Mountain Group Aquifer with a utilisable potential conservatively estimated at 100 million m^3, and the Cape Flats Aquifer which is currently under-utilised (Colvin & Saayman 2007). Artificial recharge of groundwater by transferring surface water into the aquifers during the wet Cape winter is now being promoted by DWAF (Department of Water Affairs and Forestry 2007). A similar practice has been used for 20 years in the Atlantis area (which is just north of Cape Town on the west coast), where about 40 per cent of water comes from artificial recharge with treated wastewater.[12] But this is a set of (proven) technologies with significant potential that lie outside the institutionalised 'know how' of the existing water administrations (bar the individuals who have championed and operated the Atlantis system). Instead, wealthier residents have been allowed to mine Cape Town's aquifers by sinking boreholes without being charged for using this crucial common resource (Colvin & Saayman 2007).

[12] Murray, R. (2007) — personal communication.

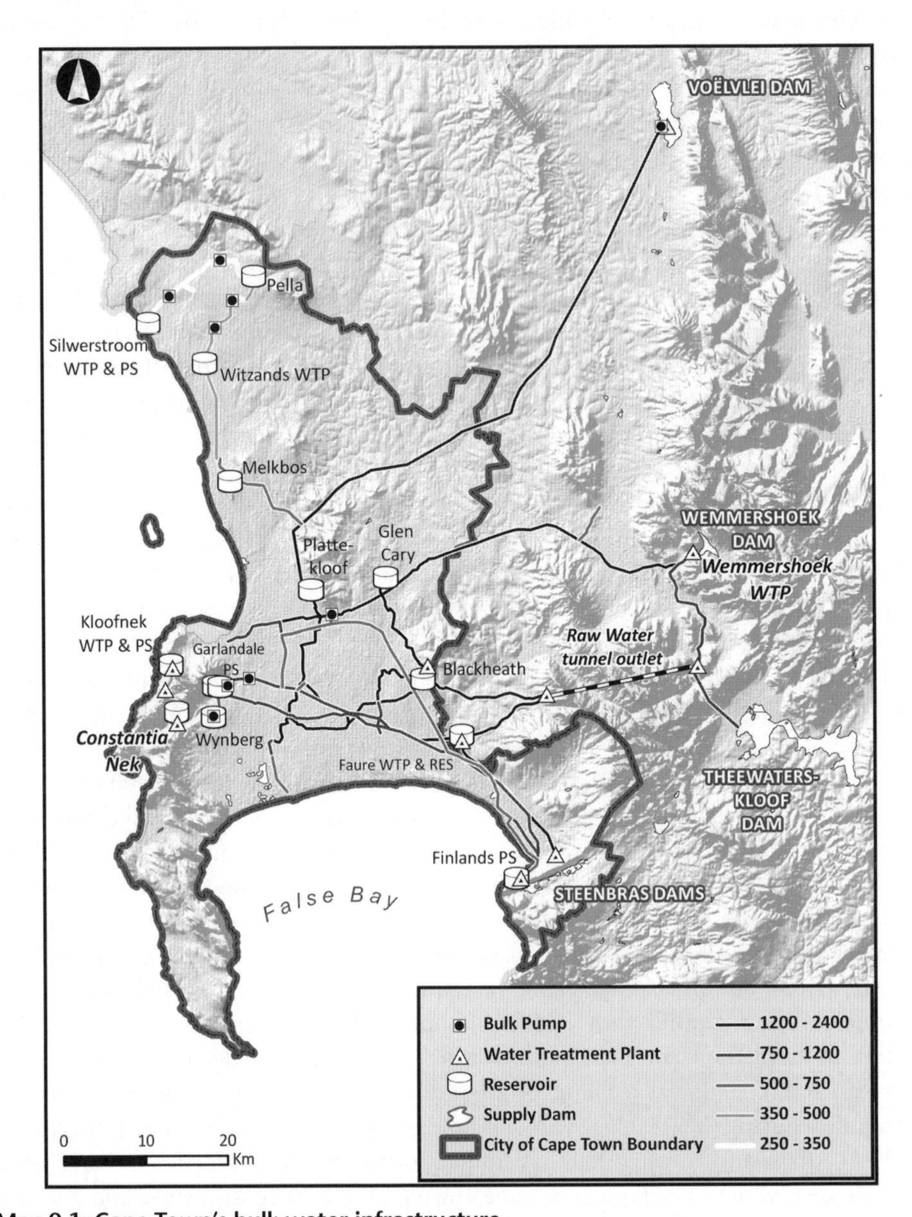

Map 9.1: Cape Town's bulk water infrastructure

(Source: Centre for Geographical Analysis, Geography & Environmental Studies, University of Stellenbosch, South Africa.)

Unlike the energy sector, domestic use makes up the bulk of Cape Town's demand for water. A total 247 million kilolitres of treated water was provided to Cape Town's water users in 2006, or 82 kl/cap/yr. This is higher than the global average of 57 kl/cap/yr and also higher than the average for several European countries (Germans use 66, UK uses 38 and the Dutch use 28), lower than developed low-density countries (USA uses

217, Australia 341), and towards the upper end of developing countries (Brazilians use 70, Chinese use 26, Indians use 38 and Mexicans use 139) (Hoekstra & Chapagain 2007). Currently 96 per cent of households have access to water and 91 per cent to sanitation. But, as with electricity, this masks substantial social inequities. In 2000, 60 per cent of domestic water was consumed by wealthy households, with 21 per cent of this used for their gardens and pools. More seriously, 61 per cent of all potable water used by Cape Town's domestic households is used for flushing toilets (Gasson 2003).

The second biggest virtual 'user' is euphemistically called 'unaccounted-for water': 23 per cent of all water that leaves the water purification works gets lost along the way.[13] The third biggest water user in the city is commerce and industry.

In many ways Cape Town's water system has already experienced the twin challenge of extending water services to the previously disenfranchised, while simultaneously reducing overall consumption to respond to water shortages. The activation of two new actors in the network (a voting majority who had never voted before with representation in the council, and significantly reduced rainfall) triggered reverse salients that forced the introduction of significant changes. Learning from this process holds lessons for what still needs to happen on a larger scale across all networked infrastructures, not just one of them.

As can be expected, actual demand steadily grew from 1994–2000 as increasing numbers of historically disenfranchised citizens were connected into Cape Town's networked infrastructures. The water administrations that managed this process were, in turn, going through complex processes of institutional amalgamation and consolidation as the governance system adjusted to the requirements of non-racialism and inclusive urbanism. However, the technologies and tariff systems inherited from the apartheid era remained basically the same—they were highly inefficient (because no incentives or penalties existed to counter wasteful use by richer households) and regressive (there was no bias in favour of the poor other than connecting them to the network). Without any major shocks to the system, an incremental progression into what is referred to in Figure 9.5 as 'unconstrained demand' appeared inevitable. However, two reverse salients triggered drastic changes.

Table 9.5: Distribution of water use

Western Cape		*Cape Town*	
Irrigation	68%	Domestic	51%
Urban	22%	Unaccounted-for water (UAW)	21%
Invasive alien plants	5%	Commercial and industrial	15%
Preliminary reserve	4%	Other	13%
Afforestation	1%		

(Source: Sustainability Institute & E-Systems 2009b)

[13] It is noted that this estimate, which is based on interviews and CCT documentation, is higher than the figure of 12 per cent for the year 2001 provided by Colvin and Saayman (2007). It is not clear whether UAW went up during the 2001–2006 period, or whether different calculation methods have been used.

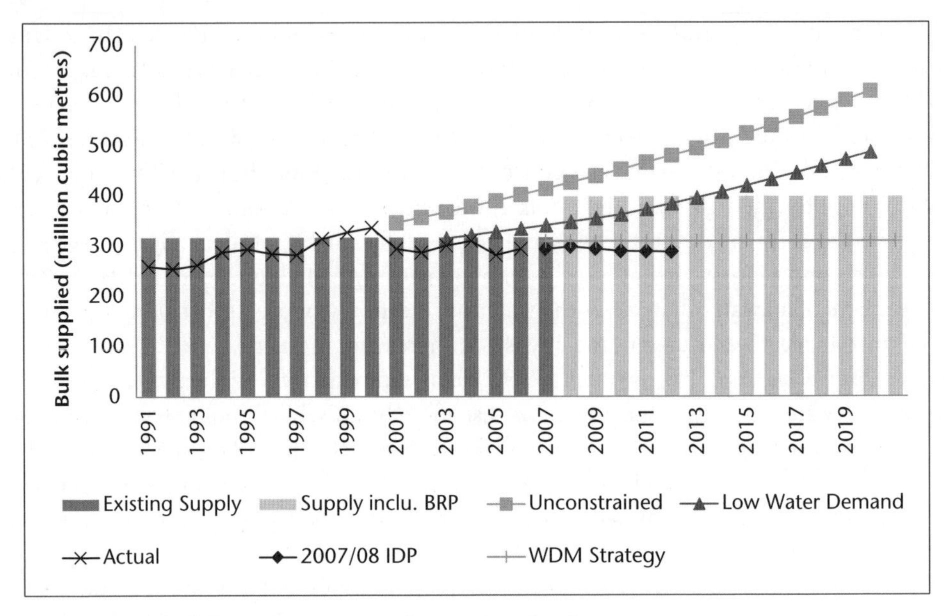

Figure 9.5: City of Cape Town water demand projections

(Source: Director, Water and Sanitation, City of Cape Town, Water Services Development Plan 2007/08.)

Note: BRP = Berg River Project.

The first was a set of intermittent droughts over a five-year period starting in 1999, which reduced some dams to dangerously low levels at certain points. Drastic conservation measures were introduced, including awareness education, severe restrictions (in particular on garden irrigation) and penalties. The second was the realisation that it was not possible to connect poor households into the networked infrastructures *and* retain a regressive tariff structure, especially if job growth was more sluggish than anticipated thus undermining the capacity of some poor households to pay service charges and levies (Jaglin 2008). Redistributive measures were introduced, including a so-called 'stepped tariff', which was supposed to make households that use more water pay a higher average charge, while poorer houses would benefit from free water quotas, plus cheaper tariffs if they kept consumption below certain levels.

The dual impact of these measures resulted in a net overall decline in the consumption of water provided via the network—this drop is reflected in Figure 9.5. So while the 2002 IDP predicted a traditional growth expectation (what is referred to in the Figure as the 'low water demand' curve), in reality the water conservation and redistributive measures resulted in a total drop in consumption by 64 million m^3 from 336 million m^3 in 2000 to 272 million m^3 in 2006. Total consumption has never returned to the 1999/2000 level despite an ongoing programme of connecting up the unserviced and meeting the needs of a growing economy. However, this good news story needs to be qualified in two respects.

Firstly, although the poor clearly benefited from the progressive tariff structure and free water, the policy failed to take into account the implications for very poor households with large, extended or multi-family units living on the same property, who inevitably consumed more than the average nuclear family. The stepped tariff structure inevitably resulted in these households paying an average unit price equal to a richer, high-consuming, but much smaller nuclear family household (Smith & Hanson 2003).

Secondly, what the figures on declining consumption mask is the fact that many wealthier households that were restricted from irrigating their gardens invested in boreholes which gave them unrestricted, free access to the aquifers below the city (for a study of this phenomenon see Colvin & Saayman 2007). However, over the long term there is nothing to stop the authorities from compelling these users to pay for the extracted water. If they do, this means the capital and operating costs of the infrastructure needed to mine the aquifers will have been paid for by private households. (Another interesting question is how much of the water extracted by private boreholes is run-off from the 'unaccounted for water'. Is this a water cycle mediated via the aquifers and private boreholes?)

Despite these two qualifications, the CCT built on the success of water demand management and in 2007 adopted the *Water Conservation and Water Demand Management Strategy* (Final Draft) — generally referred to as the 'Water Demand Management' (WDM) strategy which replaced all previous strategies and committed the CCT to investments in system efficiencies, awareness education and reuse as an alternative to the traditional response to growth in demand, in other words, massive capital projects to create additional supply infrastructures. This ambitious strategy is reflected in Figure 9.5 as the 'WDM strategy' — a target which was embedded in subsequent IDPs.

Noting that the 2000–2005 strategy resulted in a drop in daily consumption by 175 Ml/day (from 920 Ml/day in 1999 to 745 Ml/day in 2006), some officials used this precedent to justify their optimism that significant investments in water conservation, demand-side management, recovery of unpaid bills, reuse of grey water, and new sustainable technologies, could reduce consumption by another 323 Ml/day — which is to less than half the 1999 level during a period of both population and economic growth! And they may be right. Consider the suggestions in the *2005 Western Cape Reconciliation Study* to conserve water (some of which have been included in the WDM strategy): leak detection and repair; pressure management; use of water-efficient fittings; metering and plumbing repairs in low-income areas; use of grey water; use of well points and boreholes; metering; tariffs and surcharges/credit control; water-user education; rainwater tanks; exchange reclaimed wastewater for commercial irrigation; industrial reuse; reclamation to potable water standards; urban irrigation; dual reticulation in new housing (so that grey water can be supplied for toilets); and aquifer recharge (as quoted in Sustainability Institute & E-Systems 2009b: 30).

For increasing supply over the longer term, the *Reconciliation Study* made several suggestions, one of which was desalination — this is clearly an option, but it is an energy-intensive technology which makes sense (from an emissions perspective) only if powered by renewable energy. CSP-powered desalination plants must surely be a key guarantor of a more durable *liveable urbanism* over the long term.

The WDM strategy represents Cape Town's first significant investment in the sustainable and more equitable use of a finite natural resource. This is decoupling in practice: doing a lot more with less with major developmental spin-offs, ultimately leading to the kind of absolute decoupling needed to build a more liveable urbanism.

The dysfunctional state of the sewerage infrastructure in the city is often front page news. Under the heading 'Sewerage Shock' the *Cape Times* highlighted the environmental consequences and limitations to development caused by the lack of past investment in sewerage infrastructure (*Cape Times* 4 April 2007). There are three critical issues. First, is the provision of basic services to existing informal settlements and new arrivals. By 2007/08, approximately 30,000 households in Cape Town did not have access to basic sanitation, and 17,050 informal households shared a toilet with four others. Second, is the urgent need for upgrading and extending the existing wastewater treatment plants. A combination of rapid urban development and many years of under-investment in replacement and upgrade have put severe pressure on Cape Town's wastewater systems. There is a critical shortage of treatment capacity in the areas experiencing the most rapid expansion: Bellville, Klipheuwel, Gordon's Bay, Kraaifontein, Parow and Potsdam, and new housing developments which assume that bulk infrastructure will be provided by the city are not being approved. Major pipe collapses, leakages and storm-water infiltration require urgent attention. Many sewer systems are running over capacity and discharging sub-standard effluent and sewage sludge into receiving water and marine environments. The areas in which water and sewer infrastructure are severely stressed and in need of significant upgrades include:

- West Coast/Parklands development corridor
- De Grendel/N7 development node
- Northern development corridor
- Bottelary development corridor
- Fast-track housing projects (for example, N2 Gateway)
- Macassar/AECI development node.

Third, is the fact that the large bulk of the existing sewage flow is not converted into reusable inputs — methane (to generate energy), usable water and nutrients for agricultural production and greening are the key resources here.

Cape Town's bulk wastewater infrastructure consists of:

- 20 wastewater treatment works
- 3 marine outfalls,
- 27 major pump stations
- 15 major interceptor sewers
- About 120 km of bulk gravity sewers[14]
- 395 smaller pump stations and associated local-level reticulation networks.

[14] CCT, 2001: Bulk water infrastructure and statistics (internal document).

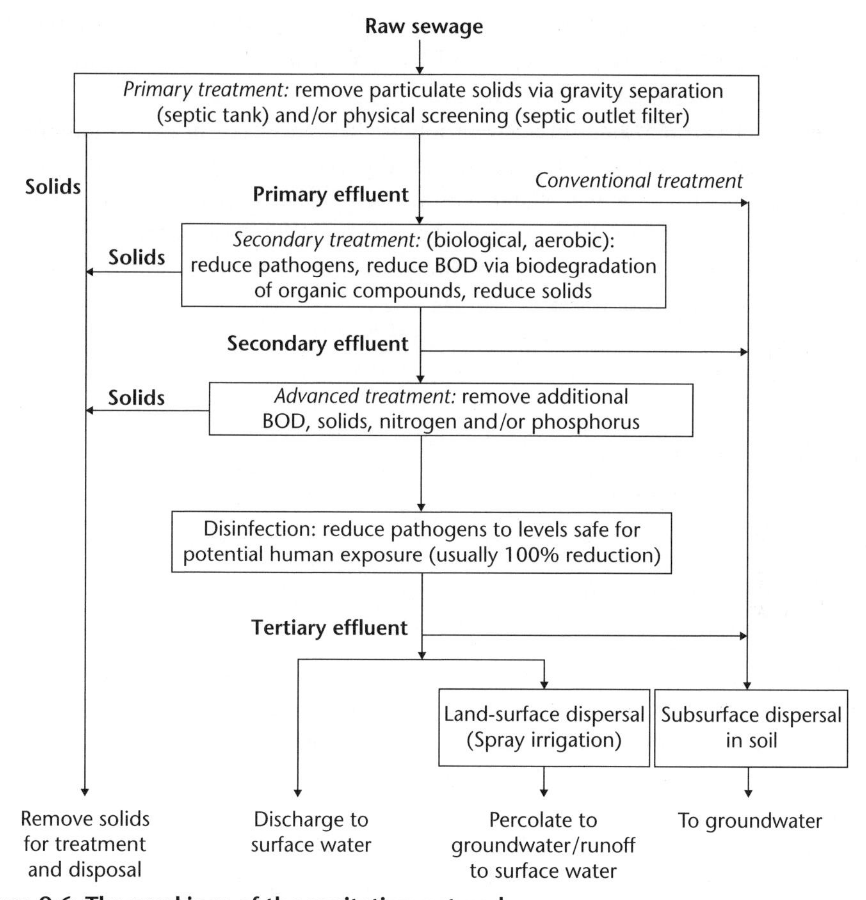

Figure 9.6: The workings of the sanitation network

The workings of the sanitation network are schematised in Figure 9.6. Most of the wastewater flow of approximately 567 Ml/day is conveyed via pump stations and treated at 20 wastewater treatment works. Of this, 31 Ml/day (5–9 per cent) is discharged directly into the sea. About 53,000 tonnes of dry sludge is produced annually (this being the residue left behind after treatment), most of which could be productively reused (if metals could be removed) but is, instead, sent to the landfill. Only 32 Ml/day (6 per cent) of treated effluent is reused.

Of the 20 wastewater treatment plants in Cape Town, 70 per cent or 14 out of 20 have less than 75 per cent compliance to DWAF standards for one or more of the listed variables. Although many wastewater practices violate the Water Act, DWAF favours a co-operative governance approach and has not enforced the law at this stage.

> *The limited financial situation in the City versus the high demand for new housing has created a scenario where the City is not in a position to maintain existing infrastructure and to provide the required bulk infrastructure for connection of new developments.* (WSDP quoted in Sustainability Institute & E-Systems 2009b)

It is not surprising, therefore, that the *City State of the Rivers*[15] report on the health of rivers in the greater Cape Town area found the ecological health of rivers downstream of the wastewater treatment works to be 'poor to bad'.

Wastewater treatment has three by-products: treated water, biogas (methane gas) and the nutrients embodied in sewage sludge. The city produces over 50,000 tonnes of sewage sludge per year, of which 57 per cent was beneficially used in 2006. The challenge is to find beneficial uses for the grey water, biogas and remaining sludge.

The following disposal processes for sludge were considered in an earlier strategic investigation for bulk wastewater.[16]

a. Incineration
b. Composting
c. High lime process
d. Drying and pelletisation
e. Co-disposal in landfills
f. Direct agricultural use
g. Manufacture of organic fertiliser
h. Brick making and allied fields
i. Co-combustion in coal fired power stations.

This list is interesting because b., f. and g. reflect a realisation that sludge comprises mainly useful nutrients. Pelletised sludge (d.) is already used by a Western Cape brick-maker, Claytile, as an alternative to the coal chips that get inserted in the wet clay brick before going into the kilns. Dry sludge can easily be incorporated as a composite into brick-making or used for co-combustion for a variety of purposes. But what about the reuse of the H_2O component of the sewage flow?

Treated effluent reuse is possibly the most under-exploited resource in Cape Town. Two-thirds of the city's water consumption ends up in more than 20 wastewater treatment works (WWTW) across the city, from which the final effluent is normally discharged back into the environment. The total average, daily summer reuse is estimated at 30 Ml per day, or 7 per cent of total wastewater treated. But more could be achieved, as reflected in the report to Council in 2007:

> *The investigation established that the potential of treated effluent use could be expanded to 170 megalitres per day (40 per cent of the total summer wastewater treated per day) at an average total supply cost of below R2/kl. This equals 30 per cent of the annual supply from the new Berg River Dam Project.*

A feasibility assessment in 2004 determined that at R2 per kilolitre, 23,000 Ml/yr can (in the short term) be recycled to a number of potential large consumers. The capital cost of R202 million would be recovered in four years.

Although there are large active biogas digesters at the Athlone WWTW generating biogas from sewage that is used to generate electricity, this represents a small proportion

[15] Cited in CCT, 2006 WSDP for 2006/07.
[16] CMC, 1999: Strategic investigation of bulk wastewater, Study Synopsis, Cape Wastewater Consultants.

of what is possible. AGAMA, a Cape Town-based renewable energy consultancy, has estimated that far more could be done. They calculated that Cape Town's WWTWs generate 4 million m^3/yr of biogas which is equal to 107,826 GJ/yr of energy. This, in turn, could generate 7 GWh/yr of electrical power. At 40 c/kWh, this is equal to R2.8 million per year of electricity. Additional income from carbon credits of about R2.1 million could be generated. Based on these calculations, AGAMA proposed reusing biogas from Athlone WWTW to generate electricity, or convert to biodiesel or methane for gas cars or sell/pipe to homes for use. A sewage-to-gas plant at Athlone, built at an estimated cost of R14 million and with an operating cost of R500,000 per annum, could earn Cape Town R2.8 million in energy savings and R2.1 million in carbon credits, resulting in a payback period of 3.2 years.

As far as the sanitation end of the urban water cycle is concerned, the evidence suggests that there has not been the same level of innovation here as in the water supply sector. Fewer unserviced households have been connected to sanitation infrastructures and there do not seem to be any really significant investments in recycling and reuse to make better use of strategic limited resources and the constituent elements of the sewage flow (methane, water and nutrients).

The problem, as already mentioned, is that inclusive urbanism during the 1994–2006 period was financed by systematically under-investing in repair and maintenance, as well as upgrade and replacement, which resulted in serious system weaknesses and even breakdowns, in particular with respect to sanitation infrastructure. To counteract these trends and protect themselves, senior managers realised that they needed to find ways to increase revenues and stop the transfers into the rates account. They turned to the popular institutional recipes of the time, namely the so-called 'new public management' inspired by neoliberalism and justified by the national government policies of the 1996–2002 period (Swilling 2008). Smith has documented this process in some detail, demonstrating how attracted senior managers were to the idea of 'corporatisation', that is, treating the water and sanitation sector as a self-contained 'business unit' which meant in practice much greater autonomy to set service charges (the ideology of 'cost recovery') and retain surpluses for reinvestment in upgrade and replacement (Smith & Hanson 2003; Smith 2004a). The problem was that 'corporatisation' — like elsewhere in the world — must of necessity go hand-in-hand with the spatial realities of splintered urbanism (discussed in Chapter 8). As Jaglin has demonstrated, there is little evidence that splintered urbanism was embraced by the city, which has been dominated by a political leadership (from both major political parties) that was reluctant to relinquish the post-apartheid necessity to forge a non-racial identity via a programme of inclusive urbanism. Nor was the financial logic of 'corporatisation' adequate for comprehending — and responding to — the challenge of sustainable resource use. For these reasons, neoliberal visions of 'corporatisation' were not nearly as hegemonic or comprehensively pursued in the water and sanitation sector as Smith's writing implies. Both cost recovery and retention of surpluses remain a dream for a dwindling pool of senior managers who still think that 'corporatisation' can be viably reconciled with inclusive urbanism.

Cape Town's solid waste system

As in most places in the world, whereas energy and water have their own powerful national departments, ministries and corresponding national legislation and budgets, up until 2008 solid waste in South Africa was managed almost entirely by local governments in accordance with a hodgepodge of regulatory provisions located within a wide range of different sectorally specific pieces of legislation (from water, to mining, to environmental management). The turning point was the passing of the National Environmental Management Waste Act in 2008 after nearly 10 years of consultation and research — a process driven by the National Department of Environmental Affairs and Tourism. Not only does this Act stipulate the first set of national standards (including — contra the energy sector perspective — respect for the constitutionally entrenched role of local government as the key role player), it also provides for a so-called '3R approach' — 'reduce, recycle, reuse'. Once again, it was in Cape Town where the challenges of implementing sustainable waste-management practices were first addressed.

The absence of a national regulatory framework for waste made it almost impossible to amalgamate the many different waste management administrations inherited from the apartheid era. Although service delivery was systematically extended to almost every area, the level and standard of service differed from area to area in accordance with historical (racially distorted) administrative practices. The passing of the Waste Act in 2008 changed all that, providing the top management of the Waste Department with the excuse they needed to draft what they (significantly) called the Integrated Waste Management (IWM) by-law which makes administrative integration possible while simultaneously introducing the '3R approach' as a key strategy to deal with the rising costs and diminishing returns of the traditional waste management technologies and systems. Unlike electricity provision where Eskom actively subverts an integrated energy planning approach by ghettoising its share of provision, the CCT's Waste Department has physical, financial and regulatory control of Cape Town's waste system, including a wide range of sub-contracting relationships with private sector operators (from collection and cleaning, to separation and recycling, to purchase for reuse).

To build the financial case for the IWM by-law, the Sustainability Institute was contracted to model the long-term financial implications of 'business-as-usual' and compare this to alternatives which offset the increased costs of a '3R approach' against savings generated by reduced landfill requirements (De Wit & Nahman 2009; De Wit 2009). A total of 2.5 million tonnes of waste was transported to landfill in 2006/07, with 40 per cent generated by households and the remainder by commerce and industry (de Wit & Nahman 2009: 5). An additional 0.4 million tonnes of waste was generated, but this was recycled. This means that Cape Town's total waste output was 2.9 million tonnes per annum.

Half of what gets thrown away by households comprises a combination of easily recyclable materials: garden waste (17 per cent), plastics (17 per cent), paper-based materials (13 per cent) and food waste (13 per cent) (De Wit & Nahman 2009: 5). In other words, 1.4 million kg/day of waste (which amounts to more than 500,000 tonnes per year) could be recycled and reused.

Approximately 50 per cent of all household waste is generated by the wealthiest households which, in turn, comprise only 14.5 per cent of Cape Town's 850,000 households. Approximately two thirds of the 500,000 tonnes of potentially recyclable domestic waste originates from these households.

Unlike water and electricity use where there has been absolute and relative decoupling over the past decade, during the period 1998–2007 waste flows grew at 7 per cent per annum. Although this was certainly due to increased waste disposal by Cape Town's households (as the middle class expanded) and by businesses as the economy grew, it is also the result of improved service delivery in poorer areas, which has reduced the quantity of uncollected waste in these areas.[17] But as Figure 9.7 reveals, in response to the introduction of limited waste recycling by the CCT and the growing volumes of building rubble recycled and reused by private sector operators since 2006/07, the absolute amount of waste to landfill declined.

Although there is evidence that an increasing proportion of industrial waste and building rubble is being recycled, the bulk of Cape Town's solid waste is managed in accordance with traditional 'throw, dump and forget' technologies. Basically this means collection of unseparated wastes from each household using large diesel trucks and manual labour; transportation to landfills which are either located near poor communities or far from residential areas thus exacerbating transport costs already pushed up by ever-rising oil prices; ending with disposal into landfills with virtually no attempt to recover usable materials (except for the informal pickers on the dumps) and methane gas. This vast

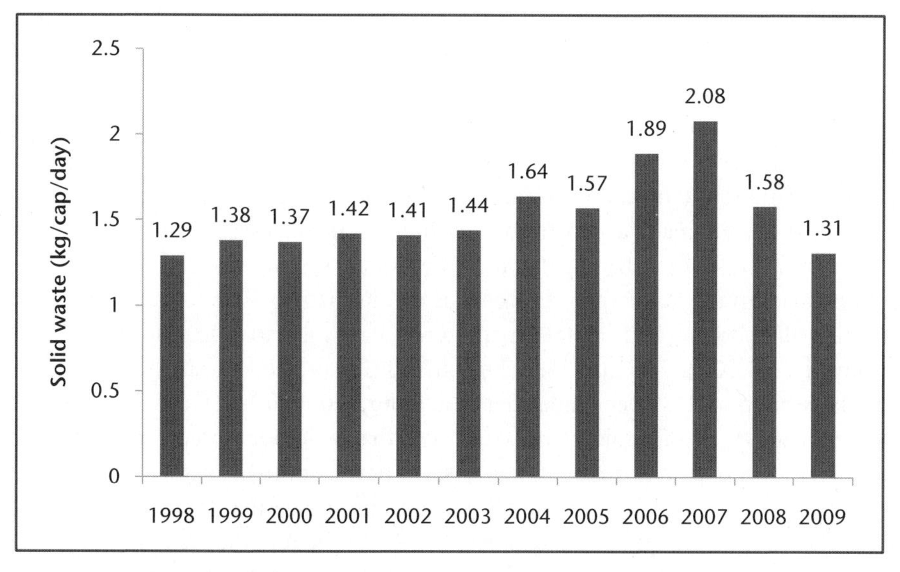

Figure 9.7: Solid waste generation in the city of Cape Town (kg/cap/day)
(Source: De Wit & Nahman 2009: 3.)

[17] However, poor households dispose 0.5 kg/pp/day compared to richer households which dispose over 2 kg — thus this factor should not be overestimated.

infrastructure cost R1 billion to operate in 2005/06 but which has received a surprisingly small annual capital budget of just below R40 million over the past five years.

By the mid-2000s the traditional system was under enormous pressure. The 'reverse salients' were the following: the alarming 7 per cent growth of waste outputs (declining from 2007) with major financial implications; the impact on operating budgets of inclusive strategies to extend service delivery networks into poorer areas (but by 2007 only 50 per cent of informal settlements were serviced), and the rising cost of area and street 'cleaning'; the rapidly rising costs of building and operating landfill sites; the fast rate that landfills were filling up; and the problem of finding suitable sites for new landfills in the face of residential opposition, while less contentious ones imply massive escalations in transport costs. In short, the senior managers of the Solid Waste Department had realised by the mid-2000s that the traditional system was clearly unviable.

All these trends would have been resolvable if extra funds had been available to extend the traditional system using existing technologies and systems. However, the commitment to funding the connections of previously unserviced households had already forced the CCT into financial stress, the already low charges for waste collection had not increased at the rate of inflation over the years, and as in most local government systems, the Solid Waste Department was not politically influential. In short, sustainability became the discursive framework for thinking about alternatives because it was a way of attracting political support for a strategy that needed more upfront funding in the short term. The problem with solid waste anywhere in the world is that switching to a '3R approach' entails costly additions to the logistics infrastructure. It is far too simplistic to assume that because less goes to landfill and more is sold to recyclers, overall costs go down. The easy part is passing a bylaw to compel separation at source. But the separated waste must then get collected by a fleet of waste collection trucks which are not equipped to do this. This means setting up an entirely new parallel system and administration complete with a separate fleet of collection trucks (or new trucks with dual compartments), waste separation depots, and mechanisms for on-selling to recyclers.

Given the overall financial constraints, setting up a parallel system for recycling could only be justified by demonstrating that if the traditional system was retained landfill costs would escalate much faster than the costs of an alternative. Significantly, landfill costs (including all operating and capital expenditures) have increased dramatically from R54 per tonne in 2005/06 to R192 per tonne in 2007/08. The projections are that by 2013 costs would have risen to R235 per tonne after which they would level off for a while (De Wit & Nahman 2009). This fivefold increase over less than a decade (driven by inflation, fuel price increases and the amortised annual costs of new landfills), together with the socio-geographical challenge of finding suitably located landfill sites that will not create massive transport costs, is clearly the most significant reverse salient in the system.

To assist the decision-making process, the Solid Waste Department commissioned a study which modelled three alternatives: 'business-as-usual' (BAU); the BAU system plus recycling for the richest households; and what was called the MaxiMin option. The MaxiMin option was selected as the preferred option because it maximised the amount of waste diverted from landfills into recycling operations at the minimum cost. More could be done, but it was calculated that the cost per tonne beyond this baseline would go up considerably.

The MaxiMin option, if implemented, would involve normal area cleaning plus waste buy-back centres in poorer areas; waste collection for all households with compulsory separation for the most affluent (132,000 households); 20 per cent recovery at landfills; and diversion of 20 per cent of building rubble and compost material into recycling activities. It was calculated that this would make it possible to divert 1.2 million tonnes of the total waste flow of 2.9 million tonnes into recycling activities. The cost of increasing the level of recycled material from below 10 per cent to over 40 per cent is estimated to be between 3–15 per cent of the total budget for solid waste management (De Wit & Nahman 2009). This compares to much more significant long-term cost increases if the BAU system is retained.

To conclude, we need to recall that the IWM strategy was not just a response to rising costs by way of a sustainability strategy—it was also about the integration of separate administrative systems that were undermining the aspiration to provide universal services for all. In other words, sustainability began to be used to reinforce an inclusive urbanism agenda but which evolved for discursive reasons into a green urbanism alternative. This evolution had a lot to do with the unique structure of the waste sector compared to the other EWWS sectors, each of which are dominated by a powerful centralised delivery agency. Solid waste has a vast number of private sector and community-based actors who play key roles in the overall value chain. This means there is flexibility in the system which, in turn, could make it possible to respond in contextually specific ways to the wide range of different communities and users. Although this could be interpreted as a move towards some sort of privatisation with a splintered urbanism agenda, this need not be the case if the overall goal remains equity and affordable services for all rather than just cost recovery.

Forward planning versus decoupling

Since 2006 there has been a growing realisation that development strategies need to address the question of sustainable resource use. This has been recognised in the CCT's policy documents (City of Cape Town 2008), and in the policy documents of the Western Cape Provincial Government (Department of Environmental Affairs and Development Planning 2007; Western Cape Provincial Government 2007; Western Cape Provincial Government 2008). The academic literature has also started to reflect similar arguments (Clark *et al.* 2007; Petrie and Ocran 2007; Sustainability Institute 2007; Swilling 2006; Swilling & Annecke 2006; Crane & Swilling 2008; Pieterse 2010; Swilling 2010; Swilling 2008; Swilling 2010). One key indication of this emerging policy and academic literature is that service delivery will not be able to address the needs of the poor if these services depend on traditional technologies and systems. It follows that job-creating growth will depend on a very different set of technologies and practices which reinforce the decoupling trends that are already evident in the EWWS sectors.

Most EWWS sectors do their forward planning by correlating growth in demand for their services to economic and population growth projections. The evidence, however, suggests that there is a decoupling of resource consumption from economic and population growth in all the EWWS sectors. This has major policy implications for capital budgets and operational expenditure.

The growth factors that are used for projections by the city are as follows:

- ***For electricity use***: 2.7 per cent in 2007/08, and 3.5 per cent thereafter (City of Cape Town 2007a). The 2008 IDP review factored a nil increase in electricity growth mainly due to the implementation of an energy savings plan, but this was only for one year (City of Cape Town 2008)
- ***For water use***: 3 per cent per annum for the foreseeable future for what was defined as unconstrained demand, with a similar rising curve for 'low water demand' (DWAF)
- ***For solid waste***: 7 per cent up until 2007/08 without waste minimisation, thereafter dropping *if* waste minimisation measures were approved and funded (City of Cape Town 2007b).

As demonstrated earlier, these growth factors are out of line with the actual growth rates of the primary resources that course through the networked infrastructures. This is especially the case for projections for water use. The average growth in water use from 2001–2006 was minus 2 per cent, and for the 10 years from 1996–2006 only 0.04 per cent, compared to an average annual economic growth rate of 3.9 per cent per annum over the same period. It is the volatility in the growth of water use that complicates planning for future use. Water-use growth peaked at 4.9 per cent in 2003/04 compared to minus 12.4 per cent year-on-year growth in 2000–2001. The obvious driver here was the drought coupled to municipal awareness campaigns, enforced restrictions and pro-poor tariff reform. This, of course, proves the point: consumption behaviour can be changed.

Electricity is a similar story. The growth in electricity use over the last few years is also lower than the current estimates which are tied to economic growth projections (excluding the single reference to 0 per cent growth in the 2008 IDP). Growth in electricity use is expected to be even lower in the immediate future because of supply constraints and much higher electricity tariffs from 2008.

Solid waste grew by an average of 8.9 per cent over the period 2001–2006, with a high of 20.4 per cent in 2003/04 and a low of 2.6 per cent in 2001/02. But as the preceding discussion on waste demonstrated, growth rates have declined since 2007 after relatively minor recycling practices were introduced suggesting that with relatively little effort a lot can be achieved.

In short, as in many cities, there is a disjuncture between the actual year-on-year growth rates in the EWWS sectors and the growth rates used by the forward planners that manage these large technical systems. Although it is common practice for the managers of large technical systems to derive budget projections from growth and population coefficients, the obvious consequence is that decoupling is excluded from the start as a strategic option, thus removing any incentive for innovation. The end result can be the perpetual subsidisation of inefficiency and, ultimately, unsustainability.

Greening slum urbanism

Over the past two decades, homeless people have been quietly encroaching — and at times noisily invading — whatever spaces they could to connect to the networked infrastructures needed for survival in Cape Town. As demonstrated in the emerging literature on Cape Town's slum urbanism (Cross 2006; Ngxabi 2001; Robins 2006; Smit 2000; Smit 2006; Swilling *et al.* 2010; Yose 1999) and the ever-present reports in the press, it was not only the planners and engineers who decided whom to connect and how. People found their own ways to tap into pipes and cables, build toilets, dispose of their waste, and access energy and transport. We do not intend to rehearse this well-known story here. What is interesting is the way slum urbanism has begun to respond to the City's resource constraints. An excellent example is provided by the urban design vision and detailed infrastructure plans developed by a Cape Town-based architecture firm, ARG Design, for the most densely populated informal settlement in Cape Town known as *Kosovo.* This design framework managed to creatively merge an incremental approach with ecological design (Goven 2010; Goven 2007).

Towards a sustainable Cape Town?

The trends addressed in this chapter suggest that inclusive urbanism will persist, but that this will need to go further than simply connecting everyone to the city's networked infrastructures. As recent writing suggests (Ewing & Mammon 2010; Hendler 2010), unless sprawl is limited, densification promoted and genuine mixed-income neighbourhoods created, the aspirations of inclusive urbanism will not be realised. That said, the fiscal constraints are enormous, thus driving decision-makers to find ways of incorporating the private sector. These same constraints also drive the quiet encroachments into the city's networked infrastructures by the poorest households, often in the process redefining the boundaries, densities and cultures of the city. But if more private sector investment means splintered urbanism via orthodox cost recovery, this will undermine the non-racial vision of the kind of inclusive urbanism that successive political leaderships have pursued since 1994 and reinforce slum urbanism. However, as developments in the solid-waste sector suggest, opening up the system to a greater number of private and community-based actors can create additional capacity and flexibility without necessarily commodifying the service.

As the underlying resource constraints undermine the conventional ways of doing things, so green urbanism (doing more with less) and liveable urbanism (restoration of life) emerge as frameworks for thinking about solutions. Since these are early days, there is a tendency to underestimate how serious the resource constraints really are, thus justifying the retention of tried-and-tested delivery technologies. So it is not surprising that green urbanism is the more attractive of the two and, of course, cheaper in the short run. The Green Building Council (GCB), which is Cape Town-based, has become the intellectual centre of learning for professionals who service the commercial sector. Every week there is one or other workshop or conference on sustainability, and business-linked bodies such as Accelerate Cape Town, Cape Town Partnership and Cambridge Programme for Industry deliver a constant flow of events and discussions that relate to the future of the city. The local universities, CCT and Western Cape Provincial Government all have ongoing programmes that are, in one way or another, about making Cape Town more sustainable. In general, however, the more progressive wing of Cape Town's green urbanism movement is focused on how to do more with less and the language used is invariably about cost savings through greater efficiencies. This is reflected in investments in making buildings energy efficient, moving more people onto buses, increasing densification, reducing water losses, using renewable energy to power traffic and street lights, installing solar water heaters, limited recycling and the greening of commercial developments. These are all vitally important steps in the right direction, but as Birkeland has argued, sustainability is not achieved by retarding collapse (Birkeland 2008).

As our analysis of the EWWS sectors revealed, there are things that a utopian realist can envisage which could reconcile urban growth with eco-system restoration. Inevitably, these are more radical, probably more costly in the short run, but more cost effective in the long term, and conceptually somewhat beyond what most actors are able to imagine. Practical suggestions have already surfaced in stakeholder discussions about alternatives, and some have even started to emerge in practice. Although it is not possible to discuss all the alternatives, options include the introduction of large-scale investments in solar power (including Concentrated Solar Power [CSP] plants and Solar PV) to generate electricity (including base-load) and heat; halving water consumption (compared to 2000 levels) via efficiency interventions; harvesting additional water supplies from the sustainable use of aquifers, and desalination plants powered by CSP; and the application of zero-waste principles to both solid waste and sewage flows, seeing both these waste streams as supplies of key resource inputs. Beyond the EWWS sectors, the big shifts would be drastic restrictions on private car use in order to boost mass transit via rail and coach; regulatory interventions to stop urban sprawl on the periphery in order to drive densification around public transit nodes and corridors (especially the inner city); approval of all building plans made conditional on adherence to green-building guidelines, including the removal of restrictions on non-conventional building materials (such as straw bale, adobe brick, sandbag and bamboo); massive increases in local food production within the city and along the peri-urban boundaries with job creation, CO_2 reduction and nutrition improvement in mind; and significant investments to improve biodiversity and eco-system services.

Conclusion

We have tried to deploy our understanding of complexity, material flows and contested urbanisms developed in previous chapters to develop an understanding of the practicalities of urban transition. We used the example of Cape Town to demonstrate the benefits of an empirical analysis that reveals the primary material flows through the city's networked urban infrastructures and how these shape and constitute a particular set of contested urbanisms. This enabled us to consider the kinds of interventions that could redirect these flows to catalyse the evolution of a more sustainable urban development trajectory, including how to go beyond the limits of green urbanism. By doing so, we have shed light on the micro-dynamics of the global challenge of managing the transition to a more sustainable socio-ecological regime at precisely the moment we have entered an urban age dominated almost entirely by cities. Unless cities are seen as key sites for imagining and practically engineering the transition from linear to circular metabolic flows, the chances of achieving a more sustainable world will be almost nil. For this to happen we will need fewer utopian techno-fantasies for the greening elites (such as Masdar, Dongtan and Treasure Island), and more utopian realism about what it will take to retrofit cities so that everyday urban living can become more sustainable over time.

Chapter Ten

Pioneering Liveable Urbanism: Reflections on an Invisible Way

> *What is of the greatest importance grows and keeps on growing ... Don't despair too much if you see beautiful things destroyed, if you see them perish. Because the best things are always growing in secret. We have discovered an invisible way. The next stage of our evolution is to be free of our visible things. Then we will become sublime forces in the universe.*
> (Ben Okri 1995. *Astonishing the Gods*)

Entry points

It is October 2010 and we are writing in our quiet office in a 500-year-old building at University of Freiburg in the heart of Germany's 'solar city'—an unlikely place to begin to make sense of the last 10 years of our work within the emerging Lynedoch EcoVillage. Bombed by the Allied Forces in the last months of World War II, Freiburg is now one of the 'sustainability capitals' of Europe resulting from massive investments in renewable energy since the 1970s, integrated transport systems, fresh food markets and a ban on cars in the inner parts of the town.

Much humbler, far messier and within a context in the most unequal country in the world, what are the emergent patterns that we think make Lynedoch EcoVillage a space in which to experiment with a just transition? It may be too soon to tell, but the initial conditions, the gritty framing of the underpinning ethics, ethos and practice have provided some stories which seem to continually rise to the surface as we relook at the joys and pains experienced in the process of transforming this particular space.

Tessie's story, below, as well as our own, provides some sense of a set of experiences which, when reduced to any other form of analysis, runs the risk of being seen through lenses of disciplinary divides and fragmentation, which reduce the complexity of our experience into the very divides we struggle daily to bridge. Narrative integrates, creates rhythm, is timeless and offers perspectives that are different from the other forms of analyses used thus far. Other ways of telling the story of Lynedoch may lie on a continuum of dualisms such as 'triple bottom line', 'development versus environment', 'minimising environmental impacts', 'social-ecological relations', 'building human-nature connections'—all in various ways using a language structure that separates and fragments what are in reality indivisible wholes.

We start from this point of indivisibility. Humans *are* nature, although Western civilisation appears to have struggled to disprove this for many centuries. An extended sense of the individualistic 'self' includes soil, soul and society—diversity within an eternal whole that is a prerequisite for all life.

Tessie van Niekerk, a seven-year-old girl, lives at the Lynedoch EcoVillage, Stellenbosch. At the time of her birth, Tessie's immediate family—her great-grandmother, grandmother, grandfather, mother and uncle—were already one of the pioneer families in South Africa's first emerging, mixed-income, sustainable neighbourhood. She lives in a 140 m^2, government-subsided home built from adobe (clay) bricks. She has a hot bath every day from a solar-water geyser; her waste is separated at source; her flushing toilet uses water recycled through the on-site waste treatment plant that processes all black and grey water by means of earth worms and an integrated wetland system; she collects and eats vegetables from the community vegetable garden and plays endlessly in the indigenous surrounds.

Tessie's neighbours are a mixed-income group, with mixed jobs (organic farmer, professor, driver, crèche teachers, etc.) and a combination of races. Tessie speaks Afrikaans, her mother-tongue, and English fluently. To her great delight, she is learning words in isiZulu from her best friend, Velile Buthelezi.

She attends the ecologically renovated Lynedoch Primary School (a Western Cape Education Department, non-fee paying school, which serves one of the poorest communities in South Africa), which is a two-minute walk from her home and part of the Lynedoch EcoVillage. Before school, she visits the crèche, which starts with gentle music and t'ai chi exercises. She sees the children organising their breakfast table, making ready for one of the two 'farm-to-fork' meals provided to all crèche children every day. Tessie chooses to help one of the younger children with his artwork, leaving when he joins a group going to harvest the organic vegetables to prepare for lunch. On the way, she stops to watch the neighbouring Grade R class, who share the Lynedoch early learning space. Altogether there are 75 farmworker children of mixed ages at the crèche, and 300 in the primary school. Stories, rhymes and a rhythmic connection with life are the threads that bind their experience of everyday life.

Tessie, along with 65 others, participates in the 'Changes' aftercare group (so named by the teenage boys after the Tupac song). This aftercare facility was the result of the Lynedoch community's decision to tackle head-on the issue of under-fathered boys—triggered by a group of teenage youth who had started to experiment with knives, alcohol and drugs. Tessie does not know of the frightening visit with the boys to the Stellenbosch prison, where they were introduced to hard-core prisoners inside dank, dark cells. She does not know of the choices with which the boys were confronted—the demands made on them by their community to choose lives of honour. Nor does she know of the demanding work that goes into creating alternative experiences for firstly those boys, and now an entire aftercare facility due to the demand for participation by so many local farm children—boys and girls. What she does know of, along with the rest of the Lynedoch community, is the extraordinary success of the Lynedoch United under-11, under-15 and under-17 soccer teams, who won the Stellenbosch LFA junior league. This was no small feat for a straggling group of bored boys, our own son included, who did not even have a team or a ball three years ago, let alone the amount of organisation and support required to compete against 40 other teams for the Stellenbosch trophy. With South Africa holding the highest reported rape statistics in the world, increasing drug dependency and gang-related violence, especially in Western Cape communities, the Lynedoch community's emphasis on creating alternative models of masculinity for teenage boys is as much in its own interest, as in that of the boys. After some of our Sustainable Development masters students took five boys night-hiking at full moon (despite much initial resistance), the boys returned having loved the experience, saying: 'we left our stress on the mountain'. These children, like Tessie, come from a context rooted in the remnants of South Africa's violent apartheid history, which has left the Western Cape rural areas (in particular) with the highest rate of Foetal Alcohol Syndrome in the world.

Tessie is able to choose how she would like to spend her afternoons: homework and remedial support if necessary; drama classes; martial arts; gardening; sustainability lessons on the various ecological aspects of the Lynedoch EcoVillage; holiday camps; cycling; and, when she is a little older, learning to nurture with nature—joyous celebrations of girls becoming young women.

Having only recently heard some stories of Lynedoch's history, Tessie chided her grandmother for not telling her earlier about what a 'terrible place' Lynedoch used to be. 'You must tell me these things, Ouma. I want to know what happened. But is it true?' *Her grandmother is struck by Tessie's total disbelief at how things used to be. While in Tessie's short lifetime Lynedoch has transformed in its entirety, it sometimes still seems only a heartbeat away from its drunken, apartheid and violent past.*

What invisible strands made Tessie's life possible?

Context

Stellenbosch is a small university town some 35 minutes' drive inland from Cape Town. It is surrounded by three mountain ranges and is home to the unique fynbos biome, the sixth and smallest of Earth's biomes. Stellenbosch is also the commercial centre of a wealthy, white-owned, agricultural community dominated by the winemaking industry. Black people have not only suffered from exclusion from economic ownership in this region since agriculture began in the late 1600s (with key exceptions made from time to time, for example, after the end of slavery in the 1830s when some free slaves were given land, which they later lost), but also from housing, education and higher learning. The infamous '*dop* system' (paying workers partially in alcohol) has caused profound social damage which will take at least a generation to heal, with children being the key to a better, safer future for this region—but only if they can break loose from the cycles of poverty, alcoholism and domestic violence.

Instead of adding to the cacophony of voices that have already done an excellent job calling for a grassroots movement of alternatives (for example, see Hawken 2007; Sen & Waterman 2007), we reflect on 10 years of doing this in practice as we imagined and co-built with others the Lynedoch EcoVillage, the first ecologically designed, socially mixed, intentional community in South Africa.

Neither of us was trained in the design or technical professions. We come from the social sciences and spent the first five years of the democratic period in South Africa teaching public and development leadership in Johannesburg.[1] During the decade leading up to 1994, both of us worked with NGOs and CBOs, which were the lifeblood of the social movements that brought apartheid to an end (Swilling 1999).

This meant we were able to bring into the Lynedoch EcoVillage experience an appreciation of empowerment through participation, an adaptive leadership approach (Heifetz 1994) and an appreciation of the importance of embedding design within robust social processes. Our technical partners, architects Alistair Rendall and Gita Goven, who set up ARG Design[2] in Cape Town, were committed to exploring what ecological design meant in practice within the South African context. It is never easy to reflect critically on one's own practice, but we concur with a well-established literature which insists that direct participation in social change is the best way to get close enough to learn something useful for the benefit of others (for the seminal contribution see Touraine 1981). For us, 'direct participation' has meant living and learning in the community of Lynedoch.

Adaptive design

While it is true to say that we started off seemingly hopelessly ill equipped for this journey — no grand plan, barrage of consultants or finances — in retrospect, it turned out that we were actually well equipped with a sensibility we have combined to create a philosophical approach or framework, which we now call 'adaptive design'. This framework was based on the work on 'adaptive leadership' (an influential theory of leadership which proceeds from the assumption that there are two basic types of challenges: technical and adaptive) pioneered in the 1990s by Professor Ron Heifetz, and the work on 'positive development' by the Australia-based Professor of Architecture, Janice Birkeland (Birkeland 2008; Heifetz 1995). To complement these perspectives, we share a radical pragmatism inspired by the bioregional ethics of place (Hattingh 2001; Norton 1991) and a commitment to sustainability-oriented innovations with a social- rather than a market-driven purpose (Seyfang & Smith 2007).

Technical challenges are what most people are comfortable with: a problem arises, tried-and-tested solutions are identified, a particular solution is then selected for application. This often means that the decision-maker enjoys relatively uncontested authority to make the decision. The underlying assumption is that most challenges are technical challenges which have been addressed in some way before. However, new and increasingly complex problems often emerge within contexts in which there is no recognisable or generally accepted authority with the requisite capacity and capabilities to generate appropriate solutions — and, in fact, where the technical solutions may still need to be figured out while titanic clashes of worldviews, paradigms and values are under way. Where leadership is so diffuse and the context beset by unfolding complexities, it can become extremely risky — and often counter-productive — for any one party (particularly if it is a strong player) to narrow things down to a technical

[1] Mark Swilling was Director of the Graduate School of Public and Development Management, University of the Witwatersrand, and Eve Annecke headed up the leadership development programme within this school until they resigned in 1997.

[2] Website details for ARG Design: http://www.argdesign.co.za

challenge for which known solutions already exist and therefore readily applied. Quite simply, it does not work. As argued in Chapter 4, this is particularly risky from a sustainability perspective, as new challenges emerge which require niche innovations that are unlikely to be generated from within established knowledge networks and technological regimes.

Heifetz argues for seeing these as 'adaptive challenges' that require 'adaptive solutions', which are the emergent outcome of (often lengthy) interactions between stakeholders prepared to explore options where no obvious technical solutions exist. The problem, of course, is that this is easy to say, but it requires a very different type of leadership if one wants participation — or, indeed, trans-disciplinary collaboration — to be more than the tokenism that it often becomes. Contrary to the conventional assumption that leaders (in particular, design professionals) 'know what is best', adaptive leadership entails new skills to help groups face their own tough challenges by working with them to figure out what is precious and what is expendable. The obstacle to change is not that people resist change as such, but rather they resist loss — the sense of loss that comes from having to let go of deeply held values and conceptions of what they know or what they assume is regarded by society (or 'science') as an acceptable norm.

As exemplified in the Appreciative Inquiry methodology (Cooperrider and Srivastva 1987), if what is precious becomes the animator of co-operative creativity, extraordinary solutions can emerge which often surprise, energise and transform with remarkable ease. Adaptive leadership entails cultivating the diversity of perspectives that make this possible, including encouraging productive conflict and utilising the power of multiple leaders to transcend preconceptions that are barriers to innovation. The primary aim is, therefore, to lead changes in values, attitudes and behaviour in order to resolve an adaptive challenge. In this context, formal authority/expertise and power is not much use. As is so familiar across Africa, the (often pompous) ritual of formal authority is secretly mocked into insignificance by those invited to applaud, leaving very little in its place (Swilling *et al.* 2003).

By contrast, adaptive leadership is about the skills needed to make (and protect) the spaces for the emergence of (often unanticipated) initiatives to come up with creative solutions that work because they are carried by the self-organised capacities of the group rather than driven by a leader who colludes with those who want him/her to 'know it all' (or worse, groups who collude in refusing to tell the 'emperor that he has no clothes'). This capacity for adaptive leadership needs to be built if niche innovations for sustainability are to be promoted across strategic sectors of the economy. Indeed, it is a form of leadership that may well be a *sine qua non* for the type of innovation-promoting developmental state advocated in Chapter 4.

Van der Ryn and Cowan's seminal book, *Ecological Design,* marked the key break away from conventional design by consolidating the synthesis of sustainability thinking and design that had emerged in practice during the 1970s and 1980s. It was this book that offered the founding statement of ecological design, namely that 'the environmental crisis is a design crisis'. Given the context, their definition of ecological design was certainly radical for the time:

> *We define ecological design as any form of design that minimises environmentally destructive impacts by integrating itself with living processes.* (Van der Ryn & Cowan 1996: 18)

This basic idea was extended into the commercial world and popularised by Hawken *et al.* who argued that real estate development that does 'less harm ... [is] even more profitable' (Hawken *et al.* 1999: 87). This notion that 'less harm is profitable' inspired the Green Building movement and what we characterised in Chapter 5 as the *green urbanism* approach. It is, however, unfair to blame Van der Ryn and Cowan for the way ecological design has been co-opted by the property development industry and the global signature architectural firms. They were partially aware of the need to go beyond 'minimising damage' when they wrote that design must also be about how to 'heal the living world' (Van der Ryn & Cowan 1996: 18).

It was, however, Janice Birkeland who took this idea further. Reflecting on the co-option of ecological design by the property development industry from the late 1990s, she argued that the environmental rating systems that have emerged focus only on the 'symptoms', such as waste and climate change, while neglecting the 'root causes' which are all about resource consumption and unsustainable material flows. The result is that ecological design is turned into a 'war with nature' in which we manage how much we extract, but we never contemplate making peace with nature by reversing the extractive process (Birkeland 2008: 4–5). From this flows her argument that the aim of design must be the restoration of life:

> *If we are to sustain the economy, then urban development must also restore and expand the ecological base of the surrounding region.* (Birkeland 2008: xix)

However, Birkeland is pessimistic about alternatives and suggests that there are very few examples of the approach she advocates. In their review of trends, Hodson and Marvin suggest that there are examples of what we have called *liveable urbanism* with a focus on restorative design. Referring in particular to the Transition Towns movement in the UK, they refer to these as 'alternative responses' to the 'bounded responses' of *green urbanism.* They depict this contrast in a useful way in Table 10.1.

When we started work on the Lynedoch EcoVillage, there were very few examples of ecological design we could look to for inspiration and guidance.[3] Indeed, Van der Ryn and Cowan had only recently published their seminal text and most architects we spoke to smiled patronisingly and made clear their tack of interest. With hindsight, the Lynedoch EcoVillage can be firmly located in the right-hand column of Table 10.1 as a work in progress. It is this experience that has enabled us to understand the distinction between 'minimising damage' and 'restoration of life', especially now that 'green buildings' and 'eco-estates' have started to pop up all around us. Looking back,

[3] Key exceptions were Auroville in India (www.auroville.org), Crystal Waters near Brisbane, which inspired the Global EcoVillage Network (http://gen.ecovillage.org) and Tlholego EcoVillage in South Africa (www.sustainable-future.com). It was only later on that we discovered older experiments like Gaviotas in Colombia, Sekem in Egypt, Hivre Bazar in India, Vauban in Germany, Bedzed in the UK, EcoYoff Living and Learning Centre in Senegal and many others that have now mushroomed all over the world.

Table 10.1: 'Alternative responses' to the 'bounded responses' of *green urbanism*

Bounded responses	*Feature*	*'Alternative' responses*
Transcendence of limits	Ecological constraints	Work within limits
Commercial—banks, developers, architects, utilities	Social interests	Community—NGOs, environmental groups, charities
Divisible	Concept of security	Collective
Productionist-scale economies	Scale of solution	Consumption—small, local
Eco-urbanism—eco-cities, regions, blocks and towns	Type of build	Retrofitting—existing and new
Product of bounded security and by-pass	Consequences	Mutual interdependencies
Dontan (Shanghai), Masdar (UAE), Treasure Island (California)	Exemplars	Transition towns, relocalisation, ecovillages, bioregionalism

(Source: Hodson & Marvin 2010: 310.)

the only really sustainable outcome has been a profound process of social learning that has built a relational, adaptive capacity for ongoing innovation.[4] And this, in the end, is what adaptive design should be all about — not a perfect technical solution, but rather a legacy of sustained capacity to resolve both adaptive and technical challenges as they come up every day, or what in a different literature is called 'resilience' (Beatley *et al.* 2009; Kay & Richardson 2007; Roux *et al.* 2008).

With hindsight, this can all boil down to some very simple things: making sure the water pumps keep working; that energy is not being used wastefully by various appliances; that worms in the sewage treatment plant don't die; that buildings get painted with the correct kind of paints; that residents realise that closing the curtains on north-facing windows during winter will increase their heating costs; that leaking water pipes are fixed; that people learn how to grow food and plant gardens; and that conflicts are resolved amicably through dialogue. Co-operating to deal with these practical, day-to-day matters (however bumpily) is what building community and sustainable living is all about.

By fusing the notions of 'adaptive leadership' and 'restorative design', we discover the purpose of adaptive design: to put in place networked infrastructures and flows that depend on — and, indeed, foster — social co-operation to keep things going. This is very different to design for individualised consumption which requires someone — at considerable financial expense — to manage the system 'at arm's length' (which is invariably kept 'out of sight', and therefore 'out of mind'). Surely this must be what liveable urbanism is about — the endless creative acts to restore life by the way we live co-operatively with one another, while reconnecting to the entire web of living beings.

[4] There is a growing body of research about Lynedoch that confirms this conclusion (Dowling 2007; Irrgang 2005; Mokheseng 2010; Spies 2010; Stuwe 2009).

Beginnings: Finding Lynedoch's heartbeat

It is old development lore to locate oneself deeply within a context in order to first find its rhythms and then to simply work with what is there, slowly connecting the processes so that they are allowed to unfold in their own creative ways. It seems seldom though that the contrived 'outcomes and goals orientation' cast in managerial speak can make space for this to occur. By searching for a restorative role in a context that seemed to present staggering challenges (no land reform; chemical- and pesticide-reliant farming; soil degradation; farmworker evictions; high alcoholism; minimal education), two things emerged as key levers for change: working with children and restoring the health of the soil.

Since the 1970s, the school that the farmworker children attended, Lynedoch Primary School, was housed in what were supposed to be temporary prefabricated buildings. But like most promises made to rural farmworker families, by the end of the 1990s no permanent structures had been built. It was probably no coincidence that on arrival, we discovered that the adjoining Spier Wine Farm, bought by businessman-cum-philanthropist Dick Enthoven in 1994, promised support to the school by way of land on the farm and some funds. After it became clear in 1999 that zoning regulations and objections from white neighbours would scupper this plan, funds were made available to a project initiated by Spier (and managed by ourselves when we were still employees of Spier).[5] The project, initially to house premises for the new school, entailed the purchase of a 7 ha property close to Spier, which is where the Lynedoch EcoVillage is now located. A R1 million donation from Spier plus a R2 million loan from a commercial bank made it possible to purchase the property in the name of a non-profit company called Lynedoch Development (LD).

The initial land purchase included the following structures:

- The old historic country hotel (Drie Gewels) in a dilapidated condition and barely functional (717 m^2)
- Three family-sized residential houses in varying degrees of disrepair
- The large, dominating dance hall-cum-restaurant, referred to as the 'main building' (4,023 m^2)
- A small brick shed (50 m^2)
- Approximately 30 prefabricated asbestos units of about 10 m^2 which had served as guest rooms for the hotel and needed to be demolished immediately.

In 2000, with little more forethought than cleaning up the site and relocating the school, renovations began on the main building to create a space of beauty for the Lynedoch Primary School, and a performing arts group sponsored by Spier. In January 2002, the school moved into its new premises. As is common practice since 1994, when something good happens within the public sphere, the South African flag is proudly displayed — in this case, the children walked ceremoniously across the road to the new school, with a child in front carrying the flag.

[5] We worked for the Spier Group from 1998–2002.

From a dilapidated, prefabricated building on the land of a local white farmer, the school now inhabited an ecologically renovated building[6] sporting wind chimneys, clay bricks, passive cooling and heating, along with 10 classrooms, a staff room, computers, a community hall with a sprung floor and huge space for a multitude of activities. The beauty of the building gives meaning to Birkeland's notion that development must be about the 'restoration of life': not only did it trigger the tree planting and gardening programme, it also became the hub for social transformation across the valley, with children the central focus at all times (Spies 2010).

Worth noting was the complex set of relationships that made this possible: the private sector with Spier's donation for the building of the new school; the Western Cape Education Department which agreed to operate the school and to enter into a long-term lease to contribute to the upkeep; and the non-profit sector in the form of the Sustainability Institute which provided all the development facilitation, design and fund-raising support.

While children were the one foundation of the Lynedoch EcoVillage, the other was the soil. In September 1999 the first seedlings were planted in what was to become a 100 ha land-reform project located a short distance away from the village. Responding to the need to assist landless farmers to secure land, Spier agreed to relinquish control of 100 ha of publicly owned land which it leased from the municipality. This land—known as 'commonage'—was land that the white-controlled Stellenbosch Municipality had leased to white landowners shortly before the democratic elections in 1994 in a bid to prevent land reform. Countering this racist move, Spier agreed that this commonage could be made available to a land-reform project serving between 14 and 20 small farmers. While the building was being renovated to house the school, the board of Lynedoch Development began considering what to do with the property as a whole in order to make it financially viable (including how to pay off the loans). Gradually, this led to the idea of building an EcoVillage around the school, initially just for farmworkers, but the vision expanded over time to building a socially mixed community.

Building vision

The Board of Lynedoch Development comprised a mixed group of local community leaders, professionals (including ourselves) and the principal of Lynedoch Primary School (who chaired the Board).

Set up in 2000, this Board was energised by the possibility of building an inclusive living and learning community that would demonstrate in practice what it means to live in sustainable ways. The Board, however, was never able to raise the resources required to cover the full costs of the innovation and social facilitation processes that were needed to close the gap between vision and implementation. This gap was addressed by the Sustainability Institute (a non-profit Trust founded in 1999, initially named the Spier

[6] ARG Design was awarded a Best Practice Award by the Council for Scientific Industrial Research (CSIR) in 2004.

Leadership Institute) based at Lynedoch,[7] which worked in partnership from 2002 with Spier and the School of Public Leadership at the University of Stellenbosch (see www.sustainabilityinstitute.net). The Sustainability Institute effectively acted as the animator of the adaptive design activities, institution building and community empowerment processes. This NGO-university alliance was able to mobilise intellectual capital, research networks and a sense of vision-in-practice that made the project credible in the eyes of the providers of senior debt[8] (the Development Bank of Southern Africa — a government-owned bank), local bankers (Nedbank), Spier, local authorities and, most importantly of all, the buyers of the properties.

Inspired by this commitment, three goals were formulated in 2000 to guide the various aspects of the planning and implementation of the project. The goals were:

- To be a socially mixed community (both in terms of race and class) organised around a child-centred learning precinct
- To strive to be a working example of a liveable ecologically designed urban system
- To be a financially and economically viable community that would not require external funding to sustain itself over time.

Above all else, the Board members were determined that the Lynedoch EcoVillage provide a safe space in which South Africans from all backgrounds could attempt to live in peace with one another and in harmony with nature. It should also, they believed, 'be a place where people from all over the world could come and share in the life of the community while they learn, think, and create works of art and knowledge that will contribute to the making of a better world. It must, in other words, be a place where all life is celebrated and beauty in all its forms treasured for this and future generations.'

By the time of writing (June 2011), the key features of the Lynedoch EcoVillage were in place:

- Village vegetable gardens and landscaped areas planted with indigenous plants
- A primary school for up to 400 children drawn mainly from the families of local farmworkers (completed December 2001)
- A pre-school for 45 children, with an upstairs roof space used by the 'Changes Youth Club' (aftercare for children and teenagers)
- A large multi-purpose hall (completed in December 2001)
- Offices and classrooms for the Sustainability Institute (completed in December 2001)
- Conversion of the old Drie Gewels Hotel and an existing residential house to provide accommodation for 25 people, mainly participants in the programmes

[7] The Sustainability Institute obtained core funding from the Ford Foundation for the period 2002–2004. This grant made it possible for the Institute to engage generally in the Lynedoch Development while simultaneously setting up a new master's programme in sustainable development that includes the learning from the Lynedoch Development in the curriculum of the various modules.

[8] Senior debt is large-scale debt that is repaid over a long period of time, with all other short-term debt (such as overdraft facilities) regarded as subordinated to the senior debt.

of the Sustainability Institute, as well as a small conference venue for general use (completed 2004)
- 42 new residential sites, with 15 earmarked for purchase at a price of R20,000 by people who qualified for a government housing subsidy (10 of which were completed by March 2006), with the remainder sold at a commercial rate ranging from R90,000 to R275,000 per erf—breaking from usual South African practice, the urban design did not spatially separate the subsidy erven from the commercially priced erven[9]
- Commercial spaces for offices, small manufacturers and crafters
- A land-reform organic farm in partnership with Spier
- Traffic restrictions: a limit on the number of cars in the village, and parking restricted to designated communal parking areas, which secures the space for children and pedestrians.

Future phases of the development have been the focus of ongoing discussion between the Lynedoch Development Company and the Lynedoch Home Owners' Association (LHOA), which is also a non-profit Section 21 Company. It is compulsory for every property owner to be a member of the LHOA. This means that after the sites were transferred to the property owners in February 2005, the LHOA was able to start operating as the democratically elected governing body of the village. The Stellenbosch municipality approved the Lynedoch EcoVillage development on condition the municipality was not expected to deliver services within a settlement it did not understand—hence the zoning conditions that required the establishment of the LHOA which was then delegated the municipal powers and functions (especially with respect to the provision of energy, water, sanitation, security, cleanliness of public areas, roads and access).

Building strategy

The Lynedoch Development Board believed that the design and development of the village should be an explicit intervention to demonstrate in practice an alternative to the inequities and injustices that characterise the communities of the Winelands region. We walked the talk by being the first middle-class family to build a house in the village.

Set up and preliminary funding

While the main building renovation for the new Lynedoch Primary School began in 2000, Lynedoch Development submitted a full development application to the authorities for approval in June 2000. Approval was only finally secured in May 2002. The delay was caused by racist objections to the proposed development from white neighbours. These objections needed to be overruled by the Western Cape Provincial

[9] An erf is a formally defined plot of land with a surveyed boundary registered on a title deed and legally authorised to be bought and sold on the market. Erven is the word used to refer to two or more of these plots of land.

Government. The approval that was obtained was for 150 housing units on plots ranging in size from 80 m^2 to over 300 m^2, with the large majority in the 120–160 m^2 range. During the participatory planning process, it became clear that a more mixed community with a diverse range of activities and incomes had a greater chance of being self-sustaining than a community that was entirely dependent on returns from workers earning very low wages. The reason for this is that the approvals from local and provincial government were based on the assumption that neither of these levels of government would contribute financially to the development. This meant that none of the costs of the energy, water and waste services would be cross-subsidised. (This later changed slightly when, under pressure from the emerging Lynedoch community and school, the provincial government agreed to provide R6 million for the upgrading of the road intersection, which was a zoning condition imposed by the municipality.)

Whereas the initial development plan presumed that deals would get made with farmers to relocate entire groups of workers off the farms and into Lynedoch, the Board—which included two people who lived on farms and played leadership roles to defend farm dwellers from forced evictions—decided that it could not be party to this because the end result would be another dumping ground for disgruntled workers, who may have preferred to remain living where they had lived for generations. Furthermore, the housing stock on the farms would probably be demolished, meaning that there would be no net gain in housing stock for the country, just a relocation of stock from the farms into the Lynedoch Development. Local community leaders and elected councillors were vociferously opposed to housing developments which reinforced the farmers' agenda to evict farm dwellers.

The initial development plan as approved by the local authorities envisaged a single-phase largely low-income housing development built around the conversion of the Main Building into the premises of the Lynedoch Primary School. However, after the school premises were completed in December 2001 and occupied by the school in January 2002, the Lynedoch Development Board hosted a strategic planning session in April 2002 to consider options for the way forward. It was at this meeting that local community stakeholders expressed reservations about the idea of building 150 low-income houses around the school. They argued that the school was aiming to become a safe, protected and beautiful learning space for the children, unlike their local communities where social problems created challenging and often threatening environments for them. An exclusive low-income housing development similar to other areas where failure levels were high, they argued, would threaten this new safe space from which the children were benefiting. All these factors triggered the strategic decision to adopt a two-phase approach to the housing: the first phase was to have a reduced number of housing units, which could only be viable if an increased number of commercially priced units were included to recover the land costs (which included mounting interest on the loan) and infrastructure costs. This would then be followed by a second phase which would be discussed once the core of the community was established. At the time of writing (2011), discussions about the next phases of the development were just beginning.

Development of infrastructure, and the immediacy of dealing with inequity, needed to be designed with future ecological trends in mind—trends that at the time we could conceptualise but not prove in quantitative terms. However, it is important to note here that, as required by the zoning conditions, the infrastructure was transferred to the Lynedoch Homeowners' Association, which became the body responsible for the operation and maintenance of Lynedoch's entire service infrastructure in terms of the legal framework provided for in the Western Cape's Land Use Planning Ordinance and the Municipal Systems Act.

By 2002/03, we had at best a very vague understanding of the technical and institutional substance of what this would entail in practice. An integrated approach to building a sense of place entailed a combination of an ethical commitment to ecological sustainability and financial viability. To make the ecological and financial case, a 20-year perspective was articulated from 2004 onwards which suggested that it was necessary to design an infrastructure and a set of buildings that took these trends into account.

Although our assumptions about long-term trends took a few more years to test via an extended research programme, they were stated up front in order to argue that it made sense to focus on reducing water consumption in each house, treating all waste water (black- and grey-water streams) on site and reusing the treated water for toilet flushing, reducing household energy consumption, eliminating the need for solid-waste removal from the site, raising densities by shrinking the average size of erven in a way that does not discriminate between rich and poor, and maximising the economic benefits of a socially mixed development. These were all, with hindsight, part of a 'minimising damage' perspective.

We were determined to be practical and realistic, which has stood us in good stead over the years. But the germ of an alternative perspective was planted by Max Lindegger from the Global EcoVillage Network who ran workshops, organised by Spier, on his work at one of the world's pioneer EcoVillages, namely Crystal Waters on Australia's East Coast (outside Brisbane). Lindegger, who had worked with Bill Mollison who pioneered the permaculture movement, passionately believed that it was possible to reconcile human settlement with the restoration of ecological systems (for an articulation of this see Jackson & Svensson 2002). However, it is one thing to aspire to this ideal, but quite another to embed it in designs which then become claims that others use to judge the project. Under-claiming in this business is a virtue, especially if you are (vaguely) aware of how little you know.

Configuring space

An adaptive design process must inevitably start with a deep engagement with the context. With hindsight, it is possible to define three key adaptive challenges that have bedevilled the Lynedoch EcoVillage from the start: How to enable a social mix without allocating poor and rich people to separate designated areas (thus replicating the worst of South Africa's apartheid past)? How to ensure that all houses are oriented northward (to benefit from solar penetration in winter and shade in summer) on a site that is situated on a northwest-southeast axis, with the best views to the south? How to ensure that the engineers, architects and urban planners work together to design a single, integrated system of resource flows (energy, water, sewage and waste) through the neighbourhood and houses in order to avoid the common mistake of designing 'green buildings' which get embedded in conventionally designed networked infrastructures. In general, the urban design that was submitted for approval did not achieve the best result with respect to northward orientation.

By contrast, the architectural team worked well with the engineers, which resulted in an urban infrastructure that broke fundamentally with conventional templates. Furthermore, by providing for small plots (by South African standards) it was possible to avoid allocating plots in advance for poorer or richer households — instead, each plot was given two prices, with people who qualified for housing subsidies entitled to purchase for an average of R120/m^2, while everyone else had to pay R700/m^2 or more. The outcome was a mix that overcomes traditional spatial divisions between richer and poorer households. Most significantly of all, the technical designs evolved as part of a much wider socio-institutional logic that was captured in a detailed set of constituting documents including the Constitution of the Home Owners' Association, design guidelines for the building of the houses, a Code of Conduct (mainly behavioral) applicable to all residents, plus a (continuously evolving) landscape design.[10] These documents were discussed at some length over the 18-month period leading up to what was called the Founding Annual General Meeting of the LHOA which took place on 8 May 2004. By May 2004, the contractors had already completed the landscaping of Phase 1 and dug the trenches for the pipes and cables. By the end of 2005, the first houses were complete and ready for occupation.

Flows and networks

An infrastructure was designed and built which has to a large extent achieved its objectives. This does not mean, however, that it is operated as originally intended.

The following system has been constructed:

Water and stormwater

The original overall aim of the water and stormwater system was to reduce the consumption of potable water supplied by Stellenbosch Municipality by 40 per cent of what a development like this would normally use. Research by Dowling in 2007 revealed that the reductions in the households may be as high as 55 per cent (Dowling 2007).

During the course of 2010 a sophisticated water monitoring and control system was installed to set up the information base that is required to test the original design assumption.[11] The data generated confirmed that 40 per cent of all the water consumed was provided by the recycled grey water, mainly for toilet flushing and irrigation. However, it was also revealed that there is more recycled grey water available than required — making it possible to reduce even further the amount purchased from the municipality thus gradually shifting from a 'minimising damage' to a 'restorative' approach. The system has the following elements:

[10] For details see http://www.sustainabilityinstitute.net/newsdocs/documents-mainmenu-31/cat_view/47-lynedoch-ecovillage

[11] Funding provided by University of California, Berkeley.

- Dual water supply consisting of the following:
  - Potable water is supplied by the municipality to the village via a single meter — the LHOA then on-sells the water to each building
  - Each household is supplied with potable water and recycled water — the recycled water is used for toilet flushing and garden irrigation
  - The recycled water supply is generated by two on-site waste water treatment plants that treat all the sewage from the site (described below)
  - Two water meters per household have been installed, one for potable water and the other for recycled water. Readings are taken by the LHOA and one invoice per household is generated with two line items — a fee for potable water which the LHOA pays to the municipality, and a fee for recycled water that goes towards the operation and maintenance of the on-site water recycling system (described below).
- Water-saving taps and shower heads have been installed where possible, as well as dual-flush or low-flush toilet systems.
- To reduce the cost of the recycled water infrastructure, the pipes that conduct the recycled water supply are the same as those required by law to supply water to the fire hydrants.
- Stormwater run-off from the ground and roofs is conveyed in open channels and pipes to a dam located at the bottom of the site. The longer-term plan is to connect this supply to the recycled water system. This has not happened to date because of a lack of funds for deepening the dam, protecting the pump from theft and some doubt about the safety of the flows given that the village is surrounded by vineyards which are sprayed with chemical pesticides.
- The open stormwater channels were originally planted with Kikuyu grass and designed to complement the natural character of the development. However, these have tended to silt up resulting in their replacement during the course of 2009/10 with fixed channels made from recycled brick.
- Stormwater run-off has been minimised by restricting hard landscaping, thereby increasing percolation into the ground-water supply which, in turn, is accessed via two functioning boreholes when required. This might be changed later if the community decides to pave the roads with impermeable surfaces thus making the dam supply more viable.
- Rainwater harvesting is optional for each household, although strongly encouraged, especially for those who will require irrigation water which may not be forthcoming in the dry summer months if the dam drops below the required levels.

Sewage

The overall aim of the ecological sanitation system was to recycle and reuse all three elements of the effluent — the H_2O (water) for flushing toilets and irrigation, the CH_4 (methane) for energy production (mainly cooking), and the nutrients (N and P) for gardening and food production. To this extent, the original conception of the sanitation system was more 'restorative' than 'minimising damage', especially in light of the fact that nutrient retention has contributed significantly to greening what was originally quite a desolate space, but also to biomass and food production over the years.

The system has been constructed so that grey and black effluent from households passes through septic tanks (one per two or three erven) where the main solids are deposited (which also means the risks of abuse that can block the pipes are confined to the user and not transferred into the village-wide system), and then gets pumped on to a constructed Vertically Integrated Wetland (VIW) at the bottom of the site via a purpose-built sump. The sewage from the septic tanks enters at the surface level of the VIW from where it gradually sinks down through a multi-layered filter comprising river sand, small stones, straw, geo-fabric and iron filings (to capture the phosphates). The surface of the VIW is planted with halophytes which are heavy nitrogen feeders (such as the arum lily or bloetriet plant). The iron filings and the halophytes help to remove the bulk of the nutrients, while the filter removes the bulk of the solids. After sinking to the bottom into an oxygen-starved anaerobic environment which helps deal with some pathogens, the treated effluent goes into a sump, from which it gets pumped into a series of storage tanks (at the bottom and the top of the site) for onward transmission into the households for toilet flushing and irrigation.

The grey and black effluent from the guest house and main building is channelled directly into a biolytic filter which is an engineered micro-ecology consisting of a peat filter inoculated with earthworms. This system effectively deals with the solids in an aerobic environment which results in treated water that has retained the primary nutrients (nitrogen and phosphorus) for reuse as a natural organic fertiliser for developing a nursery or irrigating orchards.

Both the constructed VIW and the biolytic filter are rapid, odour-free, fly-free environmentally appropriate filtration systems which do not require the use of chemical inputs, although they do use energy to drive the relatively low-voltage pumps. They are appropriate for high-density environments, with many examples of the VIW in The Netherlands and Germany. Interestingly, the key element of the biolytic filter are the worms commonly referred to as 'red wrigglers' but whose scientific name is *Eisenia foetida* (*E. Foetida*). They thrive in decaying organic matter and get their name from the pungent smell they exude when disturbed. This natural defence may explain why it is possible to judge whether the worms are thriving or not by the number of flies around: if there are none, the worms are safe and healthy. These hermaphrodites reproduce at a rate that is commensurate with their food supply, thus creating a complex self-organising system which requires no external energy or inputs and performs best when far from equilibrium. The worms consume the organic matter in the sewage flow, and what they secrete in tiny particles of nitrogen and phosphorus attach to the H_2O molecules, thus creating the precious, naturally processed nutrient flow that is returned to the soils of Lynedoch. This extraordinary cycle, which effectively captures from the sewage flow the nitrogen and phosphorus resources that would otherwise be lost via a conventional sanitation system, is probably the most significant restorative dynamic at work in the Lynedoch EcoVillage.

Where nutrient and pathogen removal is required (that is, for the treated effluent that goes into the toilets), this takes place in the oxygen-starved anaerobic environment at the base of the VIW. Once the dam is deployed, nutrient removal will be even more effective if a way can be found to introduce fish without them being stolen. Although pathogen removal through the biolytic filter is limited, it is possible to remove the pathogens by passing the effluent from the biolytic filter through a sand followed by an ultra-violet light. This energy intensive stage has been installed, but is used only when gardens need to be irrigated with the biolytix filtrate.

No stormwater or black- or grey-waste water leaves the site except via groundwater flows, leakages and evaporation (mainly from the dam). By treating all black and grey water on site, the village will never need to face the burden of capital costs for bulk sanitation or municipal charges for such a service (normally indirectly via rates).

It is worth noting that three buyers with neighbouring erven collaborated during the infrastructure construction phase to replace their septic tanks with a single biogas digester. This biogas digester (made of brick in a dome structure) collects all the grey and black water, plus kitchen organic waste, and captures the methane gas at the top of the dome where it is released via a pressure valve back into the houses for use as cooking gas which replaces the LPG gas supply. Whereas the pressure from a traditional LPG bottle that supplies a gas stove is about 2.5 Kpa, the pressure at the biogas digester itself is 9 Kpa and between 2 and 3 Kpa by the time the methane gets to the stoves that are operating in two of the houses (including ours). The effluent from the biogas digester then flows into the village-wide treatment system as described above. If we knew in 2002 what we know today about biogas digesters, there is no doubt that the entire effluent flow would have been directed into a large, village biogas digester to generate energy for productive use.

Energy

The original aim of the energy system was to reduce the amount of energy that would normally be used by a similar size residential development by 60 per cent. However, as time has passed, innovations have taken place with respect to photovoltaics and biogas that may make it possible for Lynedoch to eventually become 'energy neutral' (that is, a settlement that generates as much energy — or more — than it uses).

The electrical infrastructure was designed to ensure that each structure has an electricity supply from the national electricity grid (which, in turn, is mainly supplied by coal-fired power stations and a small percentage by nuclear power). Solar-water heaters were installed on existing and new buildings (and will be installed on all future buildings), with each solar-water heater fitted with a thermostat that switches to electricity during days when sunlight is insufficient to heat up the water (a few weeks a year in winter). The solar water heaters eliminate between 50 and 60 per cent of their normal electricity consumption.[12] No electric stoves were permitted — all cooking is done via LPG hob or biogas. Space heating and cooling has been achieved naturally via effective design, a proper north–south orientation (in most cases), correct roof overhangs (long enough to shade in summer and short enough to allow winter sun penetration during winter), the most appropriate insulation (varying with affordability levels), and the use of thermal mass (such as rock stores) or even geo-thermal systems (underground piping systems that make use of steady state temperatures that exist at around 2 m or more below the surface). This all saves money by reducing natural resource use consumption. Low energy lighting has been installed, namely 11 W compact fluorescents lights (CFLs) instead of the usual 60 W incandescent lighting.[13] Except for one experimental solar street light, street lights (at the time of writing in 2011) have not been installed, but will most likely be CFLs powered by solar panels when they are — this being a one-off capital cost to the developer and not an ongoing operating cost for the residents (except occasional replacement of the CFLs and, over a longer cycle, the battery).

The most significant aspect of the design of the electrical cabling infrastructure is that contrary to the normal design template used by electrical engineering consultants, the terms of reference specified that the cable and sub-station capacity must assume that no electric geysers, stoves or street lights will be installed in the Lynedoch EcoVillage. This significantly reduced the costs of this infrastructure. It was this saving that was then used to provide the low-income homes with a finance facility to purchase their solar water heaters and gas stoves. In other words, instead of seeing the solar hot-water heaters and stoves as privately owned consumer items payable by the home owners (via their bonds usually at above prime interest rates), they were defined as part of the village-wide (read: public) infrastructure financed via the (below prime) infrastructure loan from the DBSA. Repayments were derived from an extra charge for the electricity sold to the households by the LHOA via a pre-paid meter system. The capital and interest was paid off within three years. Of course, this subversion of the traditional distinction between public and private infrastructure opens up enormous opportunities — for example, roof structures and household electrical infrastructure of all homes could be owned (and therefore financed) by the power utility (in this case Eskom) if the rooftops are seen as minipower-stations (because they generate from solar hot-water heaters and grid-connected photovoltaic cells) with residents charged via metering systems for the use of this energy.

There is obviously significant potential for Lynedoch to generate more than it consumes. Because it is already a 'micro-grid'[14] it is an ideal site for a mixed system which could be constructed in future, building on what has already been achieved,

[12] During 2010 an energy monitoring system was installed which will generate more reliable data about actual use in future.

[13] 2 W low energy diodes (LED) would be preferable but have hitherto been too expensive.

[14] It is a micro-grid because the users are not directly supplied by Eskom — there is one meter between the national grid and the Lynedoch grid, which means that any energy generated this side of the meter is sold to other users within Lynedoch (see Marnay & Firestone 2007).

namely on-site energy generation for feeding back into the grid using solar roof tiles (photovoltaic cells embedded in roof tiles), wind power, biogas digesters for generating electricity via generators, mini-hydro systems, plus possibly a storage system using a dam at the top of the site that can be filled at night using cheap electricity and then used during peak periods when electricity is most expensive. Another alternative would be for Lynedoch to invest in a large-scale, renewable-energy generation system located elsewhere (for example a concentrated solar-power system) which could generate an equal amount to what Lynedoch consumes with payments being made directly to the supplier, thus effectively using the grid as a virtual transmission belt.[15] This would be another example of how the infrastructure costs of a neighbourhood could build into its infrastructure, an investment in a remotely located plant that secures an affordable and sustainable energy supply over the long term.

Refuse

Like all ecologically designed settlements, solid waste recycling was an objective from the start. Although seemingly obvious, it is one of the most difficult aspects of sustainable living because it is more dependent on behaviour change within each household/workplace than the energy, water and sanitation services.

Refuse collection and recycling is managed by the LHOA. All solid waste must be separated into three bins at source: organic waste, recyclables (plastic, glass, tins, paper), and non-recyclable general waste (which gets collected by the municipal waste-removal truck). The recyclables are collected by a local recycler who on-sells to manufacturers—LHOA pays him for this service. A composting depot has been established and is used to process organic waste for use in the community gardens. It is estimated that 80 per cent of Lynedoch's waste is not transported to landfill. The reuse of the organic waste clearly has a restorative outcome, while the diversion of recyclables to recyclers is more about minimising damage.

The system depends on separation at source, which is difficult to sustain over time, especially if households have teenagers or younger adults who may not live permanently in the house. The LHOA has yet to move away from voluntary separation to more punitive measures to enforce separation, for instance, by non-collection of bags.

Roads

Unfortunately, very little thought went into road design at the start. The only aspects that were incorporated into the design were access to each plot, shared parking areas to discourage parking on each residential site, cost savings by constructing gravel roads, and grassed storm-water furrows conducting run-off into the dam at the bottom on the site. No consideration was given to a design that could have located the roads on the periphery of the site providing access to the back of each household and common

[15] Eskom is setting up pilot projects of this nature already.

parking areas, thus reinforcing the public shared space in the centre with pedestrian pathways providing the main internal linkages.

Towards the end of 2010, some of the roads and the main parking area were paved with recycled bricks.

Housing

The main purpose of the housing strategy was to make sure people working within Lynedoch had decent homes and that houses were designed in ecologically sustainable ways. Whereas the former was achieved via dual pricing (already described) plus cross-subsidies for people who qualified for government housing subsidies, the latter was achieved by applying the LHOA design guidelines. Over time, it has become clear that the application of these guidelines has resulted in designs that have provided for the following features: use of materials with a low 'embodied energy' (by building with adobe bricks laid with a clay mortar, or recycled fired brick laid with a cement mortar or wood); (more or less) correct north-south orientations coupled to significant roof overhangs and appropriately located windows; maximum use of the cool southerly winds during the hot summers for ventilation; inclusion of effective insulation; use of non-toxic treatment of wood (using boron rather than CCA) and non-toxic paints (using the commercially available Pro-Nature and Breathecoat products); compulsory installation of solar hot-water heaters and gas stoves; and the optional installation of rainwater storage tanks. This is as good as it gets, given that our restorative design ambitions are constrained by what is available in the marketplace.

The 12 adobe brick houses were built for an interesting mix of people, six of whom qualified for government housing subsidies because they earned below R3,500 per month. The others earned a wide range of incomes and they came from all of South Africa's social backgrounds — most grew up in the local coloured communities, while the rest were either from African or white English-speaking backgrounds. From those who had never, in the history of their families, had their own homes and, by their own cheerful admissions were at Lynedoch '*sommer maar net om my eie huis te kry*' ('simply just to get my own house') to those who were dreaming of some form of 'eco-escape' with stringent environmental policing, we have attempted to find simple 'middle ways' where all voices are reflected and we can live together in a gentler, more normal, village life.

Building community

It is remarkable how often one comes across the refrain that many of our problems have got something to do with the 'loss of community'. Whether it is crime, psychological depression, old age care, the loss of biodiversity, teenage pregnancies, substance abuse, the failure of development or the corruption of our political leaders, the solutions offered always end up in some way referring to the need to rebuild a 'sense of community'. The problem is that a 'sense of community' cannot be built as an end-in-itself. A sense of community emerges only when there is a purpose for a 'sense of community'. In our experience, a 'sense of community' continues to emerge as we find ways to collaborate (which includes, of course, vigorous conflicts along the way) in building all aspects of the Lynedoch EcoVillage.

There are three dimensions of the community building process that we think have been important: institutionalising community governance, ensuring and maintaining a socially mixed community, and constantly following what it means to be a 'child-centred' community. We see these as embedded in nature. While connection with nature is a personal imperative for us, we have not found it helpful to beat up on ourselves and others when we still see distressing signs of damage, such as continuing litter, harm to creatures and plants and child-on-child violence. We have, instead, attempted to focus on beauty and alternatives — by planting a woodland and community vegetable gardens; by working on Eric's organic farm; and enabling the possibilities for healing.

Governance matters

The Lynedoch Development Company has acted as the developer. As has been explained, in large part this meant the Sustainability Institute provided a 'behind-the-scenes' role for a non-profit developer with minimal capacity of its own. When the local authorities approved the development application submitted by the Lynedoch Development Company, this approval was granted on condition that a 'Home Owners' Association' and a 'Special Management Zone Trust' be established. The Lynedoch Home Owners' Association (LHOA) was constituted as a non-profit company (in terms of Section 21 of the Companies Act) with a detailed Memorandum of Association, Articles of Association and several appendices, one of which is the Code of Conduct that governs daily living in the village. This association is responsible for ensuring that the community has the services it requires, such as water, refuse collection, sewage treatment, cleansing, gardens and grounds maintenance. These services are paid for from service charges and levies paid by the members of the association. The association pays the bulk service providers such as Eskom, Stellenbosch Municipality, and various private contractors (for example, for waste recycling and removal). The LHOA employs a staff to deliver these services. The association holds an Annual General Meeting at which members vote for a Board of Trustees.

The most important document is the Constitution with a Code of Conduct which defines the way in which the community would like to live on a daily basis. The Code of Conduct governs matters such as litter, waste disposal (including separating waste at source), the number of pets each owner is allowed, noise pollution, traffic control, building extensions, use of energy and water, use of common areas, planting of vegetation and food gardens, disposal of compostable organic waste, safety and security matters (especially for children), use of the community hall, conflict resolution, domestic violence, air pollution, external appearance of buildings, procedures for managing community events (for example, parties, marriages and funerals), behaviour of temporary residents (such as students), and the right to privacy in a context which is already inundated by visitors.

As far as the Special Management Zone Trust is concerned, one of the zoning conditions imposed by Stellenbosch Municipality was that the LDC establishes a Trust into which 1 per cent of all land sales must be deposited for future investment in the natural environment, such as gardens, river rehabilitation, removal of alien plants and biodiversity promotion. This remarkably progressive intervention, along with the quota for the provision of subsidy sites, confirms the important role that zoning conditions can play in directing the investments of developers.

The most significant regular event is the meeting on the third Thursday of every month of the trustees of the LHOA (who are elected annually at the AGM).

Achieving the social mix

When the Stellenbosch Municipality approved the development application of the LDC, it made the unprecedented (although largely unenforceable) decision to specify that a significant proportion of the sites *must* be developed to benefit people living in the local area who qualify for housing subsidies. This, once again, demonstrates how zoning decisions can be used in progressive ways. However, the municipality also decided that the entire development must be self-financing, thus undermining its own decision. During the course of an intensive 18-month participatory planning process during 2004/05, it became clear that those who qualified for a government housing subsidy would not be able to afford to build a decent home with the subsidies they were to receive. Due to their income level and non-standard building methods, they were also unlikely to get approval for mortgages from the bank. Instead of giving up on the ideal of a socially mixed community, institutional and financial innovations were developed in partnership with the Development Bank of Southern Africa (DBSA) to make it possible. The first was to sell the houses via an installment sale agreement in terms of the Alienation of Land Act.[16] This meant that the buyer had possession and occupation rights, while full ownership was to be transferred only once all the payments had been made (similar to what is called 'hire-purchase' in the automobile market). This gave LDC the security it needed to provide building loans. The funds for these loans were generated from sales of the remaining plots — an income that, strictly speaking, should have been put aside to repay the R3 million loan to the DBSA. Instead, the DBSA agreed to share the risk because LDC remained the legal owner of the assets. This innovative financing mechanism (which amounted to a second circuit of debt finance for infrastructure was rerouted through housing loans before repayment) was a first for the DBSA and potentially replicable in other similar projects.

A so-called 'institutional housing subsidy' was secured from the Western Cape Provincial Government which was used as a down payment that effectively reduced the size of the loan each home owner required. A separate 'building contract' was entered into between the home owner and the contractor, which governed the terms of payment, construction and post-occupation liabilities. However, because the houses were built using unconventional materials, the normal insurance provided by the government-created National Home Building Registration Council (usually a five-year policy to

[16] This Act makes it possible to craft a variety of legal mechanisms for purchasing and owning land other than via conventional freehold title.

protect home buyers against structural defects) did not apply. This meant that the LDC had to provide the same degree of coverage otherwise the government subsidies would not have been allocated. The DBSA effectively stood behind these assurances, thus once again taking on risk beyond its initial mandate.

The home owners who qualified for government subsidies effectively benefited from three additional subsidies: a lower land cost; subsidised labour cost made available by the Swiss South Africa Co-operation Initiative so that the workers who built the adobe houses could be formally trained in this building technique and the learning materials written up into a formal, replicable building course that others could use; and finally, LDC made an additional investment to finance the extra costs of building the houses — funds generated from the sale of two buildings. (As already mentioned, the sites for poorer and better-off families were not geographically separated. Each site had two prices (subsidised and non-subsidised) which meant families could choose their sites unconstrained by price. The result was a socially integrated settlement pattern.

The most significant community-building process took place in the months that led up to the founding Annual General Meeting of the LHOA in May 2004. During this three-to-four-month period, a steering committee took responsibility for translating the rather bulky Constitution of the LHOA (memorandum and articles plus appendices) into simplified English and Afrikaans versions. The steering committee members had individual meetings with the rest of the home owners in order to explain the Constitution clause by clause. By the time the 50 or so home owners met at the founding meeting, there was a high level of understanding by everyone of the document, which was duly adopted, together with a budget for the first year. After approving the Constitution and electing the first trustees, the group ceremoniously planted a tree and walked around the newly constructed gravel roads to inspect their plots.

We consider ourselves deeply fortunate to have had visionary partners in the form of Sally Wilton and Teresa Graham[17] who could 'see' the importance of restoration and connection to nature through our indigenous planting, and who — through their R1 million investment — made possible the repayment of the remainder of the DBSA loan by purchasing a significant portion of land for the community, the Lynedoch Primary School and Sustainability Institute to use as gardens, together with community vegetable allotments. While subsidised home owners who took loans continue to repay these, the loan repayments are now able to form a community fund for reinvestment.

Following the children

When one sees children playing visibly, unrestricted by fencing and imposing signposts, one realises how much has been lost by locking them up in schools that look like prisons set apart, as they are, from the economic hubs and gardens of the community. What strikes those who arrive in Lynedoch for the first time is this sense of the children — from the pre-schoolers in their ecologically designed, new crèche, built behind the guest

[17] Sally Wilton and Teresa Graham are UK-based social entrepreneurs who founded the Lexi Cinema that allocates its profits to support the work of the SI.

house, to the uniformed pupils at the Lynedoch Primary School located in the main building at the centre of the village, to the gangly teenagers playing soccer or tumbling down the stairs from the youth club-house located in the roof of the crèche building. The full significance of this was only dimly sensed when we generated an urban design way back in 2000/01 that placed the school at the centre of the village. As the head boy of the school recently remarked when asked what he likes most about his school: 'It is all the other things that happen around our school.'

In a part of the world in which the levels of violence against children and between children are among the highest in the world, it is clear that unless children are placed at the centre of the development process, social integration will remain a chimera. If they feel external to the community and economy, why should they develop any desire to be part of them? 'When the youth are not initiated into the village, they will burn it down just to feel, the heat' (African proverb).

A combination of partnerships between Lynedoch Development, the Sustainability Institute, Spier and the Western Cape Education Department (WCED) has made Lynedoch's child-centred way possible. The inspirational core is undoubtedly the crèche which caters for 0–5-year-olds and is run in accordance with a carefully crafted pedagogy that fuses Montessori education with eco-literacy. Over the years the women who manage this community-based pre-school have translated what they know into a formally accredited training programme for early childcare providers from poor communities across the country. Every few weeks trainees can be seen doing t'ai chi with the children, working in the gardens, creating their own learning materials and sitting in lectures.

Entrepreneurship and the local economy

The most destructive consequence of apartheid is that richer communities can survive within enclosed local economies that do not need anything from the poor other than their labour. The reverse, however, is not true. There are, quite simply, insufficient resources within poor communities for poor households to survive without commuting to the rich communities to work, beg or steal. The Lynedoch Development has begun to overcome this problem in various ways.

First, the assets acquired by the buyers of subsidy sites in Lynedoch are worth far more than the government subsidy and what the households have contributed — something that is not the case for poor households who buy properties in marginal, poverty-stricken, urban ghettos where a property market is non-existent. Without over-burdening middle-income buyers, it was possible to give low-income buyers a high-value asset at a discounted price due to internal cross-subsidies. The market value of this asset was, therefore, much greater than the actual capital cost simply because of the existence of a viable property market. This confirms the assumptions underlying the government housing policies that were introduced from 2002 onwards, which emphasised the importance of 'integrated human settlements', viable property markets and an 'asset-based' approach to community development.

The challenge, of course, was to make sure that the intended beneficiaries remain the beneficiaries. The Constitution of the LHOA imposes on all home owners severe restrictions on resale by making it compulsory that any seller of property must first offer the property to the LHOA, and only then offer it to a third party at a price that is not lower than the price proposed to the LHOA. This, plus a provision that allows the LHOA to approve or disapprove a potential buyer, was intended to enable the LHOA to ensure that the membership of the community remains committed to the vision and values of the village. This made it possible to make sure that the sale of a subsidy site is treated in one of two ways. Either it is sold to someone else who has qualified for a housing subsidy and can take over the financial obligations from the seller, or it is sold to a buyer who can pay a market-related price that will generate considerable profits for the seller. In the latter case, the Constitution of the LHOA requires that an agreed percentage of the selling price is contributed to the LHOA to reinvest in a way that benefits someone who qualifies for a government housing subsidy. The purpose of this provision is to make sure that the existing social housing stock is not depleted while making it possible for poorer people to benefit financially from investing in property. In this way the total stock of social housing does not get eroded, and the sellers can realise a profit on their investment if they want to sell and move elsewhere.

It needs, however, to be accepted that if the poor remain poor, the development approach has been a failure. Over 10 years, the difference between the poorest and richest members of the Lynedoch community has already narrowed quite considerably. Some households that qualified for a housing subsidy in 2004/05 were earning by 2010 well above the R3,500 level required to qualify for a housing subsidy. Even if they don't sell, the original definition of 'social mix' will change. Similarly, it will not be possible to prevent the gradual sale of subsidy sites to buyers who earn above the subsidy requirement. The objective, therefore, is not to retain the number of social housing units within Phase 2, but to make sure that transfers result in a build-up of funds for investment in social housing units in Phase 3, or even a similar development located elsewhere. Maintaining the *size* of the total social housing stock is the objective, not necessarily to ensure that this stock remains fixed in a specific location.

The LDC has worked closely with the Sustainability Institute to assist with the establishment of a Savings and Credit Co-operative (SACCO) at Lynedoch. This is essentially a non-profit community bank which provides non-secured loans to poor people based on their savings record. Affiliated to the South African Credit Union League (SACUL), the SACCO has become an important vehicle for mobilising the savings of poorer households in the rural areas surrounding Lynedoch. However, more importantly, it provides a vehicle for managing the savings and loans procedures related to the housing construction process.

Finally, it needs to be noted that the Lynedoch EcoVillage Development is connected to a land-reform project led by a farmer from a historically disadvantaged background. This project aims to supply food directly to the Lynedoch EcoVillage members, thus by-passing the intermediaries in the food chain, namely the pack-house operators who normally buy from the farmers at ridiculously low prices, and the supermarkets that require such a standardised product that wastage levels are extremely and unnecessarily high.

The developmental state in practice

Harking back to our discussion of the developmental state in Chapter 4, one of the most astounding things about the evolution of the Lynedoch EcoVillage is how important the various agencies of the state have been in making it all possible from a regulatory and financial perspective. The Stellenbosch Municipality was always encouraging — we mentioned the progressive zoning conditions earlier. The Western Cape Provincial Government eventually (after two years) overrode the objections of white neighbours and confirmed the Stellenbosch Municipality's decision. This was after the provincial Department of Education assisted with all of the contractual arrangements that led to the relocation of the school to Lynedoch. More importantly, the provincial government went out of its way to figure out how to transfer housing subsidies for the construction of non-standard buildings and it also agreed to invest R6 million in the upgrading of the intersection which was a zoning condition imposed by the Stellenbosch Municipality and Provincial Roads Department. Normally, such a cost would have had to be carried by the development and recovered in the property prices (which in this case would have excluded the poor and the school). The National Department of Water Affairs and Forestry (as it was called at the time) also went out of its way to accommodate the application for the non-standard sewage treatment system which, in turn, was a key aspect of the environmental impact assessment which was approved by the Western Cape Department of Environmental Affairs and Development Planning after quite lengthy interactions. Most important of all, the state-owned Development Bank of Southern Africa was the only financial institution that was prepared to provide the loan finance (a R3 million loan for the infrastructure and, eventually, home loans) together with a good deal of unpaid-for strategic and technical advice (including the benefits of the thorough financial due diligence that it did to assess the loan application).

Read together, this does suggest that the Lynedoch EcoVillage was made possible by a set of responses from government officials and politicians at different levels that are similar to those advocated in Chapter 4. We argued that today's developmental state should anticipate the epochal, industrial and urban transitions and see itself not primarily as the deliverer of physical infrastructure or creator of a manufacturing base, but rather as an investor in niche-level, sustainability-oriented innovations that can generate a new set of capacities and capabilities for sustainable resource use and living.

Intentions versus realities

It is striking, but not surprising, that the global discussions about the transition to more sustainable modes of production and consumption attach enormous importance to technology. The most pervasive of all assumptions is that new technologies will result in a more sustainable use of resources. While this assumption is not wrong, on its own it can result in denial of the problem of over-consumption and a failure to realise that technologies are socio-technical systems that do not adhere to the rationalities and logics of the 'techno-fixer' (who is, invariably, someone from the design professions such as an engineer or architect). Adaptive design as we have defined it will mean working with technological changes within specific contexts characterised by specific capacities and capabilities that may or may not be appropriate for the deployment of specific technologies. There are many examples that reveal the difference between what was assumed to be technically rational and therefore workable, and what happens in practice as a given technology becomes part of the daily life of a given socio-ecological system shaped by people with diverse interests and behavioural norms.

The most significant examples of the way technical designs are socially engaged and contested are reflected in the minutes of the monthly meetings of the LHOA. Many of these meetings during the period 2008–2010 have been devoted to considering the building plans submitted by home owners. Most of these plans have been drawn up by architects, while none of the trustees are trained architects and only a few have had experience with building. The most successful outcomes have been when home owners have presented their own story, including an explanation of the technical issues. The least successful have been when architects have presented plans without the home owner even being present, or possibly when the home owner is present but passive. When a home owner presents his/her story, the trustees can immediately relate to the presenter and follow the logic as it is revealed, which normally starts with the social and living arrangements of the structure. When an architect presents, the language tends to be technical and the presentation normally starts with what the architect assumes the trustees want to hear, i.e. adherence to the technical specifications contained in the guidelines about orientation, energy, water use, paint and building materials. Eventually, no matter the starting point, what then follows is a fascinating social engagement that is not driven by a desire to make sure all the houses *look the same* (which would make the exercise so much easier), but rather by the desire to make sure that the home owner and community have agreed on the best way to express what is often referred to as the *values* of the EcoVillage. The outcome, therefore, is an expression of this consensus rather than adherence to a model. If people have had the time to engage in a robust way and penetrating questions are asked, the outcome can be impressive. If the meeting is poorly attended or insufficient time has been made available by trustees to argue with a slick architect, the result will be a seriously compromised version of an ecological house.

The result is that a very diverse set of structures have been built — different aesthetic styles, a range of different building materials (adobe, wood, recycled brick), and varying degrees of commitment to adhere to the agreed plans once the building process begins and choices have to be made every day. For those with a purist interpretation of ecological design — as one incensed architect put it in an email to us — there is 'nothing ecological about how Lynedoch looks, has been built or laid out'. This may well be true if you start off with an ideal image of what an ecologically designed settlement should look like (normally a well-developed spatial and aesthetic image) and then measure all claims against this. It is probably safe to say that LHOA trustees do not share such an image—Lynedoch has not copied a development from somewhere else, nor have any of the trustees ever been professionally trained designers, nor have there ever been funds to do what rich security estates do, which is to hire an architect to represent the trustees when it comes to evaluating a plan. The outcome is, therefore, a genuine *emergent outcome* which includes all the messy contradictory elements expressed for all to see in a permanent structure which invariably makes the design purists squirm with embarrassment. One founding member of the LHOA (a successful businessman who has had to subsequently sell and move to the UK for family reasons) once defended this approach after going to visit other estates in Stellenbosch: 'What I witnessed in these other estates was a symphony of one instrument; Lynedoch must be a symphony of many instruments.' Maybe the relational sensibility of the musician is better suited to making sense of what is emerging in the Lynedoch EcoVillage.

Conclusion

Our argument in Chapter 5 was that green urbanism is not an adequate response given the challenges we face as the epochal, industrial and urban transitions unfold. We proposed liveable urbanism as an alternative with restoration of life as its defining feature, but we have resisted setting these up as opposites. They both have a place in what we have called adaptive design.

The purpose of this reflection on the Lynedoch EcoVillage experience was to hopefully deepen the exploration of the restorative alternatives to green urbanism. In many ways it is too early to really judge where Lynedoch is on the continuum between green urbanism and liveable urbanism. For this we will need the quantitative data that must still be generated by the various monitoring systems that were put in place in 2010. However, it is obvious that this data will tell us that we are doing less damage by recycling (most of) our waste, and using less grid-supplied electricity and potable water than would be the case if Lynedoch had been configured in a traditional way. But other less quantifiable information will be needed that can tell us about restorative impacts, such as improved soils, rising local food production as people develop new skills, and the generation of renewable energy. What is impossible to properly quantify is the investment in human capabilities and awareness that takes place in the pre-school, primary school, short courses and university degrees, as well as the capabilities that are evolving within the community, staff and LHOA. In our view, adaptive design must not just be about the design of buildings, spaces and infrastructure. Above all it must be about doing all this in ways that presuppose the need for social co-operation. This, above all else, is what will rebuild that 'sense of community' that we all crave.

The core message of this chapter is that liveable urbanism—living in ways that can restore life—is doable. The funds can be found, the regulatory environment exists, the technologies are available, and from pilots such as Lynedoch, important lessons are available about the technical designs, governance structures and social dynamics of building sustainable neighbourhoods. Future research will need to document these lessons as these niche innovations mature and coalesce into an alternative more sustainable regime. At this stage, the Lynedoch story is about a vision that has been realised in practice, but it remains too early to judge what exactly has been achieved with respect to the restorative agenda that is so central to the notion of liveable urbanism. At the same time similar projects are emerging in various parts of South Africa and beyond inspired by the Lynedoch experience (and similar initiatives). We hope the Lynedoch experience may assist others to imagine the impossible in practice, and help to build the courage of those who dream but may lack workable examples that point to a way forward.

Appendix 1

Phase 1 of the housing development (42 sites) was funded with a R3 million loan from the Development Bank of Southern Africa (DBSA). This was provided at a 10.4 per cent fixed interest rate repayable over 10 years, with all interest and capital repayment deferred for the first 12 months. The final unaudited financial results for Phase 1 of the development (as at February 2006) were as follows:

Revenue:	
Sales plus interest earned:	R5,900,000
Expenditure:	
Land cost:	R1,628,459
Cost of Phase 1 infrastructure (inclu. professional fees)	R2,353,091
Provision of payment of interest	R1,758,493
Total expenditure	R5,740,043
Revenue after costs	R 159,957
Market value of retained assets:	
(Guest House)	R3,000,000
Market value of undeveloped portions of the land	R6,000,000[18]

[18] Market values determined by two reports, one by an independent valuer and the other by an estate agent

Conclusion

We started and ended this book with the story of the Lynedoch EcoVillage because we believe that a *just transition* will emerge if we find ways of transcending the split between local action and global change. These exist within each other, and the transformative impact of niche innovations must not be underestimated. They may seem isolated and insignificant in light of the deepening global polycrisis, but we are convinced that adversity is creating the conditions for a rapid spread of alternatives as niche innovations coalesce into new social movements, knowledge networks and major new developments.

At the same time, we suggest that it is highly problematic to under-emphasise the severity of the global ecological and social challenges that we face. The rate of carbon emissions is not slowing down nor is eco-system degradation. There were no significant binding global agreements on these two issues in the lead-up to the so-called 'Rio+20' Earth Summit in 2012. Following spikes in oil and food prices, the global recession, which began with the financial meltdown in the USA after Lehman Brothers collapsed in 2008, pushed more people into poverty while inequalities remained unacceptably high across most regions of the world. The debt crises in the USA and Europe, sluggish global growth, deepening divides between Keynesian and free-market economic policy paradigms and unimpressive political leadership at the global level (in fora such as the G20) have clearly exacerbated the polycrisis. Our chapter on Resource Wars (Chapter 7) is meant to hold up a spectre of a possible future if nothing fundamental changes to facilitate a global, just transition.

Despite clear evidence that our problems multiply faster than our capacity to generate solutions, we argued in Chapters 3 and 4 that we should nevertheless anticipate the next long-term development cycle. On what do we base our sense of hope? To retain a balance between understanding the deep structural dimensions of the polycrisis on the one hand, and a sense of the dynamics of innovation and transition on the other, we believe it might be helpful to go back to Karl Polanyi's classic notion of a 'double movement'. Published soon after the collapse of fascism and in anticipation of the post-World War II long-term development cycle, Polanyi wanted to know how it was possible for a free-market economy, which values only competition and individual wealth, to emerge without the simultaneous destruction of the 'natural and human substance of society' (Polanyi 1946: 135). His proposition was as follows:

> *Let us return to what we have called the double movement. It can be personified as the action of two organising principles in society, each of them setting itself specific institutional aims, having the support of definite social forces and using its own distinctive methods. The one was the principle of economic liberalism, aiming at the establishment of a self-regulating market ... and using largely* laissez-faire *and free trade as its methods; the other was the principle of social protection aiming at the conservation of man and nature as well as productive organization, relying on*

> *the varying support of those most immediately affected by the deleterious action of the market ... and using protective legislation, restrictive associations, and other instruments of intervention as its methods.* (Polanyi 1946: 135—emphasis in original)

Reviewing the largely peaceful century leading up to World War I, Polanyi noticed that the movement of social protection did not arise from centralised state action motivated by a particular social theory, but rather from a vast array of accumulated, minor, legislative changes and bottom-up socio-cultural arrangements that, over decades, emerged to counteract the atomising effect of individualism and free markets.

Applied to the current polycrisis, Polanyi's notion of a 'double movement' may be helpful. The 'deleterious actions of the market' coming up against powerful social forces of cohesion has its analogy in today's growing, resource-hungry, global economy that many believe is reaching the limits of its biophysical conditions of existence. This is one movement, and followed to its logical conclusion, the end result is system collapse (Lovelock 2006). As Fischer-Kowalski put it:

> *... [W]hat drives a transition is the structural exhaustion of opportunities, and at the same time the opening of new opportunities. If only previous opportunities are exhausted, and no substantial new opportunities open up, one may rather expect system collapse.* (Fischer-Kowalski 2011: 155)

For those who remain fixated on this one movement of structural exhaustion, hope becomes a naïve distraction. The absence of alternative opportunities is not inevitable.

There are four elements of the counter-movement that in complex combinations shape the emergence of the new opportunities. Firstly, a *civil society*-based revolution is under way—what Hawken has called *Blessed Unrest* (Hawken 2007). From the interstices and capillaries of societies across the world and from all contexts, there is evidence that people read the signs and recognise the need for alternatives. For some (mainly located in the global South) this is about sheer survival (and often means protesting against the rape and destruction of natural resources), while for others (usually those with more resources) it is an ethical choice. The emergent outcome is a set of values, consumption cultures, coalitions and technologies that, over time, become increasingly mainstream (from organic food through to solar power). The second is about *market responses and technological change*—the increasingly incontrovertible evidence that resource depletion and the rising price of carbon emissions, fossil fuels and resources in general is starting to drive primarily technological and system innovations, which make possible what would have been unthinkable a few decades ago (for instance, Germany's decision to transition to renewable energy; massive advances in resource efficiency) (Von Weizsäcker *et al.* 2009). The third is about *state interventions* arising from the significance of the double bubble of 2000 and 2007/08, which marks the mid-point crisis of the Information Age. Depending on whether states can intervene in ways that clear the way for productive capital to replace financial capital as the key motor of development, and depending on whether the state can be reconceptualised as the key facilitator of sustainability-oriented innovations, the outcome could be the unleashing of the, still unexploited, potential of information and communication technologies to become the operating system for a sustainability transition. The fourth is about the radical *reconfiguration of space:* whereas in previous transitions this was about acquiring foreign territories (colonialism for the third industrial transition) or extracting cheap resources from these territories (structural adjustment in the fourth industrial transition), this time the focus will be on the endogenous process of transforming cities into incubators of sustainability-oriented innovations that address the seemingly impossible task of building a viable urban world.

The 'double movement' is never resolved, but remains a powerful dynamic which structures interests, values, conflicts and the perceptions of opportunities and possibilities. For Polanyi, it was the unresolved tension between markets and social welfare that lay at the centre of post-World War II political contestations. Since at least the Brundtland Commission Report in 1987, the unresolved tension has been between a globalising capitalist market and the limits of its biophysical conditions of existence. The two have now been irrevocably fused together to create the context for the discussion of a *just transition.*

It is not surprising that the spirit of Polanyi's 'double movement' lives on in the emerging literature on sustainability transitions which, in the words of one review, addresses 'the hopes and expectations about solving environmental problems [that] are so far removed from current reality, that there is a need for radical, large-scale and integrated socio-technical changes, well beyond traditional policy approaches' (Van den Bergh *et al.* 2011: 8). A new journal, *Environmental Innovations and Societal Transitions,* has become a focus of this work.[1] The introductory essay in the first issue identified four approaches to innovation and transition in this literature (Van den Bergh *et al.* 2011: 9): the Innovation Systems (IS) approach which addresses system failures, social learning and institutional arrangements; the Multi-Level Perspective (MLP) (as discussed in Chapters 1 and 3); the Transition Management (TM) approach which applies complexity theory to organisational change and innovation; and the Evolutionary Systems (ES) approach that has emerged out of evolutionary economics with a focus on cumulative change, multiple-selection factors, diversity and policies to escape technological 'lock-in'.

In our view, these four approaches are helpful because they grapple with the dynamics of the 'double movement', but they fall short when it comes to addressing the challenge of a *just transition.* They tend to assume the existence of democratic space, well-established markets and the capacity for technological innovation. They say little about much wider global dynamics, such as the tension between global inequality and global interdependence, the rising number of resource wars and failed states, the emergence of the 'BRIC plus' countries that has brought an end to the traditional north-south axis of economic power, the spatial implications of the second urbanisation wave, the impact of globally networked social movements, and the conceptual confusion that

[1] The journal is closely associated with the Sustainability Transitions Research Network (see www.transitionsnetwork.org)

flows from attempts to resolve global problems using a national frame of reference. We need a rich mix that goes beyond a discourse limited to markets, technologies, institutions and policy.

To contribute to this rich mix, we summarise the key arguments in this book which support our view that a *just transition* will depend on the emergence of a conceptual framework that accepts the following propositions:

- A transdisciplinary approach to problems is needed that is based on complexity thinking, transcends disciplinary apartheid in theory and in practice, and helps to redefine what progress means by salvaging it from the wreckage of modernity (Chapter 1).
- The global economy is rapidly reaching — or has, in some cases, already breached — the system boundaries of the current globalised, industrial, socio-metabolic regime which, in turn, sets up the structural (but by no means the necessary) conditions for a transition to a more sustainable socio-metabolic regime (Chapter 2).
- The 'spring and summer' phase of the next long-term development cycle will most likely fuse together the deployment period of increasingly mature information and communication technologies driven by productive capital, the re-emergence of manufacturing as the basis for a more inclusive, more equalising period of economic development, and sustainability-oriented innovations to deal with the financial consequences of resource depletion and climate change (with this more than likely driven by financial capital as part of the installation of a sixth — rather than the greening of the fifth — industrial transition) (Chapter 3).
- Given that a shift from the current period of extended economic crisis to the next long-term development cycle will depend on appropriate state interventions to enforce the necessary shifts in power, it will be necessary to reconceptualise the political ecology of developmental states with special reference to the public domains they create for the kinds of sustainability-oriented innovations that will be required to translate structural necessities into sufficient conditions for transition (Chapter 4).
- Massive investments in urban infrastructure should become the focus of sustainability-oriented innovations because they create the twin opportunities for stimulating economic development and creating the conditions for a sustainable long-term development cycle (Chapter 5).
- As the population expands from 6 to 9 billion and the diets of another billion people become more meat and dairy product-based, degradation of soils could accelerate, thus further undermining food security if alternative diets and more agro-ecological farming practices that restore the soils are not introduced (Chapter 6).
- If vested interests at the global level (such as the oil industry) retain the power to block a just transition, resource wars and failed states will spread as local elites use force to strengthen their grip on the corrupt rent-seeking systems that tend to arise in resource-rich, resource-exporting countries. Local resource wars provide the template for what collapse could look like at a global level (Chapter 7).

- South Africa is now a formal member of the BRIC club of nations and as such provides an instructive example of a democratic, developing country faced with severe inequalities and a rich resource endowment which has been significantly depleted. Its haphazard adoption of 'green economy' approaches reflects a gradual realisation that dependence on resource exploitation will, in the long run, undermine its commitment to poverty eradication and, ultimately, democratic consolidation (Chapter 8). The South African case clearly demonstrates the need for a new development paradigm with sustainability at its core.
- To build more sustainable cities, socio-metabolic flows at city level will need to become more sustainable. This can happen only if urban infrastructures are reconfigured to achieve this outcome. Cape Town is used as a case study to show how these socio-metabolic flows, the techno-infrastructures that conduct these flows through the city, and the governance arrangements that reproduce these infrastructures and flows, can be analysed and transformed (Chapter 9).
- Finally, we discuss our own 10-year endeavour to build a sustainable local community. For us, adaptive design was a creative process of imagining a liveable space for a socially mixed community that was not only embedded within the web of all life, but actively contributed to the restoration of what was a degraded and desolate landscape.

We want to end by returning to the values of generosity and restoration discussed towards the end of the introductory chapter. After all is said and done, each one of us will weigh up the evidence in the private lair of our skulls and make up our minds about the facts and the dull compulsions of historical trajectories. But this will be insufficient to evoke the deeper passions that will be needed to ignite the viral actions for change that are needed. Commenting on this, Bruno Latour argued:

> *... [W]e are trapped in a dual excess: we have an excessive fascination for the inertia of the existing socio-technical systems and an excessive fascination for the total, global and radical nature of the changes that need to be made. The result is a frenetic snails' pace. An apocalypse in slow motion ... Changing trajectories means more than a mere apocalypse, and is more demanding than a mere revolution. But where are the passions for such changes?* (Latour 2010)

Finding these passions may well mean looking for ways of thinking about generosity and restoration that transcend the old dualisms between society and nature, individual and community, us and the 'other', science and other ways of knowing that have been so central to the so-called 'modern way of thinking'. Maybe we need to look for the kind of relational worldviews expressed so clearly in the 2008 Ecuadorian and Bolivian Constitutions that have effectively transcended these dualisms by adopting the concept of *sumac kawsay* (in Quechua), *suma qamana* (in Aymara), or *buen vivir* (in Spanish) — all of which mean 'living well'. Like Bhutan's Happiness Index, *buen vivir* effectively subordinates material wealth (as measured by GDP per capita) to social justice, human dignity and ecological integrity. But the Ecuadorian

Constitution goes one step further by enshrining the Rights of Nature, or the *Pachamama.* As Escobar notes, this 'represents an unprecedented "biocentric turn", away from the anthropocentrism of modernity' (Escobar 2011: 138). He even suggests that this 'biocentric turn' effectively marks the end of the modern notion that there is only 'One World — a universe'.

> *In emphasizing the profound relationality of all life, these newer tendencies show that there are indeed relational worldviews or ontologies for which the world is always multiple — a pluriverse ... in the successful formula of the Zapatista, the pluriverse can be described as 'a world where many worlds fit'. At their best, it can be said that the rising concepts and struggles from and in defense of the pluriverse constitute a post-dualist* theory and practice of interbeing. (Escobar 2011: 139 — emphasis added)

It is this perspective that sheds light on the twin values of generosity and restoration. These belong together because they inspire the possibility of 'interbeing' in a loving pluriverse with enough space to imagine the kind of world that can arouse the passions commensurate with the task at hand. To capture the nature of these passions we can do no better than end with the beauty of Ben Okri's incantation to astonish:

> *But when a nation or an individual creates things so sublime — in a sort of permanent genius of inventiveness and delight — when they create things so miraculous that they are not seen or noticed or remarked upon by even the best minds around, then that is because they create always from the vast unknown places within them. They create always from beyond. They make the undiscovered places and infinities in them their friend. They live on the invisible fields of their hidden genius. And so their most ordinary achievements are always touched with genius. Their most ordinary achievements, however, are what the world sees, and acclaims. But their most extra-ordinary achievements are unseen, invisible, and therefore cannot be destroyed. This endures forever. Such is the dream and reality of this land. I speak with humility.* (Okri 1995. *Astonishing the Gods*: 51–52)

Bibliography

Adler, R., Claasen, M., Godfrey, L. & Turton, A.R. 2007. Water, mining and waste: A historical and economic perspective on conflict management in South Africa. *The Economics of Peace and Security Journal*, 2(2): 32–41.

African Development Bank. 2010. *African Economic Outlook 2010*. Moulineaux, Fr.: African Development Bank Group. Available: http://www.afdb.org/en/knowledge/publications/african-economic-outlook/. [Accessed: 3 November 2010].

Aghion, P. & Howitt, P. 1998. *Endogenous Growth Theory*. Cambridge, MA: MIT Press.

Airoldi, M., Biscarini, L. & Saracina, V. 2010. *The Global Infrastructure Challenge: Top Priorities for the Public and Private Sectors*. Milan: Boston Consulting Group.

Ajulu, A., Othieno, T. & Samasuw, N. 2006. Sudan: The state of transition, prospects and challenges. *Global Insight*, 56.

Akram, Q.F. 2009. Commodity prices, interest rates and the dollar. *Energy Economics*, 31: 838–851.

Aleklett, K. & Campbell, C.J. 2003. The peak and decline of world oil and gas production. *Minerals and Energy*, 18(1): 5–20.

Allan, T. 2001. *The Middle East Water Question: Hydropolitics and the Global Economy*. London: I.B. Taurus.

Allen, P. 1997. *Cities and Regions as Self-Organising Systems: Models of Complexity*. Amsterdam: Gordon and Breach.

Altieri, M. 1995. *Agroecology: The Science of Sustainable Agriculture*. Boulder, CO: Westview Press.

———2002. Agroecological principles for sustainable agriculture. In: N. Uphoff (ed.). *Agroecological Innovations: Increasing Food Production with Participatory Development*. London: Earthscan.

———2008. *Small Farms as a Planetary Ecological Asset: Five Key Reasons Why we should Support the Revitalisation of Small Farms in the Global South*. Environment and Development Series, 7. Penang, Malaysia: Third World Network.

Altieri, M.A. & Koohafkan, P. n.d. Globally important agricultural heritage systems (GIAHS): Extent, significance, and implications for development. Unpublished paper available on the GIAHS website: Globally Important Agricultural Heritage Systems, Food and Agriculture Organization. Available: http://www.fao.org/nr/giahs/en/.

Altvater, E. 2009. Postneoliberalism or postcapitalism? The failure of neoliberalism in the financial market crisis. *Development Dialogue*, January.

Amin, A. & Thrift, N. 2002. *Cities: Reimagining the Urban*. Oxford: Polity.

Ampiah, K. & Naidu, S. 2008. *Crouching Tiger, Hidden Dragon? Africa and China*. Pietermaritzburg: University of KwaZulu-Natal Press.

Amsden, A. 1989. *Asia's Next Giant: South Korea and Late Industrialization*. New York: Oxford University Press.

———1995. Like the rest: South-East Asia's 'late' industrialisation. *Journal of International Development*, 7(5): 791–799.

Anderson, B. 1991. *Imagined Communities*. London: Verson.

Ansems, A. 2009. The complex social, economic and political dynamics of the sustainable city: More than meets the eye in East Asia. Master's class assignment. Stellenbosch, South Africa: Unpublished. Available: www.sustainabilityinstitute.net/documents.

Arthur, W.B., Durlauf, S.N. & Lane, D.A. 1997. *The Economy as an Evolving Complex System.* Reading, MA: Addison-Wesley.

Ashton, P.J. & Turton, A.R. 2008. Water and security in sub-Saharan Africa: Emerging concepts and their implications for effective water resource management in the southern African region. In: H.G. Brauch, J. Grin, C. Mesjasz, H. Krummenacher, N.C. Behera, B. Chourou, U. O. Spring, P. H. Liotta and P. Kameri-Mbote. (eds). *Facing Global Environmental Change: Environmental, Human, Energy, Food, Health and Water Security Concepts — Volume IV.* Berlin: Springer-Verlag.

Association for the Study of Peak Oil and Gas — South Africa. 2007. *Peak Oil and South Africa: Impacts and Mitigation.* Cape Town: Association for the Study of Peak Oil and Gas — South Africa.

Association for the Study of Peak Oil and Gas. 2009. *Newsletter Number 100.* Available: http://aspoireland.files.wordpress.com/2009/12/newsletter100_20090.pdf. [Accessed: 3 April 2010].

Audouin, M. 2009. Modernism, environmental assessment and the sustainability argument: Moving towards a new approach to project-based decision-making in South Africa. Stellenbosch: Department of Philosophy, Stellenbosch University. Docotoral thesis. 1–254.

Ayres, R. 2008. Sustainability economics: Where do we stand? *Ecological Economics,* 67: 281–310.

Ayres, R.U., Casteneda, B., Cleveland, C.J., Costanza, R., Daly, H.E., Folke, C., Hannon, B., Harris, J., Kaufman, R., Lin, X., Norgaard, R.B., Ruth, M., Spreng, D., Stern, D. & Van den Bergh, J.C.M. 1996. *Natural Capital, Human Capital, and Sustainable Economic Growth.* Boston: Center for Energy and Environmental Studies, Boston University.

Babiker, M. 2005. *Resource-based Conflict and Mechanisms of Conflict Resolution in North Kordofan, Gedarif and Blue Nile State.* Khartoum: Sudanese Environment Conservation Society & United Nations Development Programme.

Badgley, C. & Perfecto, I. 2007. Can organic agriculture feed the world? *Renewable Agriculture and Food Systems,* 22(2): 80–85.

Badgley, C., Moghtader, J., Quintero, E., Zakem, E., Chappell, M.J., Aviles-Vazquez, K., Samulon, A. & Perfecto, I. 2007. Organic agriculture and the global food supply. *Renewable Agriculture and Food Systems,* 22(2): 86–108.

Bagchi, A.K. 2000. The past and the future of the developmental state. *Journal of World-Systems Research,* 1(2): 398–442.

Bai, Z.G., Dent, D.L., Olsson, L. & Schaepman, M.E. 2008. *Global Assessment of Land Degradation and Improvement: 1. Identification by Remote Sensing.* GLADA Report 5. Wageningen: LADA, World Soil Information & FAO. Available: www.isric.org.

Bailey, R. 2011. *Growing a Better Future: Food Justice in a Resource-constrained World.* Oxford: Oxfam. Available: www.oxfam.org/grow. [Accessed: 23 June 2011].

Ballayan, D. 2000. Soil degradation: Draft.: Rome: FAO.

Ballentine, K. & Nitzschke, H. 2003. *Beyond Greed and Grievance: Policy Lessons from Studies in the Political Economy of Armed Conflict.* Policy report. New York: International Peace Academy.

Bang, J.M. 2005. *Ecovillages: A Practical Guide to Sustainable Communities.* Edinburgh: Floris Books.

Bannon, I. & Collier, P. 2003. Natural resources and conflicts: What we can do? In: I. Bannon & P. Collier. (eds). *Natural Resources and Violent Conflict: Options and Actions.* Washington, DC: The World Bank.

Barbier, E.B. 2009. *A Global Green New Deal*: Report prepared for the Economics and Trade Branch, Division of Technology, Industry and Economics, United Nations Environment Programme.

Bardi, U. 2008. *The Universal Mining Machine.* Available: http://europe.theoildrum.com/node/3451.

Barker, D. 2007. *The Rise and Predictable Fall of Globalized Industrial Agriculture.* San Francisco: The International Forum on Globalization.

Bartelmus, P. 2003. Dematerialization and capital maintenance: Two sides of the sustainability coin. *Ecological Economics,* 46(1): 61–81.

Bayat, A. 2000. From 'dangerous classes' to 'quiet rebels': Politics of the urban subaltern in the global South. *International Sociology,* 15(3): 533–557.

Beall, J., Crankshaw, O. & Parnell, S. 2000. *Uniting a Divided City: Governance and Social Exclusion in Johannesburg.* London: Earthscan.

Beaseley-Murray, J., Cameron, M.A. & Hershberg, E. 2009. Latin America's left turns: An introduction. *Third World Quarterly,* 30(2): 319–330.

Beatley, T. 2000. *Green Urbanism.* Washington DC: Island Press.

Beatley, T., Boyer, H. & Newman, P. 2009. *Resilient Cities: Responding to Peak Oil and Climate Change.* Washington, DC: Island Press.

Beatley, T. & Newman, P. 2009. *Green Urbanism Down Under: Learning from Sustainable Communities in Australia.* Washington, DC: Island Press.

Behrens, A., Giljum, S., Kovanda, J. & Niza, S. 2007. The material basis of the global economy: Worldwide patterns of natural resource extraction and their implications for sustainable resource use policies. *Ecological Economics,* 64: 444–453.

Benington, J. & Moore, M. 2010. *In Search of Public Value: Beyond Private Choice.* London: Palgrave.

Benn, D. & Hall, K. 2000. *Globalization, a Calculus of Inequality: Perspectives from the South.* Kingston: Ian Randle.

Benyus, J. 1997. *Biomimicry: Innovation Inspired by Nature.* New York: HarperCollins.

Berkhout, P.H.G., Muskens, C. & Velthuijsen, J.W. 2000. Defining the Rebound Effect. *Energy Policy,* 28(6-7): 425–432.

Berman, M. 1988. *All that is Solid Melts into Air.* New York: Penguin.

Bindraban, P. 2009. *The Need for Agro-Ecological Intelligence to Limit Trade-offs between Global Food, Feed and Fuel.* Conference Proceedings of International Fertilizer Society Conference, 11 December, Cambridge. York: International Fertilizer Society.

Bioregional. 2002. *BEDZED – Beddington Zero Energy Development, Sutton.* General Information Report. 89. Garston, Watford: UK Government Energy Efficiency Best Practice Programme.

Birkeland, J. 2002. *Design for Sustainability.* London: Earthscan.

———2008. *Positive Development: From Vicious Circles to Virtuous Cycles through Built Environment Design.* London: Earthscan.

Bond, P. 1999. Global economic crisis: View from South Africa. *Journal of World-Systems Research,* V(2): 413–455.

———2002. *Unsustainable South Africa: Environment, Development and Social Protest.* London: Pluto Press.

———2006. *Looting Africa.* London: Zed Press.

Borja, J. & Castells, M. 1997. *Local and Global: Management of Cities in the Information Age.* London: Earthscan.

Borlaug, N. 2000. Ending world hunger: The promise of biotechnology and the threat of antiscience zealotry. *Plant Physiology,* 124: 487–490.

Bosshard, P. & Hildyard, N. 2005. *A Critical Juncture for Peace, Democracy and the Environment: Sudan and the Merowe/Hamadab Dam Project*. Berkeley & London: International Rivers Network & The Corner House.

Botha, J. 2007. Rewards from rubbish. *Urban Green File,* 12(5). December.

Boulding, K. 1991. What is evolutionary economics? *Journal of Evolutionary Economics,* 1: 9–17.

BP. 2010. *Statistical Review of World Energy*. London: BP. Available: http://www.bp.com/genericsection.do?categoryId=92&contentId=7005893. [Accessed: 14 October 2010].

Bridge, G. & Watson, S. 2000. *A Companion to the City.* Oxford: Blackwell.

Bringezu, S. & Bleischwitz, R. 2009. *Sustainable Resource Management: Global Trends, Visions and Policies*. Germany: Wuppertal Institute.

Bringezu, S., Schutz, H. & Moll, S. 2003. Rationale for and interpretation of economy-wide material flow analysis and derived indicators. *Journal of Industrial Ecology*, 7(2): 43–67.

Bringezu, S., Schutz, H., Steger, S. & Baudisch, J. 2004. International comparison of resource use and its relation to economic growth: The development of total material requirement, direct material inputs and hidden flows and the structure of TMR. *Ecological Economics*, 51(1/2): 97–124.

Brown, G. 2009. We must act now. *Newsweek*. 28 September: 25–30.

———2011. Take back the future. *Newsweek.* 15 May.

Brown, L. 2006. *Plan B 2.0: Rescuing a Planet under Stress and a Civilisation in Trouble.* New York: Norton.

———2008. *Plan B 3.0: Mobilizing to save Civilization.* New York and London: W.W. Norton & Company.

———2011a. The Great Food Crisis of 2011. *Foreign Policy*, 1–4.

———2011b. *World on the Edge: How to Prevent Environmental and Economic Collapse.* London & New York: W.W. Norton & Company.

Buck, L. & Scherr, S. 2011. Moving ecoagriculture into the mainstream. In: The Worldwatch Institute. (ed.). *State of the World 2011: Innovations that Nourish the Planet*. Washington, DC: The Worldwatch Institute.

Buhrs, T. 2007. Environmental Space as a Basis for Enhancing the Legitimacy of Global Governance. CSGR/GARNET Conference on Pathways to Legitimacy? The Future for Global and Regional Governance: Warwick University, UK. 17–19 September.

Burgess, S. 2010. *Sustainability of Strategic Minerals in Southern Africa and Potential Conflicts and Partnerships*. USAFA, CO, 80840 Unclassified report. Maxwell, AL: U.S. Air Force Academy, Institute for National Security Studies.

Burns, M., Audouin, M. & Weaver, A. 2006. Advancing sustainability science in South Africa. *South African Journal of Science*, 102.

Burns, M. & Weaver, A. 2008. *Exploring Sustainability Science: A Southern African Perspective.* Stellenbosch: Sun Media.

Byrne, D. 1998. *Complexity Theory and the Social Sciences.* London: Routledge.

———2001. *Understanding the Urban.* New York: Palgrave.

Caas, F. 2004. *Environment and the Joint Assessment Mission (JAM) in Sudan: Streamlining Environmental Concerns into the JAM Report and Sudan's Reconstruction and Development Phase.* Geneva: United Nations Environment Programme.

Cabannes, Y., Yafai, S.G. & Johnson, C. 2010. *How People Face Evictions.* London: Development Planning Unit, University College London.

Cabitza, M. 2011. *Will Bolivia make the Breakthrough on Food Security and the Environment?* Poverty Matters Blog: guardian.co.uk. Available: http://www.guardian.co.uk/global-development/poverty-matters/2011/jun/20/bolivia-food-security-prices-agriculture. [Accessed: 29 June 2011].

Campbell, C.J. 1997. *The Coming Oil Crisis.* Brentwood, Essex: Multi-Science Publishing Company and Petroconsultants.

Campbell, H. 2008. China in Africa: Challenging US global hegemony. *Third World Quarterly,* 29(1): 89–105.

Campbell, T. n.d. IPPUC: The untold secret of Curitiba—In-house technical capacity for sustainable environmental planning. *Urban Age Magazine.* San Rafael, CA: Urban Age Institute. Available: www.UrbanAge.org/magazine.php.

Capra, F. 1996. *The Web of Life.* New York: Anchor Books.

———2002. *Hidden Connections.* London: HarperCollins.

Carley, M. & Spapens, P. 1998. *Sharing the World: Sustainable Living and Global Equity in the 21st Century.* London: Earthscan.

Castells, M. 1997. *The Information Age Volumes 1, 2 and 3.* Oxford: Blackwell.

———1999. The informational city is a dual city: Can it be reversed? In: D. Schon, B. Sanyal & W. Mitchell. (eds). *High Technology and Low Income Communities.* Cambridge, MA: MIT Press.

Centre for Urban Policy Research. 2000. *Towards a Comprehensive Geographical Perspective on Urban Sustainability.* Final report of the 1998 National Science Foundation Workshop on Urban Sustainability. New Jersey: Centre for Urban Policy Research, Rutgers University.

Chambers, N., Simmons, C. & Wackernagle, M. 2001. *Sharing Nature's Interest: Ecological Footprints as an Indicator of Sustainability.* London: Earthscan.

Chang, H. 2002. The real lessons for developing countries from the history of the developed world. *Policy Innovations, Carnegie Council.* [Online: retrieved 3 April 2009]. Carnegie Council: New York. Available from: http://www.policyinnovations.org/ideas/policy_library/data/01137.

Chang, H.J. 2002. *Kicking Away the Ladder: Policies and Institutions for Development in Historical Perspective.* London: Anthem.

———2007. *Bad Samaritans: Rich Nations, Poor Policies and the Threat to the Developing World.* London: Random House.

Chen, S. & Revallion, M. 2004. *How Have the World's Poorest Fared since the Early 1980s?* WPS 3341. Washington, DC: World Bank. Policy Research Working Paper.

Chibber, V. 2002. Bureaucratic rationality and the developmental state. *The American Journal of Sociology,* 107(4): 951–989.

———2003. *Locked in Place: State-Building and Late Industrialization in India.* Princeton, N.J.: Princeton University Press.

China Council for International Cooperation on Environment and Development. 2007. *Strategic Shift of Environment and Economy in China.* Special report. Beijing: China Council for International Cooperation on Environment and Development.

Cilliers, P. 1998. *Complexity and Postmodernism: Understanding Complex Systems.* London: Routledge.

Cisse, L. & Mrabet, T. 2004. World phosphate production: Overview and prospects. *Phosphorus Research Bulletin,* 15: 21–25.

Cities Alliance. 2008. *Alagados: The Story of Integrated Slum Upgrading in Salvador (Bahia), Brazil.* Washington, DC: Cities Alliance.

City of Cape Town. 2003. *Equitable Services Policy Framework: Report to Executive Committee 1 April 2003*. 2/4/2/P. Cape Town: City of Cape Town.

——2006. *Integrated Development Plan*. Cape Town: City of Cape Town.

——2008. *5-Year Plan for Cape Town: Integrated Development Plan (IDP) 2007/8–2010/12*. Cape Town: City of Cape Town.

Clark, G., Dexter, P. & Parnell, S. 2007. *Rethinking Regional Development in the Western Cape*. Integrated Development Planning Provincial Conference: Cape Town.

Clark, W.C., Crutzen, P.J. & Schellnhuber, H.J. 2005. *Science for Global Sustainability: Toward a New Paradigm*. Harvard University, John. F. Kennedy School of Government Working Paper Series. rwp05-032. Cambridge, MA: Harvard University, John. F. Kennedy School of Government.

Collier, P. 2010. The political economy of natural resources. *Social Research*, 77(4): 1105–1132.

Collins, R.O. 1990. *The Waters of the Nile: Hydropolitics and the Jonglei Canal*. Oxford: Clarendon Press.

Colvin, C. & Saayman, I. 2007. Challenges to groundwater governance: A case study of groundwater governance in Cape Town, South Africa. *Water Policy*, 9(2): 127–148.

Commission on Growth and Development. 2008. *The Growth Report: Strategies for Sustained Growth and Inclusive Development*. Washington, DC: World Bank.

Community Organization Development Institute. 2008. *Community Upgrading Projects*. CODI Update. Bangkok: Community Organization Development Institute.

Cooperrider, D.L. & Srivastva, S. 1987. Appreciative inquiry in organizational life. *Research in Organizational Change and Development*, 1. Available from: appreciative-inquiry.org/AI-Life.htm.

Cordell, D., Drangert, J. & White, S. 2009. The story of phosphorus: Global food security and food for thought. *Global Environmental Change*, 19(2): 292–305.

Costanza, R. 2000. The dynamics of the ecological footprint concept. *Ecological Economics*, 32: 341–345.

——2003. A vision of the future of science: Reintegrating the study of humans and the rest of nature. *Futures*, 35: 651–671.

——2008. Toward an ecological economy: Creating a sustainable and desirable future. In S. Kates. *Macroeconomic Theory and its Failings: Alternative Perspectives*. Cheltenham, UK: Edward Elgar Publishing.

——2009. Science and ecological economics: Integrating of the study of humans and the rest of nature. *Bulletin of Science, Technology and Society*, 29(5): 358–373.

Costanza, R., Cumberland, J.C., Daly, H.E., Goodland, R. & Norgaard, R.B. 1997a. *An Introduction to Ecological Economics*. Boca Raton: St. Lucie Press.

Costanza, R., D'Arge, R., Groot, R.D., Farber, S., Grasso, M., Hannon, B., Limburg, K., Naeem, S., O'Neill, R.V., Paruelo, J., Raskin, R.G., Sutton, P. & Belt, M.v.d. 1997b. The value of the world's ecosystem services and natural capital. *Nature*, 387: 253–260.

Costanza, R., Wainger, L., Folke, C. & Maler, K.G. 1993. Modelling complex ecological economic systems: Toward an evolutionary, dynamic understanding of people and nature. *BioScience*, 43: 545–555.

Cotula, L., Vermeulen, S., Leonard, R. & Keeley, J. 2009. *Land Grab or Development Opportunity? Agricultural Investment and International Land Deals in Africa*. London: International Institute for Environment and Development, FAO, International Fund for Agricultural Development.

Cousins, B. n.d. *What is a 'Smallholder'? Class-Analytic Perspectives on Small-Scale Farming and Agrarian Reform in South Africa.* Working Paper, 16. Cape Town: Programme for Land and Agrarian Studies, University of the Western Cape.

Cousins, B. & Scoones, I. 2010. Contested paradigms of 'viability' in redistributive land reform: Perspectives from southern Africa. *The Journal of Peasant Studies*, 37(1): 31–66.

Coutard, O., Hanley, R.E. & Zimmerman, R. 2005. *Sustaining Urban Networks: The Social Diffusion of Large Technical Systems.* London: Routledge.

Crane, A. & Matten, D. 2004. *Business Ethics: European Perspective.* Oxford: Oxford University Press.

Crane, W. & Swilling, M. 2008. Environment, sustainable resource use and the Cape Town functional region: An overview. *Urban Forum*, 19: 263–287.

Craswell, E.T., Vlek, P.L.G. & Tiessen, H. 2010. Peak phosphorus: Implications for productivity and global food security. 19th World Conference for Soil Solutions for a Changing World: Brisbane, Australia.

Crawford-Browne, S. 2007. African leaders lean on Sudan. *Cape Times.* 8 February.

Cross, C. 2006. Local governance and social conflict: Implications for piloting South Africa's new housing plan in Cape Town's informal settlements. In: M. Huchzermeyer & A. Karam. (eds). *Informal Settlements: A Perpetual Challenge.* Cape Town: University of Cape Town Press.

Crutzen, P.J. 2002. The Anthropocene: Geology and mankind. *Nature*, 415: 23.

Dag Hammarskjold Foundation. 2007. *Development Dialogue: Special Edition on Global Civil Society.* Upsalla, Sweden: Dag Hammarskjold Foundation.

Dalby, S. 2002. Demographic change, natural resources and violence: The current debate. *Journal of International Affairs*, 56(1).

Daly, H.E. 1996. *Beyond Growth: The Economics of Sustainable Development.* Boston: Beacon Press.

Dangl, B. 2007. *The Price of Fire: Resource Wars and Social Movements in Bolivia.* Oakland, CA: AK Press.

Darley, J. 2005. *High Noon for Natural Gas.* Vermont: Chelsea Green Publishing.

Davis, M. 2005. *Planet of Slums.* London: Verso.

Dawkins, R. 1976. *The Selfish Gene.* Oxford: Oxford University Press.

De Cruz, C. & Satterthwaite, D. 2005. The current and potential role of community-driven initiatives to significantly improve the lives of 'slum' dwellers at local, city-wide and national levels. London: International Institute for Environment and Development. Background paper prepared for Millennium Project Taskforce on Improving the Lives of Slum Dwellers.

Deffreys, K.S. 2001. *Hubbert's Peak: The Impending World Oil Shortage.* Princeton, NJ: Princeton University Press.

Den Biggelaar, C., Lal, R., Wiebe, K. & Breneman, V. 2004. The global impact of soil erosion on productivity I: Absolute and relative erosion-induced yield losses. *Advances in Agronomy*, 81: 1–48.

Deng, F.M. 1995. *War of Visions: Conflict of Identities in the Sudan.* Washington, DC: Brookings Institute.

Department for International Development. 2004. *Evaluation of the Conflict Prevention Tools: Sudan.* Evaluation Report EV 647. London: Department for International Development.

Department of Energy, Republic of South Africa. 2011. Integrated resource plan for electricity, 2010–2030. *Government Gazette.* Vol. 551, No. 34263. Pretoria: Government Printer.

Department of Environmental Affairs and Development Planning. 2007. Sustainable development implementation plan - Appendices (Final draft — 16 August 2007). Cape Town: Western Cape Provincial Government.

Department of Environmental Affairs and Tourism. 2007. *State of the Environment.* Pretoria: Department of Environmental Affairs and Tourism.

Department of Minerals and Energy. 2006. *RSA Energy Balance 2006.* Pretoria: Department of Minerals and Energy.

——2007. *Energy Security Master Plan — Liquid Fuels.* Pretoria: Department of Minerals and Energy.

——Department of Water Affairs and Forestry. 2004. *National Water Resource Strategy.* Pretoria: Department of Water Affairs and Forestry. Available: http://www.dwaf.gov.za/Documents/Policies/NWRS/Default.htm.

——2005. *Groundwater Resource Assessment Phase II.* Pretoria: Department of Water Affairs and Forestry.

——2006. *South Africa's Water Sources.* Pretoria: Department of Water Affairs and Forestry. Available: http://www.dwaf.gov.za. [Accessed: October 2006].

——2007. *Artificial Recharge Strategy Version 1.3.* Pretoria: Department of Water Affairs and Forestry.

Dery, P. & Anderson, B. 2007. Peak phosphorus. *Energy Bulletin.* August 13.

De Soysa, I. 2002. Ecoviolence: Shrinking pie, or honey pot? *Global Environmental Politics*, 2(4).

De Wit, M. 2009. *Costing the Integrated Waste Management Bylaw:* Final Report (Updated). Report commissioned by Solid Waste Management Department, City of Cape Town and Sustainability Institute, University of Stellenbosch. Cape Town: De Wit Sustainable Options (Pty) Ltd.

De Wit, M., Musango, J. & Swilling, M. 2008. *Natural Resources, Poverty, Municipal Finances and Sustainable Service Provision: Results of a Participatory Systems Dynamics Scoping Model in Cape Town.* Report prepared for Sustainability Institute, Stellenbosch. Report No. DWSO 8004. Brackenfell, Cape Town: De Wit Sustainable Options (Pty) Ltd.

De Wit, M. & Nahman, A. 2009. *Costing the Integrated Waste Management Bylaw with Specific Reference to Airspace Savings: Draft Report (Phase II).* Report prepared for City of Cape Town and School of Public Management and Planning, Stellenbosch University. Cape Town: De Wit Sustainable Options (Pty) Ltd.

Diamond, J. 1997. *Guns, Germs and Steel.* New York: Norton.

——2005. *Collapse.* New York: Viking.

Dittmar, M. 2011. The end of cheap uranium. *Physics and Society.* 18 June: 1–13.

Dlamini, R.M. 2007. Investigation of sustainable indigenous agricultural practices: A systems approach. Stellenbosch: University of Stellenbosch. MPhil in Sustainable Development Planning and Management.

Dodman, D. 2009. Blaming cities for climate change? An analysis of urban greenhouse gas emissions inventories. *Environment and Urbanization*, 21(1): 185–201.

Dore, R. 2008. Financialization of the global economy. *Industrial and Corporate Change*, 17(6): 1097–1112.

Doshi, V., Schulam, G. & Gabaldon, D. 2007. Lights! Water! Motion! *Strategy and Business*, 47: 39–53.

Douglass, M. & Friedman, J. 1998. *Cities for Citizens: Planning and the Rise of Civil Society in a Global Age.* New York: John Wiley and Sons.

Douthwaite, R. 1999. *The Growth Illusion.* Gabriola Island, Canada: New Society Publishers.

Dowling, T. 2007. Lynedoch's water and sanitation system. Stellenbosch: School of Public Management and Planning, University of Stellenbosch. Masters' thesis.

Dresner, S. 2002. *The Principles of Sustainability.* London: Earthscan.

Driver, M., Smith, T. & Maze, K. 2005. Specialist review paper on biodiversity. Cape Town: Unpublished paper commissioned by the Department of Environmental Affairs and Tourism.

Economist, The. 2009. *The Long Climb.* Special report. October 2009. London: *The Economist.*

Edigheji, O. 2010. *Constructing a Developmental State in South Africa.* Cape Town: HSRC Press.

Edwards, B. 2002. *Rough Guide to Sustainability.* London: RIBA.

Ehrlich, P. 2002. *Human Natures: Genes, Cultures and the Human Prospect.* London: Penguin.

——2008. Key issues for attention from ecological economists. *Environment and Development Economics,* 13: 1–20.

Ehrlich, P. & Ehrlich, A.H. 1990. *The Population Explosion.* New York: Simon and Schuster.

Ekins, P. 2000. *Economic Growth and Environmental Sustainability.* New York: Routledge.

Elkington, J. 1998. *Cannibals with Forks: The Triple Bottom Line of 21st Century Business.* Gabriola Is., Canada: New Society Publishers.

Energy Information Administration. 2008. *Country Analysis Briefs: Sudan — Oil.* Department of Energy, US Government, 15 January 2009. Available from: http://www.eia.doe.gov/emeu/cabs/Sudan/Oil.html.

———2009. *International Energy Statistics*: Published by Energy Information Administration. Available: http://www.eia.doe.gov/emeu/international/contents.html. [Accessed: 8 September 2010].

Epstein, G. 2005. *Financialization and the World Economy.* Cheltenham, UK: Edward Elgar Publishing.

Erb, K.H., Haberl, H., Krausmann, F., Lauk, C., Plutzar, C., Steinberger, J.K., Muller, C., Bondeau, A., Waha, K. & Pollack, G. 2009. *Eating the Planet: Feeding and Fuelling the World Sustainably, Fairly and Humanely - A Scoping Study.* Social Ecology Working Paper, No. 116. Vienna: Institute for Social Ecology.

Escobar, A. 1995. *Encountering Development: The Making and Unmaking of the Third World.* Princeton, NJ: Princeton University Press.

Escobar, A. 2011. Sustainability: Design for the pluriverse. *Development,* 54(2): 137–140.

European Commission. 2008. *European Commission Proposes New Strategy to Address EU Critical Needs for Raw Materials.* IP/08/1628. 4 November. Brussels: European Commission.

Evans, B., Joas, M., Sundback, S. & Theobald, K. 2005. *Governing Sustainable Cities.* London: Earthscan.

Evans, P. 1995. *Embedded Autonomy: States and Industrial Transformation.* Princeton, NJ: Princeton University Press.

——2002. *Liveable Cities: Urban Struggles for Livelihood and Sustainability.* Berkeley: University of California Press.

——2005. The challenges of the institutional turn: New interdisciplinary opportunities in development theory. In: V. Nee & R. Swedberg. (eds). *The Economic Sociology of Capitalist Institutions.* Princeton, NJ: Princeton University Press.

——2006. What will the 21st century developmental state look like? Implications of contemporary development theory for the state's role. Conference on The Changing Role of the Government in Hong Kong, Chinese University of Hong Kong: Hong Kong. Unpublished paper.

——2010. Constructing the 21st century developmental state: Potentialities and pitfalls. In: O. Edighej. (ed.). *Constructing a Democratic Developmental State in South Africa: Potentials and Challenges.* Cape Town: HSRC Press.

Ewing, K. & Mammon, N. 2010. Cape Town den[city]: Towards sustainable urban form. In: M. Swilling & B. Nussbaum. (eds). *Imagining Sustainable Cape Town*. Stellenbosch: Sun Media.

FAO. 2006. *World Agriculture: Towards 2030/50 – Interim Report. Prospects for Food, Nutrition, Agriculture and Major Commodity Groups*. Rome: FAO.

——2008. *Soaring Food Prices: Facts, Perspectives, Impacts and Action Required*. High Level Conference on World Food Security: The Challenges of Climate Change and Bioenergy HLC/08/INF/1. Rome: FAO. 3–5 June.

Fernandes, E., Pell, A. & Uphoff, N. 2002. Rethinking agriculture for new opportunities. In: N. Uphoff. (ed.). *Agroecological Innovations: Increasing Food Production with Participatory Development*. London: Earthscan.

Field, S.L. 2004. The internal and external contexts of oil politics in Sudan: The role of actors. In: K.G. Adar, J.G. Nyuot-Yoh & E. Maloka. (eds). *Sudan Peace Process: Challenges and Future Prospects*. Pretoria: Africa Institute.

Fig, D. 2007. Technological choices in South Africa: Ecology, democracy and development. In: S. Buhlungu, J. Daniel, R. Southall & J. Lutchman. (eds). *State of the Nation: South Africa 2006–2007*. Cape Town: HSRC Press.

Fine, B. & Rustomjee, Z. 1996. *The Political Economy of South Africa: From Minerals-Energy Complex to Industrialization*. Boulder, CO.: Westview Press.

Fischer-Kowalski, M. 1998. Society's metabolism: The intellectual history of materials flow analysis, Part I, 1860–1970. *Journal of Industrial Ecology*, 2(1): 61–78.

——1999. Society's metabolism: The intellectual history of materials flow analysis, Part II, 1970–1998. *Journal of Industrial Ecology*, 2(4): 107–136.

——2011. Analysing sustainability transitions as a shift between socio-metabolic regimes. *Environmental Innovations and Societal Transitions*, 1: 152–159.

Fischer-Kowalski, M. & Haberl, H. 2007. *Socioecological Transitions and Global Change: Trajectories of Social Metabolism and Land Use*. Cheltenham, UK: Edward Elgar Publishing.

Fischer-Kowalski, M. & Swilling, M. 2011. *Decoupling Natural Resource Use and Environmental Impacts from Economic Growth*. Report for the International Resource Panel. Paris: United Nations Environment Programme.

Fisk, R. 2005. *The Great War for Civilisation: The Conquest of the Middle East*. London: Fourth Estate.

Foucault, M. 1998. *The History of Sexuality Vol. 1: The Will to Knowledge*. London: Penguin.

Frankel, J.A. & Rose, A.K. 2009. *Determinants of Agricultural and Mineral Commodity Prices*. Westfalische Wilhelm University. 15 June 2009. Muenster.

Frankel, J.A. *et al.* 2006. *South Africa: Macroeconomic Challenges after a Decade of Success*. Centre for International Development. Available from: http://www.cid.harvard.edu/cidwp/pdf/133.pdf. [Accessed: 28 June 2007].

Freund, B. 2010. Development dilemmas in post-apartheid South Africa: An introduction. In: B. Freund & H. Witt. (eds). *Development Dilemmas in Post-Apartheid South Africa*. Pietermaritzburg: University of KwaZulu-Natal Press.

Freund, B. & Witt, H. 2010. *Development Dilemmas in Post-Apartheid South Africa*. Pietermaritzburg: University of KwaZulu-Natal Press.

Friedman, H. 2003. Eating in the Gardens of Gaia: Envisioning Polycultural communities. In: J. Adams. (ed.). *Fighting for the Farm: Rural America Transformed*. Philadelphia: University of Pennsylvania Press.

Fukuyama, F. 1992. *The End of History and the Last Man*. New York: Free Press.

——2008. The fall of America, Inc. *Newsweek.* 13 October.

Fund for Peace. 2008. *The Fund for Peace.* Available from: www.fundforpeace.org. [Accessed: 14 January 2008].

Fund for Peace & Carnegie Endowment. 2005. The Failed States Index. *Foreign Policy,* July/August.

Funes, F., Garcia, L., Bourque, M., Perez, N. & Rosset, P. 2002. *Sustainable Agriculture and Resistance: Transforming Food Production in Cuba.* Oakland, CA: Food First.

Gallopin, G. 2003. *A Systems Approach to Sustainability and Sustainable Development.* Project NET/00/063. Santiago: Economic Commission for Latin America.

Gasson, B. 2002. The ecological footprint of Cape Town: Unsustainable resource use and planning implications. National Conference of the South African Planning Institute, 18–20 September. Durban.

Gedicks, A. 1993. *The New Resource Wars: Native and Environmental Struggles Against Multinational Corporations.* Cambridge, MA: South End Press.

Gelb, S. 2005. An overview of the South African economy. In: J. Daniel, R. Southall & J. Lutchman. (eds). *State of the Nation: South Africa, 2004–2005.* Cape Town: HSRC Press.

—— 2006. A South African developmental state: What is possible? Harold Wolpe Memorial Trust's Tenth Anniversary Colloquium, 'Engaging silences and unresolved issues in the political economy of South Africa', 21–23 September. Cape Town. Unpublished.

——2010. Macroeconomic policy and development: From Crisis to Crisis. In: B. Freund & H. Witt. (eds). *Development Dilemmas in Post-Apartheid South Africa.* Pietermaritzburg,: University of KwaZulu-Natal Press.

Ghani, A. & Lockhart, C. 2008. *Fixing Failed States.* Oxford: Oxford University Press.

Gibbons, M., Limoges, C., Nowotny, H., Schwartzman, S., Scott, P. & Trow, M. 1994. *The New Production of Knowledge: The Dynamics of Science and Research in Contemporary Societies.* London: Sage.

Gilbert, N. 2009. The disappearing nutrient. *Nature,* 461: 716–404.

Giljum, S., Behrens, A., Hinterberger, F., Lutz, C. & Meyer, B. 2007. *Modelling Scenarios Towards a Sustainable use of Natural Resources in Europe.* SERI Working Papers, No. 4. Vienna.

Girardet, H. 1996. *The Gaia Atlas of Cities.* London: Gaia Books.

——2004. *Cities People Planet: Liveable Cities for a Sustainable World.* Washington DC: Academy Press.

Glaister, B.J. & Mudd, G.M. 2010. The environmental costs of platinum-PGM mining and sustainability: Is the glass half-full or half-empty? *Minerals Engineering,* 23: 438–450.

Godfrey, L., Oelofse, S., Phiri, A., Nahman, A. & Hall, J. 2007. *Mineral Waste: The Required Governance Environment to Enable Re-use.* CSIR/NRE/PW/IR/2007/0080. Pretoria: Council for Scientific and Industrial Research. Available: http://researchspace.csir.co.za/dspace. [Accessed: 5 September 2010].

Goldsmith, P., Abura, L. & Switzer, J. 2002. Oil and water in Sudan. In: J. Lind & K. Sturman. (eds). *Scarcity and Surfeit: The Ecology of Africa's Conflicts.* Pretoria: Institute for Security Studies.

Goodall, C. 2008. *Ten Technologies to Save the Planet.* London: Green Profile.

Goodstein, D. 2005. *Out of Gas: The End of the Age of Oil.* New York: Norton.

Gore, C. 2010. Global recession of 2009 in a long-term development perspective. *Journal of International Development,* 22: 714–738.

Goven, G. 2007. *Green Urbanism—Kosovo Informal Settlement Upgrade: Case Study.* Holcim Forum for Sustainable Urbanism: Shanghai. Holcim Foundation. April.

Goven, G. 2010. Kosovo informal settlement upgrade: Sustainability towards dignified communities. In: E. Pieterse. (ed.) *Counter Currents: Experiments in Sustainability in the Cape Town Region*. Cape Town: Jacana and African Centre for Cities.

Government Communication and Information System. 2010. *South Africa Yearbook 2009/10*. Pretoria: Government Communication and Information System. Available: http://www.gcis.gov.za/resource_centre/sa_info/yearbook/index.html. [Accessed: 14 October 2010].

Gowan, P. 2009. Crisis in the heartland. *New Left Review*, 55: 5–29.

Graham, S. 2010. *Disrupted Cities: When Infrastructure Fails*. London: Routledge.

Graham, S. & Marvin, S. 2001. *Splintering Urbanism: Networked Infrastructures, Technological Mobilities and the Urban Condition*. London: Routledge.

Grantham, J. 2011. *Time to Wake Up: Days of Abundant Resources and Falling Prices are Over Forever*. GMO Quarterly Letter. April Issue: GMO. Available: www.gmo.com/websitecontent/JGLetterALL_1Q11.pdf. [Accessed: 16 June 2011].

Greenwood, D. & Holt, P.F. 2008. Institutional and ecological economics: The role of technology and institutions in economic development. *Journal of Economic Issues*, XLII(2): 445–452.

Grin, J., Rotmans, J., Schot, J., Geels, F. & Loorbach, D. 2010. *Transitions to Sustainable Development: New Directions in the Study of Long-Term Transformative Change*. New York: Routledge.

Gruhn, P., Goletti, F. & Yudelman, M. 2000. *Integrated Nutrient Management, Soil Fertility and Sustainable Agriculture: Current Issues and Future Challenges*. Food, Agriculture, and the Environment Discussion Paper. Washington DC: International Food Policy Research Institute. [Accessed: 11 April 2008].

Gunderson, L.H. & Holling, C.S. 2002. *Understanding Transformations in Human and Natural Systems*. Cheltenham, UK: Edward Elgar Publishing.

Guy, S. & Martin, S. 2001. Urban environmental flows: Towards a new way of seeing. In: S. Guy, S. Marvin & T. Moss. (eds). *Urban Infrastructure in Transition*. London: Earthscan.

Guy, S., Marvin, S. & Moss, T. 2001. *Urban Infrastructure in Transition*. London: Earthscan.

Haberl, H., Fischer-Kowalski, M., Krausman, F., Weisz, H. & Winiwarter, V. 2004. Progress towards sustainability? What the conceptual framework of Material and Energy Flow Accounting (MEFA) can offer. *Land Use Policy*, 21: 199–213.

Haferburg, C. & Obenbrugge, J. 2003. *Ambiguous Restructurings of Post-Apartheid Cape Town*. New Brunswick: Transaction Publishers.

Haggard, S. & Kaufman, R. 1992. *The Politics of Economic Adjustment: International Constraints, Distributive Politics, and the State*. Princeton, NJ: Princeton University Press.

Haims, M.C., Gompert, D.C., Treverton, G.F. & Stearns, B.K. 2008. *Breaking the Failed-State Cycle*. Washington, DC: Rand Corporation.

Halberg, N., Alroe, H.F., Knudsen, M.T. & Kristensen, E.S. 2005. *Global Development and Organic Agriculture: Challenges and Promises*. Wallingford, UK: CAB International.

Hall, S., Held, D., Hubert, D. & Thompson, K. 1996. *Modernity: An Introduction to Modern Societies*. Oxford: Blackwell.

Halweil, B. 2006. Can organic farming feed us all? *World Watch*, 19(3): 18–24.

Hamdok, A. 2004. The future of democracy in post-war Sudan. In: K.G. Adar, J.G. Nyuot & E. Maloka. (eds). *Sudan Peace Process: Challenges and Future Prospects*. Pretoria: Africa Institute.

Hamilton, R. 2011a. Sudan prevents food, health care from reaching Darfur, Aid group says. *The Washington Post*. 25 March.

——2011b. Trouble in Khartoum. *Foreign Policy,* 27 June.

Hamman, R. 2006. Can business make decisive contributions to development? Towards a research agenda on corporate citizenship and beyond. *Development Southern Africa*, 23(2).

Hammond, G. 2006. 'People, planet and prosperity': The determinants of humanity's environmental footprint. *Natural Resources Forum*, 30: 27–36.

Harriss-White, B. & Hariss, E. 2006. Unsustainable capitalism: The politics of renewable energy in the UK. In: C. Leys & L. Panitch. (eds). *Coming to Terms With Nature*. London: Merlin Press.

Hardikar, J. 2006. *The Rising Import of 'Suicides'. India Together.* India Together: Bangalore. Available from: www.indiatogether.org/2006/jun/opi-cotton.htm. [Accessed: 29 September 2010].

Harding, S. 2006. *Animate Earth.* Totnes: Green Books.

Hardoy, J., Mitlin, D. & Satterthwaite, D. 1996. *Sustainability, Environment and Urbanization.* London: Earthscan.

Hartnady, C. 2009. South Africa's gold production and reserves. *South African Journal of Science,* 105: 328–320.

——2010. South Africa's Diminishing Coal Reserves. *South African Journal of Science,* 106(9/10): 1–5.

Haskins, C. 2006. *Household Numbers in Cape Town — Discussion Document.* City of Cape Town: Information and Knowledge Management Department.

Hattingh, J. 2001. Conceptualising ecological sustainability and ecologically sustainable development in ethical terms: Issues and challenges. *Annales,* 2.

Hawken, P. 2007. *Blessed Unrest: How the Largest Social Movement in History is Restoring Grace, Justice and Beauty in the World.* London: Penguin.

Hawken, P., Lovins, A. & Lovins, L. 1999. *Natural Capitalism: Creating the Next Industrial Revolution.* New York: Little Brown.

Haysom, G. 2010. Urban agriculture in the City of Cape Town. In: M. Swilling. (ed.). *Sustaining Cape Town: Towards a Liveable City.* Stellenbosch: Sun Media.

Hayward, T. 2006. Global justice and the distribution of natural resources. *Political Studies,* 54(2): 349–369.

Heifetz, R. 1995. *Leadership without Easy Answers.* London: Belknap Press.

Heinberg, R. 2003. *The Party's Over: Oil, War and the Fate of Industrial Societies.* Vancouver: New Society Publishers.

——2009. Temporary recession or the end of growth? *The Oil Drum.* Available from: www.oildrum.com. [Accessed: 14 July 2010].

——2010. *Peak Everything: Waking Up to the Century of Declines.* Gabriola Island, Canada: New Society Publishers.

Held, D. 2004. *Globalisation: The Dangers and the Answers.* openDemocracy: free thinking for the world. [Online: retrieved October 2004]. openDemocracy: University of Sussex. Available from: www.openDemocracy.net.

Henao, J. & Baanante, C. 1999. *Nutrient Depletion in the Agricultural Soils of Africa.* 2020 Vision Brief, No. 62. Washington, DC: International Food Policy Research Institute.

Hendler, P. 2010. Housing and sustainable neighbourhoods. In: M. Swilling (ed). *Imagining Sustaining Cape Town.* Stellenbosch: Sun Media.

Heun, M., Van Niekerk, J.L., Swilling, M., Meyer, A.J., Brent, A. & Fluri, T.P. 2010. *Learnable Lessons on Sustainability from the Provision of Electricity in South Africa.* Proceedings of

ASME 2010 4th International Conference on Energy Sustainability: Phoenix, Arizona. 17–22 May.

Heynen, N. 2006. Justice of eating in the city. In: N. Heynen, M. Kaika & E. Swyngedouw. (eds). *In the Nature of Cities*. London: Routledge.

Heynen, N., Kaika, M. & Swyngedouw, E. 2006. *In the Nature of Cities: Urban Political Ecology and the Politics of Urban Metabolism*. London: Routledge.

Hirsch, R.L. 2008. Mitigation of maximum world oil production: Shortage scenarios. *Energy Policy*, 36: 881–889.

Hobson, J.M. & Ramesh, M. 2002. Globalisation makes of states what states make of it: Between agency and structure in the state/globalisation debate. *New Political Economy*, 7(1): 5–22.

Hodson, M. & Marvin, S. 2009a. Cities mediating technological transitions: Understanding visions, intermediation and consequences. *Technology Analysis & Strategic Management*, 21(4): 515–534.

———2009b. Urban ecological security: A new urban paradigm? *International Journal of Urban and Regional Research*, 33(1): 193–215.

———2010a. Urbanism in the Anthropocene: Ecological urbanism or premium ecological enclaves? *City*, 14(3): 299–313.

———2010b. Can cities shape socio-technical transitions and how would we know if they were? *Research Policy*, 39: 477–485.

Hoekstra, A.Y. & Chapagain, S. 2007. Water footprints of nations: Water use by people as a function of their consumption pattern. *Water and Resources Management*, 21: 35–48.

Hoff, K. & Stiglitz, J. 2001. Modern economic theory and development. In: G. Meier & J. Stiglitz. (eds). *Frontiers of Development Economics: The Future in Perspective*. New York: Oxford University Press for World Bank.

Homer-Dixon, T.F. 2000. *The Ingenuity Gap*. London: Jonathan Cape.

Holt, P.M. & Daly, M.W. 2000. *A History of Sudan: From the Coming of Islam to the Present Day*. Harlow: Pearson Education Limited.

Howell, P.P., Lock, M. & Cobb, S. 1988. *The Jonglei Canal: Impact and Opportunity*. Cambridge: Cambridge University Press.

Hubacek, K. & Giljum, S. 2003. Applying physical input-output analysis to estimate land appropriation: Ecological footprints of international trade activities. *Ecological Economics*, 44: 137–151.

Hubert, B., Rosegrant, M., Van Boekel, M.A.J.S. & Ortiz, R. 2010. The future of food: Scenarios for 2050. *Crop Science*, 50: 33–50.

Hurrell, A. & Woods, N. 1995. Globalisation and Inequality. *Millennium*, 24(3): 447–470.

Intergovernmental Panel on Climate Change. 2007. *Climate Change 2007*. Geneva: United Nations Environment Programme.

International Crisis Group. 2002. *God, Oil and Country: Changing the Logic of War in Sudan*. Brussels: ICG Press.

International Energy Agency. 2008. *World Energy Outlook*. Paris: International Energy Agency.

International Fertilizer Development Centre. 2010. *World Phosphate Rock Reserves and Resources*. Muscle Shoals, Alabama: International Fertilizer Development Centre.

International Labour Organisation. 2001. *Construction Industry in the Twenty-First Century: Its Image, Employment Prospects and Skills Requirements*. Geneva: International Labour Organization.

International Monetary Fund. 2010. *International Financial Statistics.* Paris: International Monetary Fund.

Irrgang, B. 2005. A study of the efficiency and potential of the eco-village as an alternative urban model. Stellenbosch: School of Public Management and Planning, University of Stellenbosch. Masters' thesis.

Isaac, T. & Franke, R. 2000. *Local Democracy and Development: People's Campaign for Decentralised Planning in Kerala.* New Delhi: Leftwood.

Jackson, H. & Svensson, K. 2002. *Ecovillage Living.* Totnes, UK: Green Books.

Jackson, T. 2009. *Prosperity without Growth? The Transition to a Sustainable Economy.* United Kingdom: Sustainable Development Commission.

Jacobson, A. & Kammen, D. 2005. Science and engineering research that values the planet. *The Bridge*, Winter: 11–17.

Jaglin, S. 2004. Water delivery and institution building in Cape Town: The problems of urban integration. *Urban Forum*, 15(3): 231–253.

——2008. Differentiating networked services in Cape Town: Echoes of splintering urbanism?. *Geoforum*, 39: 1897–1906.

Jarzombek, M. 2008. A green masterplan is still a masterplan. In: I. Ruby & A. Ruby. (eds). *Urban Transformation.* Amsterdam: Ruby Press.

Jayasuriya, K. 2001. Globalization and the changing architecture of the state: The regulatory state and the politics of negative co-ordination. *Journal of European Public Policy*, 8(1): 101–123.

Jenks, M. & Dempsey, N. 2005. *Future Forms and Design for Sustainable Cities.* Oxford: Architectural Press.

Johnson, C. 1982. *MITI and the Japanese Miracle.* Stanford, CA: Stanford University Press.

Jok, J.M. 2009. *Sudan: Race, Religion and Violence.* Oxford: Oneworld.

Juma, C. 2011. *The New Harvest: Agricultural Innovation in Africa.* Oxford: Oxford University Press.

Kahl, C. 2002. Demographic change, natural resources and violence: The current debate. *Journal of International Affairs*, 56(1).

Kanbur, R. 2009. *The Co-Evolution of the Washington Consensus and the Economic Development Discourse.* Working Paper. Ithaca, New York: Department of Applied Economics and Management, Cornell University.

Kate, T. 2010. *From Industrial Agriculture to Agro Ecological Farming: A South African Perspective.* Working Paper Series, No. 10. East London: Eastern Cape Socio-Economic Consultative Council.

Kates, R.W., Clark, W.C., Corell, R., Hall, J.M., Jaeger, C.C., Lowe, I., *et al.* 2001. Sustainability science. *Science*, 292: 641–642.

Kauffman, S. 1995. *At Home in the Universe.* Oxford: Oxford University Press.

Kauffman, S.A. 2008. *Reinventing the Sacred.* New York: Basic Books.

Kay, J., Regier, H., Boyle, M. & Francis, G. 1999. An ecosystem approach for sustainability: Addressing the challenge of complexity. *Futures*, 31(7): 721–742.

Kay, R. & Richardson, K.A. 2007. *Building and Sustaining Resilience in Complex Organizations: Pre-Proceedings of the 1st International Workshop on Complexity and Organization Resilience.* Mansfield, MA: ISCE Publishing.

Keil, R. & Boudreau, J.A. 2006. Metropolitics and metabolics: Rolling out environmentalism in Toronto. In: N. Heynen, M. Kaika & E. Swyngedouw. (eds). *In the Nature of Cities.* London: Routledge.

Kellard, N. & Wohar, M.E. 2006. On the prevalence of trends in primary commodity prices. *Journal of Development Economics*, 79: 146–167.
Keller, E.F. 2000. *The Century of the Gene.* Cambridge, MA: Harvard University Press.
Kelly, C. 2009. Lower external input farming methods as a more sustainable solution for small-scale farmers. Stellenbosch: University of Stellenbosch. MPhil in Sustainable Development Planning and Management.
Khan, F. 2008. Political economy of housing policy in South Africa, 1994–2004. Presentation to MPhil students, Stellenbosch University.
Khan, M. 2004. State failure in developing countries and institutional reform strategies. In: B. Tungodden, N. Stern, & I. Kolstad. (Ed.). *Toward Pro-Poor Policies: Aid, Institutions and Globalization.* New York: Oxford University Press.
Khosa, M.M. 2000. *Empowerment through Service Delivery.* Pretoria: Human Sciences Research Council.
King, L. & Levine, R. 1994. Capital fundamentalism, economic development, and economic growth. *Carnegie-Rochester Conference Series on Public Policy*, 40: 259–292.
Klare, M.T. 2001a. The new geography of conflict. *Foreign Affairs*, May/June.
———2001b. *Resource Wars: The New Landscape of Global Conflict.* New York: Metropolitan Books.
Kondratieff, N.D. 1935. The long waves in economic life. *Review of Economics and Statistics*, 27(1): 105–115.
Kooiman, J. 1993. *Modern Governance: New Government-Society Interactions.* London: Sage.
Korhonen, J. 2008. Reconsidering the Economics Logic of Ecological Modernization. *Environment and Planning A*, 40: 1331–1346.
Korten, D. 1995. *The Post-Corporate World.* San Francisco: Berret-Koehler.
———2006. *The Great Turning: From Empire to Earth Community.* San Francisco, CA: Berret-Koehler.
Krausman, F., Eisenmenger, N., Braunsteiner-Berger, J., Deju, M., Huber, V., Kohl, L., *et al.* 2005. *MFA for South Africa.* Vienna: Institute for Social Ecology.
Krausman, F., Gingrich, S., Eisenmenger, K.H., Haberl, H. & Fischer-Kowalski, M. 2009. Growth in global materials use, GDP and population during the 20th century. *Ecological Economics*, 68(10): 2696–2705.
Krausman, F., Schandl, H., Fischer-Kowalski, M. & Eisenmenger, N. 2008. The global socio-metabolic transition: Past and present metabolic profiles and their future trajectories. *Journal of Industrial Ecology*, 12: 637–656.
Krugman, P. 2008. *The Return of Depression Economics and the Crisis of 2008.* London: Penguin.
———2010a. Building a green economy. *New York Times Magazine.* 5 April.
———2010b. The finite world. *The International Herald Tribune.* 28 December.
Kuhn, T.S. 1962. *The Structure of Scientific Revolutions.* Chicago: University of Chicago Press.
Kunstler, J.H. 2005. *The Long Emergency.* London: Atlantic Books.
Lahiff, E. & Cousins, B. 2005. Smallholder agriculture and land reform in South Africa. *IDS Bulletin*, 36(2): 127–131.
Lambin, E.F. & Meyfroidt, P. 2011. Global land use change, economic globalization, and the looming land scarcity. *Proceedings of the National Academy of Science*, 108(9): 3465–3472.
Lampkin, N.H. & Padel, S. 1994. *The Economics of Organic Farming: An International Perspective.* Wallingford, UK: CAB International.

Langdon, D. 2004. *World Construction Review: Outlook 2004/5*. London: David Langdon Management.

Latour, B. 2004. Why has critique run out of steam? From matters of fact to matters of concern. *Critical Inquiry*, 30: 225–248.

——2008. 'It's development, stupid!" Or how to modernize modernization? In: J. Proctor. (ed.). *Postenvironmentalism*. Boston: MIT Press.

Latour, B. 2010. Where are the passions Commensurate with the stakes? *IDDRI SciencePo: Annual Report 2010*: 3. Paris: www.iddri.org.

Law, J. 2004. *After Method: Mess in Social Science Research*. New York: Routledge.

Le Billon, P. 2001. The political ecology of war: Natural Resources and armed conflicts. *Political Geography*, 20(5): 561–584.

——2004. The Geopolitical Economy of 'Resource Wars'. *Geopolitics*, 9(1).

——2005. *Geopolitics of Resource Wars: Resource Dependence, Governance and Violence*. London and New York: Frank Cass.

Lee, C.K. 2007. *Against the Law: Labor Protests in China's Rustbelt and Sunbelt*. Berkeley: University of California Press.

Leftwich, A. 1995. Bringing politics back in: Towards a model of the developmental state. *Journal of Development Studies*, 31(3).

——2000. *States of Development: On the Primacy of Politics in Development*. Cambridge: Polity Press.

Lind, J. & Sturman, K. 2002. *Scarcity and Surfeit: The Ecology of Africa's Conflicts*. Pretoria: Institute for Security Studies.

Liu, M. & Vlaskamp, M. 2010. Smart, young and broke: White collar workers are China's newest underclass. *Newsweek*, 28 June–5 July: 36–37.

Logan, J. & Molotch, H. 1987. *Urban Fortunes: The Political Economy of Place*. Berkeley: University of California Press.

Lovelock, J. 1979. *Gaia: A New Look at Life on Earth*. Oxford: Oxford University Press.

——2006. *The Revenge of Gaia*. New York: Basic Books.

Lovins, A.B. 2005. More profit with less carbon. *Scientific American*, 293: 74–82.

Lovins, A.B., Datta, E.K., Feiler, T., Rabago, K. & Rabago, K.R. 2002. *Small is Profitable: The Hidden Economic Benefits of Making Electrical Resources the Right Size*. Snowmass, CO.: Rocky Mountain Institute.

Low, N., Gleeson, B., Elander, I. & Lidskog, R. 2000. *Consuming Cities: The Urban Environment in the Global Economy After the Rio Declaration*. New York: Routledge.

Luhmann, N. 1996. *Social Systems*. Stanford CA: Stanford University Press.

Lundvall, B.A. 2007. *National Innovation System: Analytical Focusing Device and Policy Learning Tool*. Working Paper. Ostersund, Sweden: Swedish Institute for Growth Policy Studies.

Lynch, M.C. 2003. The new pessimism about petroleum resources: Debunking the Hubbert Model (and Hubbert Modelers). *Minerals and Energy*, 18(1): 21–32.

MacNeill, J., Winsemius, P. & Yakushiii, T. 1991. *Beyond Interdependence: The Meshing of the World's Economy and the Earth's Ecology*. Oxford: Oxford University Press.

Madeley, J. 2002. *Food for all: The Need for a New Agriculture*. London: Zed Books.

Mail & Guardian Correspondent. 2008. India digs deep. Johannesburg: *Mail & Guardian*, 7–13 March.

Maguire, S. & McKelvey, B. 1999. Complexity and management: Moving from fad to foundations. *Emergence*, 1(2): 19–61.

Malik, A. 2001. After modernity: Contemporary non-Western cities and architecture. *Futures*, 33: 873–882.

Mamdani, M. 2002. *When Victims Become Killers: Colonialism, Nativism, and the Genocide in Rwanda*. Princeton, NJ: Princeton University Press.

Manger, L. 2005. Understanding resource management in Western Sudan. In: Q. Gausset, M.A. Whyte & T. Birch-Thomsen. (eds). *Beyond Territory and Scarcity: Exploring Conflicts over Natural Resource Management*. Stockholm: Nordiska Afrika Institutet.

Manolis, J., Chan, K., Finkelstein, M., Stephens, S., Nelson, C., Grant, J. & Dombeck, M. 2008. Leadership: A new frontier in conservation science. *Conservation Biology*, 23(4): 879–886.

Marais, H. 1998. *South Africa: Limits to Change*. Cape Town: University of Cape Town Press.

———2011. *South Africa Pushed to the Limit: The Political Economy of Change*. Cape Town: University of Cape Town Press.

Margulis, L. & Sagan, D. 1997. *Microcosmos: Four Billion Years of Microbial Evolution*. Berkeley: University of California Press.

Marinova, D., Annandale, D. & Phillimore, J. 2006. *The International Handbook on Environmental Technology Management*. Cheltenham, UK: Edward Elgar Publishing.

Marnay, C. & Firestone, R. 2007. *Microgrids: An Emerging Paradigm for Meeting Building Electricity and Heat Requirements Efficiently and with Appropriate Energy Quality*. European Council for an Energy Efficient Economy. 2007 Summer Study: La Colle sur Loup, France. Ernest Orland Lawrence Berkeley National Laboratory. April.

Marsden, C. 2000. The new corporate citizenship of big business: Part of the solution to sustainability? *Business and Society Review*, 105(1).

Marvin, S. & Medd, W. 2006. Metabolisms of Obe-City: Flows of Fat through Bodies, Cities and Sewers. In: N. Heynen, M. Kaika & E. Swyngedouw. (eds). *In the Nature of Cities*. London: Routledge.

Maturana, H. & Varela, F.J. 1987. *Tree of Knowledge*. Boston, MA: Shambhala Publications.

Max-Neef, M. 1991. *Human Scale Development*. New York and London: Apex Press.

———2005. Foundations of transdisciplinarity. *Ecological Economics*, 53: 5–16.

Mayer, J. & Fajarnes, P. 2005. *Tripling Africa's Primary Exports: What? How? Where?* UNCTAD Discussion Papers, No. 180. Geneva: UNCTAD.

Mbembe, A. 2002. African modes of self-writing. *Public Culture*, 14: 239–73.

———2004. Writing the words from an African metropolis. *Public Culture*, 16(3): 347–372.

McCarthy, H., Miller, P. & Skidmore, P. 2004. *Network Logic: Who Governs in an Interconnected World?* London: Demos.

McDonald, D. & Ruiters, G. 2005. *The Age of Commodity*. London: Earthscan.

McDonald, D. & Smith, L. 2002. *Privatizing Cape Town: Service Delivery and Policy Reforms since 1996*. Johannesburg: Municipal Service Project, Occasional Paper Series, No. 7.

McDonough, W. & Braunart, M. 2002. *Cradle to Cradle: Remaking the Way We Make Things*. New York: North Point Press.

McLaren, D. 2003. Environmental space, equity and the ecological debt. In: J. Agyeman, R. D. Bullard & B. Evans. (eds). *Just Sustainabilities: Development in an Unequal World*. London: Earthscan.

McKinsey Global Institute. 2010. *Lions on the Move: The Progress and Potential of African Economies*: McKinsey Global Institute. Available: www.mckinsey.com/mgi.

McMichael, P. 2009a. A food regime analysis of the 'world food crisis'. *Agriculture and Human Values*, 26: 281–295.

——2009b. A food regime genealogy. *The Journal of Peasant Studies*, 36(1): 139–169.

Mebratu, D. 1998. Sustainability and sustainable development: Historical and conceptual review. *Environment Impact Assessment Review*, 18: 493–520.

Menegat, R. 2002. Participatory democracy and sustainable development: Integrated urban environmental management in Porto Alegre. *Environment and Urbanization*, 14(2): 181–206.

Middleton, B.J. & Bailey, A.K. 2008. *Water Resources in South Africa, 2005*. Water Research Commission Report, No. TT 381/08. Pretoria: Water Resource Commission.

Miraftab, F. 2004. Neoliberalism and casualization of public sector services: The case of waste collection services in Cape Town, South Africa. *International Journal of Urban and Regional Research*, 28(4): 847–892.

Mitlin, D. 2008. With and beyond the state: Co-production as a route to political influence, power and transformation for grassroots organizations. *Environment and Urbanization*, 20: 339–360.

Mitlin, D. & Satterthwaite, D. 2004. *Empowering Squatter Citizen: Local Government, Civil Society and Urban Poverty Reduction*. London: Earthscan.

Mkandawire, T. 2001. Thinking about developmental states in Africa. *Cambridge Journal of Economics*, 25: 289–313.

Moghraby, A.E. 2003. State of the environment in Sudan. In: M. McCabe & B. Sadler. (eds). *UNEP Studies of EIA Practice in Developing Countries*. Geneva: United Nations Environment Programme, Division for Trade, Industry and Economics.

Mohamed, S. 2010. The state of the South African economy. In: J. Daniel, P. Naidoo, D. Pillay & R. Southall. (eds). *New South African Review 1: 2010—Development or Decline?* Johannesburg: Wits University Press.

Mohr, S.H. & Evans, G.M. 2009. Forecasting coal production until 2100. *Fuel*, 88: 2059–2067.

Mokheseng, B. 2010. Solar roof tiles: A macro-economic analysis. School of Public Management and Planning, University of Stellenbosch. MPhil in Sustainable Development Planning and Management.

Monbiot, G. 2006. *Heat: How to Stop the Planet Burning*. London: Allen Lane.

Montalvo, C. 2008. Sustainable systems of innovation: How level is the playing field for developing and developing economies? Paper presented at Globelics Conference: Mexico City. 22–24 September.

Moore, M. 1995. *Creating Public Value*. Boston: Harvard University Press.

Morin, E. 1992. *Method: Towards the Study of Humankind Volume 1: The Nature of Nature*. New York: Peter Land Publishers

——1999. *Homeland Earth*. Cresskill, NJ: Hampton Press.

——2005. *Restricted Complexity, General Complexity*. Intelligence de la Complexité: Epistemologie et Pragmatique: Cerisy-La-Salle, France. 26 June.

Morse, D. 2005. *War of the Future: Oil Drives the Genocide in Darfur*. Global Policy Forum. 18 August. Available from: www.globalpolicy.org/security/issues/sudan/2005/0818darfuroil.htm.

Moss, T. 2001. Battle of the systems? Changing styles of water recycling in Berlin. In: S. Guy, S. Marvin & T. Moss. (eds). *Urban Infrastructure In Transition: Networks, Buildings, Plans*. London: Earthscan.

Motsi, I. 2006. *The Motives and Interests of International Presence in Sudan*. Electronic Briefing Paper 46. Pretoria: Centre for International Political Studies, University of Pretoria. Available: www.up.ac.za/academic/cips.

Mougeot, L.J.A. 2006. *Growing Better Cities: Urban Agriculture for Sustainable Development.* Ottawa: International Development Research Centre.

Muller, M. 2007. Adapting to climate change: Water management for urban resilience. *Environment and Urbanization*, 19(1): 99–113.

Nass, P. & Hoyer, K.G. 2009. The emperor's green clothes: Growth, decoupling, and capitalism. *Capitalism Nature Socialism*, 20(3): 74–95.

National Planning Commission. 2011. *Diagnostic Overview.* Pretoria: National Planning Commission, Republic of South Africa.

National Research Council of the National Academies. 2003. *Cities Transformed.* Washington, DC: National Academies Press.

National Research Council of the National Academies. 2003. *Materials Count: The Case for Material Flow Analysis.* Washington, DC: The National Academies Press.

Natsios, A.S. & Abramowitz, M. 2011. Sudan's secession crisis: Can the South Part from the North without war? *Foreign Affairs*, 90(1): 19.

Nederveen-Pieterse, J. 2002. Global inequality: Bringing politics back in. *Third World Quarterly*, 23(6): 1023–1046.

Nellemann, C., MacDevette, M., Manders, T., Eickhout, B., Svihus, B., Gerdien-Prins, A. & Kaltenborn, B.P. 2009. *The Environmental Food Crisis: The Environment's Role in Averting Future Food Crises - A UNEP Rapid Response Assessment.* Nairobi: United Nations Environment Programme. Available: www.grida.no.

Netherlands Environmental Assessment Agency. 2009. *Growing within Limits.* Bilthoven, Ne: Netherlands Environmental Assessment Agency.

Newman, P. & Kenworthy, J. 2007. Greening urban transportation. In: Worldwatch Institute. (ed.). *State of the World: Our Urban Future.* New York and London: W.W. Norton & Company.

Ngxabi, N. 2001 Homes or houses? Strategies of homemaking among some amaXhosa in the Western Cape. University of Cape Town. Masters' thesis.

Nicolescu, B. 2002. *Manifesto of Transdisciplinarity.* New York: State University of New York Press.

Norton, B. 1991. *Toward Unity among Environmentalists.* Oxford: Oxford University Press.

OECD & FAO. 2010. *OECD-FAO Agricultural Outlook 2010–2019: Highlights.* Paris & Rome: OECD & FAO.

Oelofse, S.H.H. 2008a. *Mine Water Pollution - Acid Mine Decant, Effluent and Treatment: Consideration of Key Emerging Issues that may Impact on the State of Environment.* Pretoria: Department of Environmental Affairs and Tourism. Available: http://soer.deat.gov.za/docport.aspx?m=97&d=28.

——2008b. Protecting a vulnerable groundwater resource from impacts of waste disposal: A South African waste governance perspective. *International Journal of Water Resources Development*, 24(3): 477–490.

Oelofse, S.H.H., Hobbs, P.J., Rascher, J. & Cobbing, J. 2007. The pollution and destruction threat of gold mining waste on the Witwatersrand: A West Rand case study. 10th International Symposium on Environmental Issues and Waste Management in Energy and Mineral Production: Bangkok. 11–13 December.

O' Keeffe, J., Uys, M. & Bruton, M.N. 1992. Freshwater systems. In: R.F. Fuggle & M.A. Rabie. (eds). *Environmental Management in South Africa.* Cape Town: Juta & Company.

Okri, B. 1995. *Astonishing the Gods.* London: Phoenix House.

——1996. *Dangerous Love.* London: Phoenix House.

——1999. *A Way of Being Free.* London: Phoenix House.

Oldeman, L.R. 1994. The global extent of soil degradation. In: D.J. Greenland & T. Szaboles. (eds). *Soil Resilience and Sustainable Land Use.* Wallingford, UK: Commonwealth Agricultural Bureau International.

Oliver, S. 2002. Demand management in cells. *Nature,* 418: 33–34.

Onis, Z. 1991. The logic of the developmental state. *Comparative Politics,* 24(1): 109–126.

Organisation for Economic Co-Operation and Development. 2002. *Indicators to Measure Decoupling of Environmental Pressure and Economic Growth.* Paris: Organisation for Economic Co-Operation and Development (OECD).

Ostrom, E. 1999. Revisiting the commons: Local lessons, global challenges. *Science,* 284: 278–282.

Oswald, F. & Baccini, P. 2003. *Netzstadt.* Basel: Birkhauser.

Parker, F. 2011. Climate change can be an 'opportunity', says Stiglitz. *Mail & Guardian* Online. 30 January.

Parnell, S. & Pieterse, E. 2007. Developmental local government. In: S. Parnell, E. Pieterse, M. Swilling & D. Wooldridge. (eds). *Democratising Local Government: The South African Experiment.* Cape Town: University of Cape Town Press.

Parnell, S., Pieterse, E., Swilling, M. & Wooldridge, D. (eds). 2002. *Democratising Local Government: The South African Experience.* Cape Town: University of Cape Town Press.

Patel, R. 2008. *Stuffed and Starved.* New York: Melville House.

——2009. *The Value of Nothing: How to Reshape Market Society and Redefine Democracy.* London: Portobello.

Patel, R. & McMichael, P. 2008. A political economy of the food riot. *Review,* XXXII(1): 9–35.

Patzek, T.W. & Croft, G.D. 2010. A global coal production forecast with Multi-Hubbert Cycle Analysis. *Energy,* 35: 3109–3122.

Peat, D. 2002. *From Certainty to Uncertainty: The Story of Science and Ideas in the Twentieth Century.* Washington, DC: Joseph Henry Press.

Peet, R. & Watts, M. 2000. *Liberation Ecologies: Environment, Development, Social Movements.* London: Routledge.

Peralta, A. 2003. The globalization of inequality in the Information Age. In: P.D. van Seters. (ed.). *Globalization and its New Divides.* Amsterdam: Dutch University Press in association with Purdue University Press.

Pereira, P.R. 1994. New technologies: Opportunities and threats. In: F. Sagasti, J.J. Salomon & C. Sachs-Jeanter. (eds). *The Uncertain Quest: Science, Technology and Development.* New York: United Nations University Press.

Perez, C. 2002. *Technological Revolutions and Financial Capital: The Dynamics of Bubbles and Golden Ages.* Cheltenham, UK: Edward Elgar Publishing.

——2007. *Great Surges of Development and Alternative Forms of Globalization.* Working Papers in Technology Governance and Economic Dynamics. Norway and Estonia: The Other Canon Foundation (Norway) and Tallinin University of Technology (Tallinin).

——2009. The double bubble at the turn of the century: Technological roots and structural implications. *Cambridge Journal of Economics,* 33: 779–805.

Perez, C. 2010. *The Financial Crisis and the Future of Innovation: A View of Technical Change with the Aid of History.* Working Papers in Technology Governance and Economic Dynamics, 28. Norway and Tallinin: The Other Canon Foundation & Tallinin University of Technology.

Perkins, J. 2004. *Confessions of an Economic Hit Man.* San Francisco: Berrett-Koehler.

Pervaiz, A., Rahman, R. & Hasan, A. 2008. *Lessons from Karachi: The Role of Demonstration, Documentation, Mapping and Relationship Building in Advocacy for Improved Urban Sanitation and Water Services*. Human Settlements Programme. London: International Institute for Environment and Development.

Peter, C., Swilling, M., Du Plooy, P., De Wit, M., Wakeford, J., Lanz, J., *et al.* 2010. *Greening the South African Growth Path: Challenges, Prospects and Trajectories*. Stellenbosch: Sustainability Institute.

Petterson, D. 1999. *Inside Sudan: Political Islam, Conflict and Catastrophe*. Boulder, CO: Westview Press.

Pezzoli, K. 1997. Sustainable Development: A Transdisciplinary Overview of the Literature. *Journal of Environmental Planning and Management*, 40(5): 549–574.

Pickering, M. 2008. The distribution of power: Local government and electricity distribution industry reforms. In: M. van Donk, M. Swilling, E. Pieterse & S. Parnell. (eds). *Consolidating Developmental Local Government: Lessons from the South African Experience*. Cape Town: University of Cape Town Press.

Pieterse, E. 2008. *City Futures*. Cape Town: University of Cape Town Press.

Pieterse, E. 2010. *Counter Currents: Experiments in Sustainability in the Cape Town Region*. Cape Town & Johannesburg: Jacana & African Centre for Cities.

Pile, S. & Thrift, N. 1996. Mapping the subject. In: S. Pile & N. Thrift. (eds). *Mapping the Subject: Geographies of Cultural Transformation*. London: Routledge.

Pillay, D. 2007. The stunted growth of South Africa's developmental State discourse. *Africanus*, November: 176–191.

Pimentel, D., Hepperly, P., Hanson, J., Douds, D. & Seidel, R. 2005. Environmental, energetic and economic comparisons of organic and conventional farming systems. *BioScience*, 55(7): 573–582.

Polanyi, K. 1946. *Origins of our Time: The Great Transformation*. London: Victor Gollancz Ltd.

Portney, K.E. 2003. *Taking Sustainable Cities Seriously*. Cambridge, MA: MIT Press.

Pretty, J. 2006. Agroecological approaches to agricultural development. RIMISP-Latin American Center for Rural Development Reports for World Bank, *World Development Report 2008*. Santiago: RIMISP-Latin American Center for Rural Development (www.rimisp.org).

Pretty, J.N., Morison, J.I.L. & Hine, R.E. 2003. Reducing food poverty by increasing agricultural sustainability in developing countries. *Agriculture, Ecosystems and Environment*, 95: 217–234.

Pugh, C. 1996. *Sustainability, Environment and Urbanization*. London: Earthscan.

——1999. *Sustainability in Cities in Developing Countries: Theories and Practice at the Millennium*. London: Earthscan.

Pulzelli, R. & Tiezzi, E. 2009. *City Out of Chaos: Urban Self-Organization and Sustainability*. Southampton: WIT.

Qiu, J. 2010. Phosphate fertilizer warning for China. *Nature News*: 21 October.

Rabie, M.A. & Day, J.A. 1992. Rivers. In: R.F. Fuggle & M.A. Rabie. (eds). *Environmental Management in South Africa*. Cape Town: Juta & Company.

Rastomjee, Z. 2011. NPC dialogue – RSA's low carbon economy. Powerpoint presentation to National Planning Commission Workshop on the Low Carbon Economy: National Planning Commission, Pretoria. 29 February.

Ravetz, J. 2000. *City Region 2020: Integrated Planning for a Sustainable Environment*. London: Earthscan.

Reardon, T., Timmer, C.P., Barrett, C.B. & Berdegue, J. 2003. The rise of supermarkets in Africa, Asia, and Latin America. *Americal Journal of Agricultural Economics*, 85(5): 1140–1146.

Reed, C. 2007. *Mayor Reed's Green Vision for San Jose.* San Jose: City of San Jose. Available from: http://www.sanjoseca.gov/mayor/goals/environment/GreenVision/GreenVision.asp. [Accessed: 6 August 2009].

Reed, D. 2001. *Economic Change, Natural Resource Wealth and Governance: The Political Economy of Change in Southern Africa.* London: Earthscan.

Renner, M. 2004. Anatomy of resource wars. The Hague Conference on Environment, Security and Sustainable Development: Institute for Environment and Society. 9–12 May. www.envirosecurity.org.

Republic of South Africa. Department of Environmental Affairs and Tourism. 2007. *People, Planet, Prosperity. A Framework for Sustainable Development in South Africa.* Department of Environmental Affairs and Tourism: Pretoria: 24 February. Available from: www.sustainabilityinstitute.net. [Accessed: 24 February 2008].

Republic of South Africa. Department of Water Affairs and Forestry. 2004. *National Water Resource Strategy.* Pretoria: Department of Water Affairs and Forestry. [Accessed: 4 November 2006].

Republic of South Africa. National Planning Commission. 2010. *South Africa: Development Indicators.* Pretoria: National Planning Commission.

Republic of South Africa. Western Cape Provincial Government. 2007. *iKapa Provincial Growth and Develoment Strategy.* Cape Town: Department of the Premier, Western Cape Provincial Government.

Revi, A., Prakash, S., Mehrotra, R., Bhat, G.K., Gupta, K. & Gore, R. 2006. Goa 2100: The transition to a sustainable *urban* design. *Environment and Urbanization*, 18(1).

Ritzen, J., Easterly, W. & Woolcock, M. 2000. *On 'Good' Politicians and 'Bad' Policies: Social Cohesion, Institutions and Growth.* World Bank Working Paper. Washington, DC: World Bank.

Roberts, P. 2005. *The End of Oil: The Decline of the Petroleum Economy and the Rise of a New Energy Order.* London: Bloomsbury.

Robertson, R. 2007. A brighter shade of green: Rebooting environmentalism for the 21st century. *What is Enlightenment?* 38, October–December.

Robins, S. 2006. When shacks ain't chic! Planning for 'difference' in post-apartheid Cape Town. In: S. Bekker & A. Leilde. (eds). *Reflections on Identity in Four African Cities.* Stellenbosch: African Minds.

Rockstrom, J. 2009. A safe operating space for humanity. *Nature*, 461(24): 472–475.

Rockstrom, J., Steffen, W., Noone, K., Persson, A., Chapin, F.S., Lambin, E.F., *et al.* 2009. Planetary boundaries: Exploring the safe operating space for humanity. *Ecology and Society*, In press. 14 September.

Rodrik, D. 2000. *Growth and Poverty Reduction: What are the Real Questions?* Finance and Development. Report, Harvard University. [Accessed: 25 March 2008].

———2006. Understanding South Africa's economic puzzles. Paper prepared for the Harvard Center for International Development Project on South Africa. Cambridge, MA: John F. Kennedy School of Government, Harvard University.

———2007. World too complex for one-size-fits-all models. *Post-Autistic Economics Review*, 44: 73–74.

Rodrik, D., Subramanian, A. & Trebbi, F. 2004. Institutions rule: The primacy of institutions over geography and integration in economic development. *Journal of Economic Growth*, 9: 131–165.

Roelofs, J. 1996. *Greening Cities: Building Just and Sustainable Communities.* New York: The Bootstrap Press.

Romanovsky, P. 2006. *Executive Summary: Population Projection for Cape Town 2001–2021*. Cape Town: City of Cape Town: Information and Knowledge Management Department.

Ross, F. 2005. Model communities and respectable residents? Home and housing in a low-income residential estate in the Western Cape, South Africa. *Journal of Southern African Studies*, 31(3).

Ross, M. 2003. The natural resource curse: How wealth can make you poor. In: I. Bannon & P. Collier. (eds). *Natural Resources and Violent Conflicts: Options and Actions*. Washington, DC: The World Bank.

Rotman, D. 2009. Can technology save the economy? *Technology Review*: May/June.

Rotmans, J. 2006. A complex systems approach for sustainable cities. In: M. Ruth. (ed.). *Smart Growth and Climate Change: Regional Development, Infrastructure and Adaptation*. Cheltenham, UK: Edward Elgar Publishing.

Rotmans, J. & Loorbach, D. 2009. Complexity and transition management. *Journal of Industrial Ecology*, 13(2): 184–196.

Roux, D.J., Murray, K. & Van Wyk, E. 2008. Learning to learn for social-ecological resilience. In: M. Burns & A. Weaver. (eds). *Exploring Sustainability Science: A Southern African Perspective*. Stellenbosch: Sun Media.

Rutledge, D. 2011. Estimating long-term world coal production from S-Curve Fits. *International Journal of Coal Geology*, 85: 23–33.

Sachs, J.D. & Warner, A.M. 2001. The curse of natural resources. *European Economic Review*, 45(4–6): 827–838.

Sachs, W. 2002. *The Joburg Memo: Fairness in a Fragile World*. Berlin: Heinrich Boll Foundation.

Salopek, P. 2003. Shattered Sudan: Drilling for oil, hoping for peace. *National Geographic* February.

Samuels, J. 2005. *Removing Unfreedoms: Citizens as Agents of Change in Urban Development*. London: ITDG.

Sandercock, L. 2003. Dreaming the sustainable city: Organising hope, negotiating fear, mediating memory. In: B. Eckstein & J. Throgmorton. (eds). *Storytelling and Sustainability: Planning, Practice and Possibility for American Cities*. Massachusetts: MIT Press.

Sangina, P.C., Waters-Bayer, A., Kaaria, S., Njuki, J. & Wettasinha, C. 2009. *Innovation Africa: Enriching Farmers' Livelihoods*. London: Earthscan.

Sassen, S. 2000. *Cities in a World Economy*. London: Pine Forge.

Satterthwaite, D. 2001. *Sustainable Cities*. London: Earthscan.

——2006. Review of planet of slums and shadow cities. *Environment and Urbanization*, 18(2): 543–546.

——2007. *The Transition to a Predominantly Urban World and its Underpinnings*. Human Settlements Discussion Paper. London: International Institute for Environment and Development.

——2009. The implications of population growth and urbanization for climate change. *Environment and Urbanization*, 21(2): 545–567.

Scenario Building Team. 2007. *Long-Term Mitigation Scenarios: Strategic Options for South Africa*. Pretoria: Republic of South Africa. Department of Environmental Affairs and Tourism.

Schaffer, P. 2008. New thinking on poverty: Implications for globalisation and poverty reduction strategies. *Real World Economics Review*, 47: 192–231.

Scheffer, M. 2009. *Critical Transitions in Nature and Society*. Princeton, NJ: Princeton University Press.

Scherr, S. 1999. *Soil Degradation: A Threat to Developing-Country Food Security by 2020?* Food, Agriculture and the Environment Discussion Paper No. 63. Washington, DC: International Food Policy Research Institute.

Schmitter, P., O'Donnell, G. & Whitehead, L. 1986. *Transitions from Authoritarian Rule: Prospects for Democracy Volumes I–IV*. Baltimore: Johns Hopkins University Press.

Schumpeter, J.A. 1939. *Business Cycles: A Theoretical, Historical and Statistical Analysis of the Capitalist Process.* New York: McGraw Hill.

Schwartz, H. 2004. *Urban Renewal, Municipal Revitalisation: The Case of Curitiba, Brazil.* Alexandra, VA: Hugh Schwartz.

Seekings, J. & Nattrass, N. 2005. *Race, Class and Inequality in South Africa.* New Haven: Yale University Press.

Sen, A. 1999. *Development as Freedom.* New York: Knopf.

———2009. Capitalism beyond the crisis. *The New York Review of Books,* 56(5). 26 March: 1–8.

Sen, J. & Waterman, P. 2007. *World Social Forum: Challenging Empires.* Montreal: Black Rose Books.

SERI Global & Friends of the Earth Europe. 2009. *Overconsumption? Our use of the World's Natural Resources.* Vienna/Brussels: SERI Global. Available: www.seri.at/resource-report.

Seyfang, G. & Smith, A. 2007. Grassroots innovations for sustainable development: Towards a new research and policy agenda. *Environmental Politics,* 16(4): 584–603.

Shaffer, P. 2008. New thinking on poverty: Implications for globalisation and poverty reduction strategies. *Real World Economics Review,* 47: 192–231.

Shand, M. 2005. Overview of Cape Town water supply system. Unpublished report compiled for South African Academy of Engineering. Cape Town: Ninham Shand.

Shiva, V. 2002. Power Shift. *Resurgence,* August: 32–34.

———2005. *Earth Democracy: Justice, Sustainability and Peace.* London: South End Press.

Simone, A. 2001. Between ghetto and globe: Remaking urban life in Africa. In: A. Tostensen, I. Tvedten & M. Vaa. (eds). *Associational Life in African Cities: Popular Responses to the Urban Crisis.* Stockholm: Elanders Gotab.

———2004a. People as infrastructure: Intersecting fragments in Johannesburg. *Public Culture,* 16(3): 407–429.

———2004b. *For the City Yet to Come: Changing African Life in Four Cities.* Durham, NC: Duke University Press.

Smit, W. 2000. The impact of the transition from informal housing to formalized housing in low-income housing projects in South Africa. Nordic Africa Institute Conference on the Formal and Informal City—What Happens at the Interface? Copenhagen. 15–18 June.

———2006. Understanding the complexities of informal settlements: Insights from Cape Town. In: M. Huchzermeyer & A. Karam. (eds). *Informal Settlements: A Perpetual Challenge?* Cape Town: University of Cape Town Press.

Smith, A., Stirling, A. & Berkhout, F. 2005. The governance of sustainable socio-technical transitions. *Research Policy,* 34: 1491–1510.

Smith, A., Voss, J.P. & Grin, J. 2010. Innovation studies and sustainability transitions: The allure of the multi-level perspective and its challenges. *Research Policy,* 39(4): 435–448.

Smith, L. 2004. The murky waters of the second wave of neoliberalism: Corporatization as a service delivery model in Cape Town. *GeoForum,* 35: 375–393.

Smith, L. & Hanson, S. 2003. Access to Water for the Urban Poor in Cape Town: Where Equity Meets Cost Recovery. *Urban Studies,* 40(8): 1517–1548.

Smith, M., Hargroves, K., Stasinopoulos, P., Stephens, R., Desha, C. & Hargroves, S. 2007. *Energy Transformed: Sustainable Energy Solutions for Climate Change Mitigation*. Brisbane: The Natural Edge Project. [Accessed: 13 February 2008].

Smith, V. 2000. *Feeding the World — A Challenge for the 21st Century.* Cambridge: MIT Press.

Sneddon, C., Howarth, R.B. & Norgaard, R.B. 2006. Sustainable Develoment in a Post-Brundtland World. *Ecological Economics*, 57: 253–268.

Snow, C.P. 1993. *The Two Cultures.* Cambridge: Cambridge University Press.

Soderbaum, P. 2009. A financial crisis on top of the ecological crisis: Ending the monopoly of neoclassical economics. *Real World Economics Review*, 49: 8–19.

Soil Association. 2010. *Telling Porkies: The Big Fat Lie about Doubling Food Production*. Bristol, UK: Soil Association.

South African Reserve Bank. 2010. *Quarterly Bulletin, June 2010.* Pretoria: South African Reserve Bank.

Southall, R. 2006. Introduction: Can South Africa be a developmental state? In: S. Buhlungu, J. Daniel, R. Southall & J. Lutchman. (eds). *State of the Nation: South Africa, 2005–2006.* Cape Town: HSRC Press.

——2009. Scrambling for Africa? Continuities and discontinuities with formal imperialism. In: R. Southall & H. Melber. (eds). *A New Scramble for Africa? Imperialism, Investment and Development.* Pietermaritzburg: University of KwaZulu-Natal Press.

Southall, R. & Melber, H. 2009. *A New Scramble for Africa? Imperialism, Investment and Development.* Pietermaritzburg: University of KwaZulu-Natal Press.

Spencer, F. 2010. Sustainable energy. In: M. Swilling. (ed). *Towards Sustaining Cape Town.* Stellenbosch: Sun Media.

Spies, M. 2010. Children and Lynedoch EcoVillage. Stellenbosch: University of Stellenbosch. Master's thesis.

Srinivasan, S. 2008. A marriage less convenient: China, Sudan and Darfur. In: K. Ampiah & S. Naidu. (eds). *Crouching Tiger, Hidden Dragon? Africa and China.* Pietermaritzburg: University of KwaZulu-Natal Press.

Stamm, A., Dantas, E., Fischer, D., Ganguly, S. & Rennkamp, B. 2009. *Sustainability-Oriented Innovation Systems: Toward Decoupling Economic Growth from Environmental Pressures?* Bonn: German Development Institute.

Stanhill, G. 1990. The comparative productivity of organic agriculture. *Agriculture, Ecosystems and Environment*, 30: 1–26.

Statistics South Africa. 2010. *Mining and Production Sales.* Statistical Release P2041. Pretoria: Statistics South Africa.

Steel, C. 2008. *Hungry City: How Food Shapes our Lives.* London: Chatto & Windus.

Steffen, W.A., Sanderson, A., Tyson, P.D. *et.al.* 2004. *Global Change and the Earth System: A Planet Under Pressure.* Berlin: Springer.

Stengers, I. n.d. The challenge of complexity: Unfolding the ethics of science — in Memoriam Ilya Prigogine. Talk delivered in memoriam of Ilya Prigogine. Unpublished mimeo.

Stern, N. 2007. *Stern Review: Economics of Climate Change*. Cambridge: Cambridge University Press.

Sterner, T. 2003. *Policy Instruments for Environmental and Natural Resource Management.* Washington, DC: RFF Press.

Stewart, I. 1998. *Life's Other Secret: The New Mathematics of the Living World.* London: Penguin.

Stiglitz, J. 2009. Sharing the burden of saving the planet: Global social justice for sustainable development. Paper presented at the meetings of the International Economic Association: Istanbul. June.

——2010a. *Freefall: Free Markets and the Sinking of the Global Economy*. London: Allen Lane.

——2010b. *The Stiglitz Report: Reforming the International Monetary and Financial Systems in the Wake of the Global Crisis*. New York: New Press.

Stiglitz, J., Sen, A. & Fitoussi, J.P. 2009. *Report by the Commission on the Measurment of Economic Performance and Social Progress*. Report Commissioned by President Sarkozy, French Government. Available: www.stiglitz-sen-fitoussi.fre

Strahan, D. 2007. *The Last Oil Shock*. London: John Murray.

Stuwe, C. 2009. Diffusion of sustainability innovations around Lynedoch EcoVillage. Oldenburg, Germany: Department of Sustainability Economics and Management, Carl von Ossietzky Universitat Oldenburg. Master's thesis.

Suliman, M. 1994. *Civil War in Sudan: The Impact of Ecological Degradation*. Zurich & Bern: Centre for Security Studies and Conflict Research & Swiss Peace Foundation, Environment and Conflicts Project.

Summerton, J. 1994. *Changing Large Technical Systems*. Boulder, CO: Westview Press.

Sustainability Institute & E-Systems. 2009a. *Solid Waste: Integrated Analysis Baseline Report* (Final draft). Report commissioned by the Sustainability Institute for Integrated Resources Management for Urban Development, UNDP Project No. 00038512. Stellenbosch: Sustainability Institute. Available: www.sustainabilityinstitute.net.

——2009b. *Water and Sanitation in the City of Cape Town: Integrated Analysis Baseline Report* (Final draft). Report commissioned by the Sustainability Institute for Integrated Resources Management for Urban Development, UNDP Project No. 00038512. Stellenbosch: Sustainability Institute. Available: www.sustainabilityinstitute.net.

——2009c. *Energy: City of Cape Town Integrated Analysis Baseline Report (Final Draft)*. Report commissioned by the Sustainability Institute, Integrated Resources Management for Urban Development, UNDP Project No. 00038512. Stellenbosch: Sustainability Institute. Available: www.sustainabilityinstitute.net.

Swartz, M. 2010. Pre-colonial Xhosa Cultural Aesthetic Perspectives on Human-Nature Relations. School of Public Leadership, University of Stellenbosch. Doctoral thesis.

Swilling, M. 1984. 'The buses smell of blood': The 1983 East London bus boycott. *South African Labour Bulletin*, 9(5): 45–75.

——1999a. Changing conceptions of governance. *Africanus*, 29.

——1999b. Rival futures. In: H. Judin & I. Vladislavic. (eds). *Blank: Architecture, Apartheid and After*. Rotterdam: NAI Publishers.

——2005. 'Hear the forest grow': Evaluation of Shackdwellers International—Case studies of Kenya, Malawi and South Africa. Cape Town: Unpublished report commissioned by Shackdwellers International.

——2006. Sustainability and infrastructure Planning in South Africa: A Cape Town case study. *Environment and Urbanization*, 18(1): 23–0.

——2007. *Growth, Dematerialisation and Sustainability: Resource Options for 2019*. Stellenbosch: Paper commissioned by The Presidency, South African Government.

——2008. Tracking South Africa's elusive developmental state. *Administratio Publica*, 16(1).

——2010a. Greening public value: The sustainability challenge. In: J. Bennington & M. Moore. (eds). *In Search of Public Value: Beyond Private Choice*. London: Palgrave.

——2010. *Sustaining Cape Town: Imagining a Liveable City*. Stellenbosch: Sun Media.

Swilling, M. & De Wit, M. 2010. Municipal finance, service delivery and the prospects for sustainable resource use in Cape Town. In: M. Swilling. (ed.). *Sustaining Cape Town: Imagining a Liveable City*. Cape Town: University of Cape Town Press.

Swilling, M., Frankel, P. & Pines, N. 1988. *State, Resistance and Change*. London: Croom Helm.

Swilling, M., Khan, F. & Simone, A. 2003. 'My soul I can see': The limits of governing African cities in a context of globalisation and complexity. In: P. McCarney & R. Stren. (eds). *Governance on the Ground: Innovations and Discontinuities in Cities of the Developing World*. Baltimore and Washington DC: Woodrow Wilson Centre Press and Johns Hopkins University Press.

Swilling, M., Khan, F., Van Zyl, A. & Van Breda, J. 2008. Contextualising social giving: An analysis of state fiscal expenditure and poverty in South Africa, 1994–2004. In: A. Habib. (ed.). *Social Giving in South Africa*. Cape Town: HSRC Press.

Swilling, M. & Phillips, M. 1989. The emergency state: Its structure, power and limits. In: G. Moss & I. Obery. (eds). *South African Review 5*. Johannesburg: Ravan Press.

Swilling, M., Roux, P. & Guyot, A. 2010. Agonistic engagements: Difference, meaning and deliberation in South African cities. In: P. Cilliers & R. Allen. (eds). *Complexity, Difference and Identity*. Heidelberg/New York: Springer.

Swyngedouw, E. 2006. Metabolic urbanization: The making of cyborg cities. In: N. Heynen, M. Kaika & E. Swyngedouw. (eds). *In the Nature of Cities*. London: Routledge.

Swyngedouw, E. & Kaika, M. 2000. The environment of the city ... Or the urbanization of nature. In: G. Bridge & S. Watson. (eds). *A Companion to the City*. Oxford: Blackwell.

Talberth, J. 2008. A new bottom line for progress. In: Worldwatch Institute. (ed.). *2008 State of the World: Innovations for a Sustainable Economy*. New York: W.W. Norton and Company.

Talberth, J., Cobb, C. & Slattery, N. 2007. *Sustainable Development and the Genuine Progress Indicator: An Updated Methodology and Application in Policy Settings*. Oakland, CA: Redefining Progress.

Tallis, R. 1997. *Enemies of Hope: A Critique of Contemporary Pessimism*. New York: St Martin's Press.

Tan, Z.X., Lal, R. & Wiebe, K.D. 2005. Global soil nutrient depletion and yield reduction. *Journal of Sustainable Agriculture*, 26(1): 123–146.

Touraine, A. 1981. *The Voice and the Eye: An Analysis of Social Movements*. Cambridge, UK: Cambridge University Press.

Trewavas, A. 2002. Malthus foiled again and again. *Nature*, 418: 678–684.

Tull, D.M. 2008. Postconflict reconstruction in Africa: Flawed ideas about failed states. *International Security*, 32(4): 106–139.

Turton, A.R. 2008. Three strategic water quality challenges that decision-makers need to know about and how the CSIR should respond. Keynote address to CSIR Conference on Science Real and Relevant. Centre for Scientific and Industrial Research, Pretoria, 18 November.

Tyler, E. & Winkler, H. 2009. Economics of climate change: Context and concepts related to mitigation. Unpublished report. Cape Town: Energy Policy Research Unit/Energy Research Centre, University of Cape Town.

Ulrich, A., Malley, D. & Voora, V. 2009. *Peak Phosphorus: Opportunity in the Making*. Winnipeg: Water Innovation Centre and International Institute for Sustainable Development.

UN-HABITAT. 2008a. *The State of African Cities 2008: A Framework for Addressing Urban Challenges in Africa*. Nairobi: United Nations Human Settlements Programme (UN-HABITAT).

——2008b. *State of the World's Cities 2008/2009: Harmonious Cities*. London: Earthscan.
United Nations. 2005. *Millennium Ecosystem Assessment*. New York: United Nations.
——2006. *State of the World's Cities 2006/07*. London: Earthscan and UN-HABITAT.
United Nations Centre for Human Settlements. 2003. *The Challenge of Slums: Global Report on Human Settlements*. London: Earthscan.
United Nations Development Programme. 1998. *Human Development Report 1998*. New York: United Nations Development Programme. 5 November 2006. Available from: http://hdr.undp.org/reports/global/1998/en/. [Accessed: 22 October 2006].
——2004. *Inclusive Resource Management for Livelihood Security and Sustainable Development: Workshop Report*. Renk: United Nations Development Programme.
——2005. *Reduction of Natural Resource-Based Conflict among Pastoralists and Farmers: UNDP Workshop Proceedings and Recommendations*. Khartoum: United Nations Development Programme.
——2007. *Human Development Report 2007/08 Fighting Climate Change: Human Security in a Divided World*. New York: United Nations Development Programme. 5 November 2006. Available from: http://hdr.undp.org/reports/global/1998/en/.
United Nations Development Programme—RSA. 2004. *Ten Days in Johannesburg*. Pretoria: Department of Environmental Affairs and Tourism and the United Nations Development Programme.
United Nations Environment Programme. 2006. *Global Environmental Outlook 3*. Nairobi: United Nations Environment Programme.
——2008. *Organic Agriculture and Food Security in Africa*. New York and Geneva: United Nations.
——2009. *From Conflict to Peacebuilding: The Role of Natural Resources and the Environment*. Nairobi: United Nations Environment Programme.
United Nations Human Settlements Programme. 2003. *The Challenge of Slums: Global Report on Human Settlements*. London: Earthscan.
United Nations Research Institute for Social Development. 2002. *The Greening of Business in Developing Countries*. London: Zed Books.
United States Geological Survey. 2010. *Mineral Commodity Summaries: Phosphate Rock*. Washington, DC: United States Geological Survey. Available: http://minerals.usgs.gov/minerals/pubs/commodity/phosphate_rock/. [Accessed: 4 November 2010].
Uphoff, N. 2002a. The agricultural development challenges we face. In: N. Uphoff. (ed.). *Agroecological Innovations: Increasing Food Production with Participatory Development*. London: Earthscan.
——2002b. *Agroecological Innovations: Increasing Food Production with Participatory Development*. London: Earthscan.
Uprichard, E. & Byrne, D. 2006. Representing complex places: A narrative approach. *Environment and Planning A*, 38: 665–676.
Urry, J. 2000. *Sociology Beyond Societies: Mobilities for the Twenty-First Century*. London: Routledge.
Vaccari, D.A. 2009. Phosphorus: A looming crisis. *Scientific American*, June: 54–59.
Van den Bergh, J.C.M., Truffer, B. & Kallis, G. 2011. Environmental innovation and societal transitions: Introduction and overview. *Environmental Innovation and Societal Transitions*, 1: 1–23.
Van der Ryn, S. & Cowan, S. 1996. *Ecological Design*. Washington, DC: Island Press.
Van Vuuren, D.P., Bouwman, A.F. & Beusen, A.H.W. 2010. Phosphorus demand for the 1970–2100 period: A scenario analysis of resource depletion. *Global Environmental Change*, 20(3): 428–439.

Von Blottnitz, H. 2005. Solid waste. Cape Town: Unpublished paper commissioned by the Department of Environmental Affairs and Tourism.

Von Braun, J. 2007. *The World Food Situation: New Driving Forces and Required Actions.* Food Policy Report No. 27. Washington: International Food Policy Research Institute.

Von Weizsäcker, E., Hargroves, K.C., Smith, M.H., Desha, C. & Stasinopoulos, P. 2009. *Factor Five.* London: Earthscan.

Von Weizsäcker, E., Lovins, A.B. & Lovins, L.H. 1997. *Factor Four: Doubling Wealth, Halving Resource Use.* London: Earthscan and James & James.

Wackernagle, M. & Rees, W. 2004. *Our Ecological Footprint.* Gabriola Island, Canada: New Society Publishers.

Wakeford, J. 2007. *Peak Oil and South Africa: Impacts and Mitigation.* Cape Town: Association for the Study of Peak Oil and Gas – South Africa.

Wassung, N. 2009. Moving towards urban sustainability. Assignment for MPhil in Sustainable Development Planning and Management. Stellenbosch, South Africa: Unpublished paper. Available: www.sustainabilityinstitute.net/documents.

Watson, R.T., Wakhungu, J. & Herren, H.R. 2008. *Agriculture at a Crossroads.* International Assessment of Agricultural Knowledge, Science and Technology for Development (IAASTD). http://agassessment.org/reports/IAASTD/EN/agriculture%20at%20a%20Crossroads-Synthesis%20Report%20(English).p [Accessed: 21 April 2008].

Watson, V. 2002. *Change and Continuity in Spatial Planning: Metropolitan Planning in Cape Town Under Political Transition.* London: Routledge.

Weart, S.R. 2008. *The Discovery of Global Warming.* Cambridge, MA: Harvard University Press.

Weber, S. 2004. *The Success of Open Source.* Cambridge, MA: Harvard University Press.

Wescott, A. 2006. China's Africa adventures: Beijing woos Africa's contemptibles. *Business Day.* 18 July.

Western Cape Provincial Government. 2005. *Micro-Economic Development Strategy Synthesis Report: Version 1.* Cape Town: Western Cape Provincial Government: Department of Economic Development.

Western Cape Provincial Government. Department of Environmental Affairs and Development Planning. 2008. *A Climate Change Strategy and Action Plan for the Western Cape.* Cape Town: Western Cape Provincial Government.

Westley, F., Olssen, P., Folke, C., Homer-Dixon, T., Vredenburg, H., Loorbach, D., *et al.* 2011. *Tipping Towards Sustainability: Emerging Pathways of Transformation.* Third Nobel Laureates Symposium on Global Sustainability: Transforming the world in an era of global change. Stockholm. 16–19 May.

Wheeler, S. & Beatley, T. 2004. *The Sustainable Urban Development Reader.* London: Routledge.

Wiggins, S. 1995. Change in African farming systems between the mid-1970s and the mid-1980s. *Journal of International Development,* 7(6): 807–848.

Wilde-Ramsing, J. & Potter, B. 2008. Blazing the green path: Renewable energy and state-society relations in Costa Rica. *The Journal of Energy and Development,* 32(1): 68–91.

Wilkinson, P. 2004. Renegotiating local governance in a post-apartheid city: The case of Cape Town. *Urban Forum,* 15(3): 213–230.

Wilson, C. & Tisdell, C. 2001. Why farmers continue to use pesticides despite environmental, health and sustainability costs. *Ecological Economics,* 39: 449–462.

Wilson, E.O. 1998. *Consilience: The Unity of Knowledge.* New York: Knopf.

Woke, L. 2011. Organic food a growing biz for the health-aware. *China Daily.* 27 June.

World Bank. 2002. *Globalization, Growth and Poverty*. Washington, DC: World Bank.

———2006. *Where is the Wealth of Nations? Measuring Capital for the Twenty-First Century*. Washington, DC: World Bank.

———2008. *Rising Food and Fuel Prices: Addressing the Risks to Future Generations*. Washington, DC: World Bank. Available: http://siteresources.worldbank.org/DEVCOMMEXT/Resources/Food-Fuel.pdf?resourceurlname-Food-Fuel.pdf.

———2009. *Global Economic Prospects: Commodities at the Crossroads*. Washington, DC: World Bank. Available: http://siteresources.worldbank.org/INTGEP2009/Resources/10363_WebPDF-w47.pdf.

World Commission on Environment and Development. 1987. *Our Common Future*. Oxford: Oxford University Press.

World Information Technology and Services Alliance. 2010. 17th World Congress on Information Technology 2010: Challenges of Change. Amsterdam. World Information Technology and Services Alliance. 25–27 May.

World Nuclear Association. 2010a. *Nuclear Power in South Africa*. London: World Nuclear Association. Available: http://www.world-nuclear.org/info/inf88.html. [Accessed: 15 October 2010].

———2010b. *World Uranium Mining*. London: Word Nuclear Association. Available: http://www.world-nuclear.org/info/inf23.html. [Accessed: 15 October 2010].

World Resources Institute. 2002. *Decisions for the Earth: Balance, Voice and Power.* Washington DC: World Resources Institute.

Worldwatch Institute. 2007. *State of the World 2007: Our Urban Future*. Washington, DC: Worldwatch Institute.

———2008. *Green Jobs: Towards Decent Work in a Sustainable, Low-Carbon World*. Green Jobs Initiative. Nairobi: United Nations Environment Programme.

———2011. *State of the World Report 2011: Innovations that Nourish the Planet*. Washington, DC: The Worldwatch Institute.

World Wildlife Fund. 2008. *Living Planet Report 2008*. Gland, Switzerland: World Wildlife Fund.

World Wildlife Fund, Zoological Society of London & Global Footprint Network. 2006. *Living Planet Report 2006*. Gland, Switzerland: WWF, ZSL and Global Footprint Network.

Wright, J. 2008. *Sustainable Agriculture and Food Security in an Era of Oil Scarcity*. London: Earthscan.

Yong, R. 2007. The circular economy in China. *Journal of Material Cycles and Waste Management*, 9: 121–129.

Yose, C. 1999. From shacks to houses: Space usage and social change in a Western Cape shanty town. University of Cape Town. Master's thesis.

Index

Z